Excel

SCIENCE STUDY GUIDE

Year 8

Get the Results You Want!

Geoffrey Thickett & Jim Stamell

PASCAL
PRESS

Reprinted 2015, 2017, 2020, 2022, 2025

ISBN 978 1 74125 392 4

Pascal Press
PO Box 250
Glebe NSW 2037
www.pascalpress.com.au

Publisher: Vivienne Joannou
Project editor: Mark Dixon
Edited by Leanne Poll
Cover and page design by DiZign Pty Ltd
Typset by Precision Typesetting (Barbara Nilsson)
Printed by Vivar Printing/Green Giant Press
Photos by Dreamstime and iStockphoto except Figure 3.26, p. 98,
© David Houlder; Figure 4.6, p. 114, © Leanne Poll, Letterati Publishing Services;
Figure 5.9, p. 147, thanks to Dr Stuart Thickett and the Chemistry School,
Sydney University for permission.

Contents

How to use this book

The Australian Curriculum

This study guide covers the complete course of the Year 8 Science Australian Curriculum including:

- all core knowledge
- all required problem-solving skills
- all aspects of the scientific method.

The book is divided into sections based on the Australian Curriculum's three learning strands and substrands. The three learning strands in the Australian Curriculum are:

- Science Understandings (Biological Sciences, Chemical Sciences, Earth and Space Sciences, Physical Sciences)
- Science as a Human Endeavour
- Science Inquiry Skills.

The Science as a Human Endeavour strand is covered in Chapters 1 to 4, plus each substrand of the Science Understandings strand is treated as follows:

- Chapter 1—Biological Sciences
- Chapter 2—Chemical Sciences
- Chapter 3—Earth and Space Sciences
- Chapter 4—Physical Sciences.

Chapter 5 covers the Science Inquiry Skills strand.

Tips for tests and examinations

Preparation

In order to prepare for your school exams, you will need to examine the Contents section at the front of the book and then identify the relevant sections of this book before you begin to revise and practise exam-style questions.

Each topic in your school programs for Year 8 may have integrated content from more than one of these different areas of study. It is unlikely that you will have studied the content in this order.

You need to allow sufficient time for this thorough revision—at least one week for a class test and at least three weeks for a major examination. Revise the content of the chapter or section and attempt all questions. Use the supplied answers to determine which areas you need further work in.

Test/exam questions

Multiple-choice questions

In order to answer these sorts of questions, make sure that you do the following.

- Read the stem of the question thoroughly.
- Look carefully at any diagrams, flow charts or tables, and interpret them thoroughly.
- Choose the letter of the best response and not just a correct independent statement; if you don't know the answer, make the most logical choice you can.

Free-response questions

In order to answer this type of question effectively, follow the guidelines below.

- Highlight the verbs and key words in the question and respond accordingly.
- Don't waste time by restating the question; keep your answers concise.
- Label any diagrams that you draw with a pencil and ruler.
- Line graphs should occupy more than 80% of the available grid space.
- All graphs must have a title and the axes should have linear scales and be appropriately labelled with titles and units.
- Any experimental methods must be written as a series of numbered sentences in the present or past tense.
- Repeating the experiment at least five times or more can improve the reliability of experimental results.
- For questions involving the scientific method, ensure that you state the dependent and independent variables, the variables you controlled (kept the same) and the experiment that you used as a control.
- When drawing a table of data, ensure that the table is fully bounded by lines to create columns and rows. The column headings and units should occupy the first row.

Features of this book

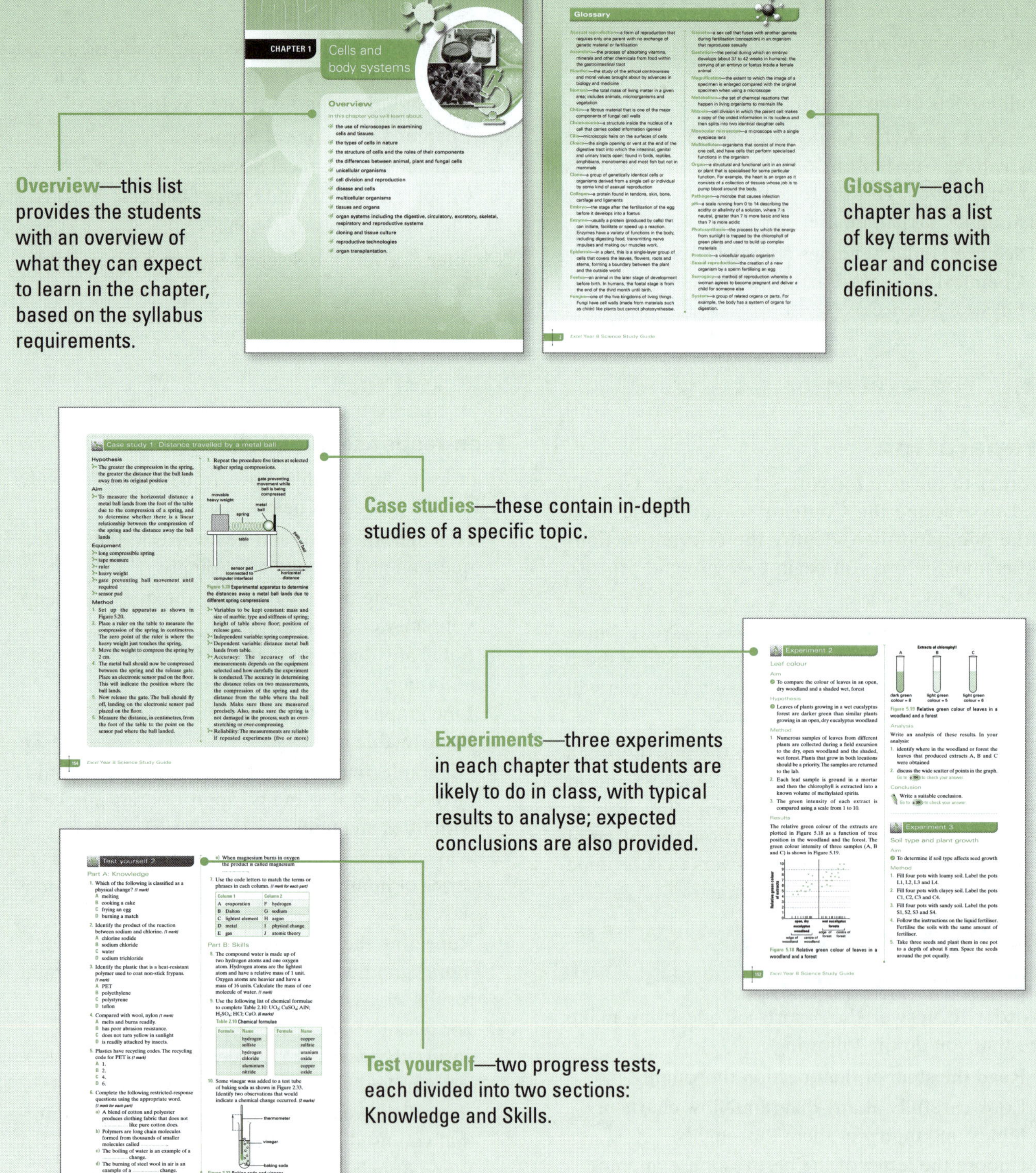

Overview—this list provides the students with an overview of what they can expect to learn in the chapter, based on the syllabus requirements.

Glossary—each chapter has a list of key terms with clear and concise definitions.

Case studies—these contain in-depth studies of a specific topic.

Experiments—three experiments in each chapter that students are likely to do in class, with typical results to analyse; expected conclusions are also provided.

Test yourself—two progress tests, each divided into two sections: Knowledge and Skills.

You will be thoroughly prepared for examinations and tests when you use this study guide. This guide is an effective revision and study program for exams and class tests in Year 8.

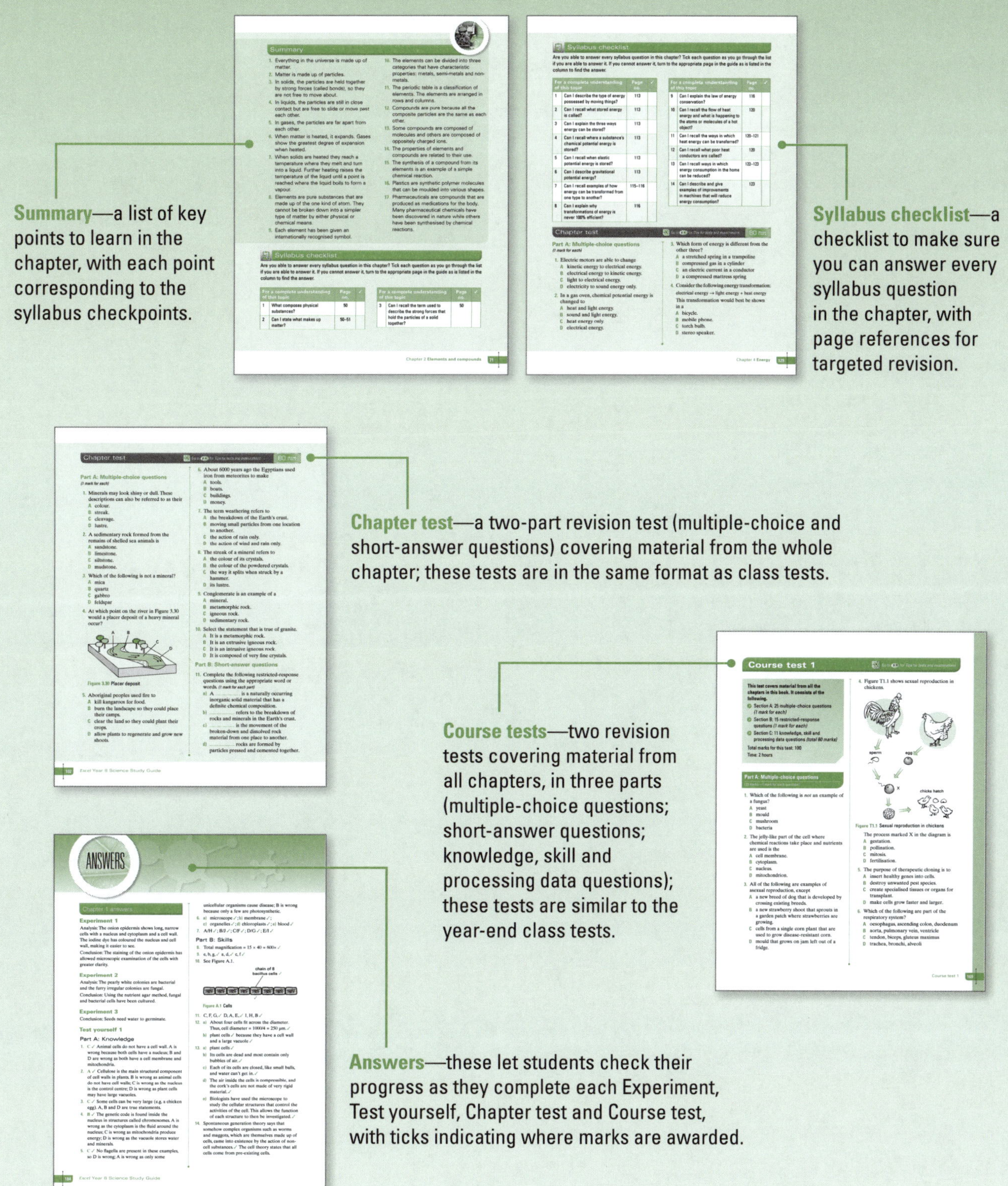

Summary—a list of key points to learn in the chapter, with each point corresponding to the syllabus checkpoints.

Syllabus checklist—a checklist to make sure you can answer every syllabus question in the chapter, with page references for targeted revision.

Chapter test—a two-part revision test (multiple-choice and short-answer questions) covering material from the whole chapter; these tests are in the same format as class tests.

Course tests—two revision tests covering material from all chapters, in three parts (multiple-choice questions; short-answer questions; knowledge, skill and processing data questions); these tests are similar to the year-end class tests.

Answers—these let students check their progress as they complete each Experiment, Test yourself, Chapter test and Course test, with ticks indicating where marks are awarded.

CHAPTER 1

Cells and body systems

Overview

In this chapter you will learn about:

- the use of microscopes in examining cells and tissues
- the types of cells in nature
- the structure of cells and the roles of their components
- the differences between animal, plant and fungal cells
- unicellular organisms
- cell division and reproduction
- disease and cells
- multicellular organisms
- tissues and organs
- organ systems including the digestive, circulatory, excretory, skeletal, respiratory and reproductive systems
- cloning and tissue culture
- reproductive technologies
- organ transplantation.

Glossary

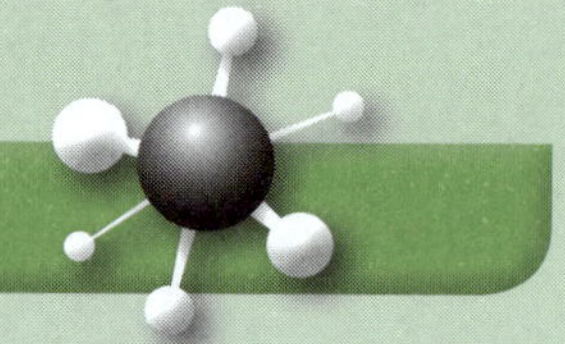

Asexual reproduction—a form of reproduction that requires only one parent with no exchange of genetic material or fertilisation

Assimilate—the process of absorbing vitamins, minerals and other chemicals from food within the gastrointestinal tract

Bioethics—the study of the ethical controversies and moral values brought about by advances in biology and medicine

Biomass—the total mass of living matter in a given area; includes animals, microorganisms and vegetation

Chitin—a fibrous material that is one of the major components of fungal cell walls

Chromosome—a structure inside the nucleus of a cell that carries coded information (genes)

Cilia—microscopic hairs on the surfaces of cells

Cloaca—the single opening or vent at the end of the digestive tract into which the intestinal, genital and urinary tracts open; found in birds, reptiles, amphibians, monotremes and most fish but not in mammals

Clone—a group of genetically identical cells or organisms derived from a single cell or individual by some kind of asexual reproduction

Collagen—a protein found in tendons, skin, bone, cartilage and ligaments

Embryo—the stage after the fertilisation of the egg before it develops into a foetus

Enzyme—usually a protein (produced by cells) that can initiate, facilitate or speed up a reaction. Enzymes have a variety of functions in the body, including digesting food, transmitting nerve impulses and making our muscles work.

Epidermis—in a plant, this is a single-layer group of cells that covers the leaves, flowers, roots and stems, forming a boundary between the plant and the outside world

Foetus—an animal in the later stage of development before birth. In humans, the foetal stage is from the end of the third month until birth.

Fungus—one of the five kingdoms of living things. Fungi have cell walls (made from materials such as chitin) like plants but cannot photosynthesise.

Gamete—a sex cell that fuses with another gamete during fertilisation (conception) in an organism that reproduces sexually

Gestation—the period during which an embryo develops (about 37 to 42 weeks in humans); the carrying of an embryo or foetus inside a female animal

Magnification—the extent to which the image of a specimen is enlarged compared with the original specimen when using a microscope

Metabolism—the set of chemical reactions that happen in living organisms to maintain life

Mitosis—cell division in which the parent cell makes a copy of the coded information in its nucleus and then splits into two identical daughter cells

Monocular microscope—a microscope with a single eyepiece lens

Multicellular—organisms that consist of more than one cell, and have cells that perform specialised functions in the organism

Organ—a structural and functional unit in an animal or plant that is specialised for some particular function. For example, the heart is an organ as it consists of a collection of tissues whose job is to pump blood around the body.

Pathogen—a microbe that causes infection

pH—a scale running from 0 to 14 describing the acidity or alkalinity of a solution, where 7 is neutral, greater than 7 is more basic and less than 7 is more acidic

Photosynthesis—the process by which the energy from sunlight is trapped by the chlorophyll of green plants and used to build up complex materials

Protozoa—a unicellular aquatic organism

Sexual reproduction—the creation of a new organism by a sperm fertilising an egg

Surrogacy—a method of reproduction whereby a woman agrees to become pregnant and deliver a child for someone else

System—a group of related organs or parts. For example, the body has a system of organs for digestion.

Glossary

Tissue—a part of an organism consisting of a collection of cells having a similar structure and function; a group of cells in an organism that work together such as the muscles in an animal or the outer surface of leaves in a plant

Toxic—the extent to which a substance can damage or cause harm to an organism. Toxic substances include chemicals as well as materials produced by organisms.

Toxin—a poisonous substance produced by living cells or organisms

Transplant—a tissue or organ transplanted from a donor to a recipient; an operation that moves an organ from one organism (the donor) to another (the recipient). In some cases the patient can be both donor and recipient.

Unicellular—an organism made of one cell only

Xylem—cells that transport water in plants

Zygote—a fertilised egg cell

1.1 Microscopes

Our knowledge of tiny organisms such as bacteria and protozoa has been made possible by the invention of the magnifying glass and the microscope. Leonardo da Vinci used a magnifying glass to study small objects in 1485. But the Dutch spectacle makers Hans and Zacharias Janssen did not invent the first microscope until about 1590. This crude microscope could magnify up to nine times the original.

In 1665, Englishman Robert Hooke invented a two-lens microscope. Hooke used his microscope to examine thin slices of cork. He observed structures that resembled the boxes of a honeycomb, as shown in Figure 1.1. He called these boxes 'cells' because they reminded him of the tiny cells or rooms in which monks lived in a monastery. This term has been used ever since. Hooke's microscope could only magnify 30 times.

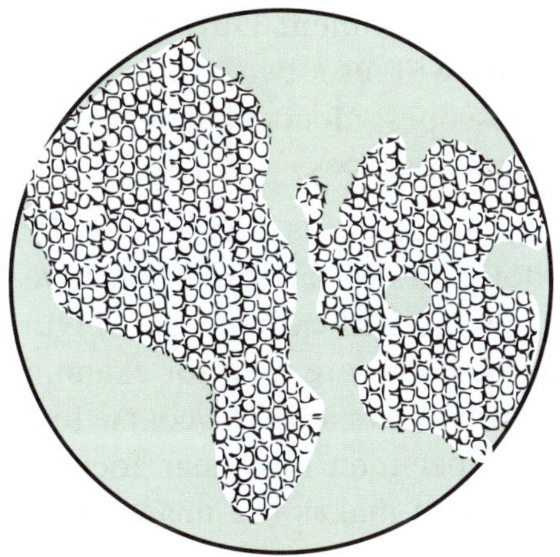

Figure 1.1 Hooke's drawings of cork 'cells'

Another Dutchman, Antoine van Leeuwenhoek, developed a single-lens microscope in 1676 that could magnify 270 times (see Figure 1.2). His hobby was to grind perfect lenses. These lenses allowed him to see the microscopic world with wonderful clarity. He examined a whole range of materials, including blood and pond water. He discovered the existence of many different types of protozoa in pond water and many different types of bacteria in his own saliva.

Improvements in microscope design occurred over the next two centuries.

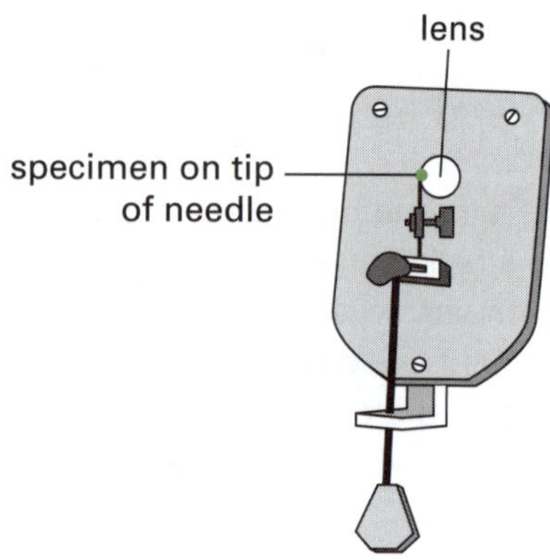

Figure 1.2 Van Leeuwenhoek's microscope

Light microscopes

Your school microscopes are called light microscopes because they use visible light to observe the specimen. During your science course you will use two common types of light microscopes: monocular microscopes and binocular microscopes.

Monocular microscopes have one eyepiece lens and a choice of two or three objective lenses of different magnifications (see Figure 1.3). These microscopes are used for examining thin specimens on glass slides. The coarse focus knob brings the specimen into near focus and the fine focus knob makes the image clearer. The specimens can be magnified up to 400 times their normal size. As the light passes through the specimen, an image is formed on the retina of your eye.

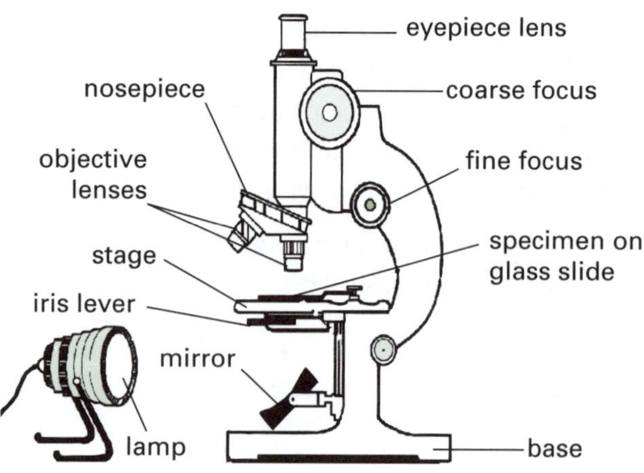

Figure 1.3 Monocular microscope

Binocular, or stereo microscopes, have two eyepiece lenses so that you can see a specimen with both eyes open. This gives a more three-dimensional effect. These microscopes are used to examine larger specimens (such as flower parts, insects and minerals). Their magnification is usually much less than a monocular microscope. Light reflects off these specimens before it enters the lens system of the microscope.

Using a monocular microscope

1. Place the microscope on a flat bench so that the mirror faces towards the lamp.
2. Rotate the nosepiece to select the lowest power objective lens.
3. Adjust the angle of the lamp and the mirror so that the light reflects off the mirror into the lens and up the tube. You should see a bright circle of light as you look through the eyepiece lens.
4. Adjust the light intensity by using the iris lever.
5. Place the specimen (which will be on a glass slide) on the stage and secure it with the clips.
6. Looking from the side, lower the objective lens until it is just above the slide surface.
7. Looking through the eyepiece lens, use the coarse focus knob to raise the tube upwards (away from the specimen) until it comes into focus.
8. Complete the focusing using the fine focus knob. Never focus downwards towards the slide as you might break the glass and the lens.
9. View the specimen at higher magnification by rotating a new objective lens into place. Use the fine focus knob to refocus.

Using a binocular microscope

1. Select the lowest power objective lens.
2. Place the specimen on a tile or glass dish and shine the lamp to illuminate the specimen.
3. Lower the tube by looking from the side. Then use the focus knob to raise the tube and focus while looking through the eyepiece lenses.

Magnification

In a microscope, the image of the specimen is enlarged or magnified. The eyepiece (or ocular) lens and the objective lens together determine the final magnification of the microscope. The typical and total magnification of each lens is as follows:

- eyepiece lens = 10×
- low power objective lens = 10×
- high power objective lens = 40×
- total magnification on low power = 10 × 10 = 100×
- total magnification on high power = 10 × 40 = 400×.

As the magnification gets larger, the field of view gets smaller.

Experiment 1

Preparation of a wet mount

A 'wet mount' is a specimen of plant or animal tissue that will be placed in a drop of water on a microscope slide to be examined microscopically.

Aim

- To prepare a stained, wet mount of the epidermal cells of an onion bulb

Method

1. Use a safety blade and tweezers to strip several thin pieces of onion skin (epidermis) from an inner surface of the bulb.
2. Place one piece of the onion tissue on a glass slide and add a drop of water.
3. Gently lower a cover slip over the specimen using a probe as shown in Figure 1.4. Avoid trapping air bubbles.
4. Examine the specimen under low power and note the shape of the cells. View at a higher power.
5. Place a second piece of onion tissue in a Petri dish containing some strong iodine solution. Leave the tissue to absorb the stain (about 5 to 15 minutes, depending on the thickness of the tissue and the strength of the solution).
6. Mount the tissue in a drop of water as before and examine under low power and then higher power. Some parts of the onion cell will now be clearer as they become stained by the iodine.

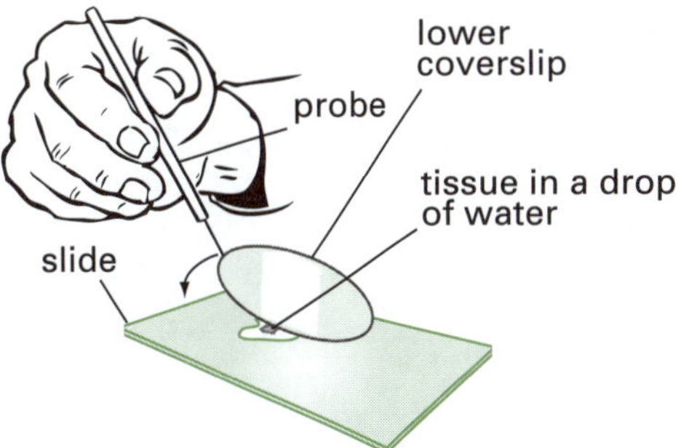

Figure 1.4 Preparing a wet mount of onion epidermis

Results

The results of this experiment are shown in Figure 1.5.

Figure 1.5 Lower power view of onion epidermis

Analysis

Describe what you see in this slide of onion epidermis and explain how the iodine dye has helped to view the specimen. Go to p. 184 to check your answer.

Conclusion

Write a suitable conclusion.

Go to p. 184 to check your answer.

Using digital technology to view cells

Digital microscopes allow us to capture and record magnified images and videos on computers (see Figure 1.6). The images of the specimen can be viewed on TV screens or shared via email or social networking websites.

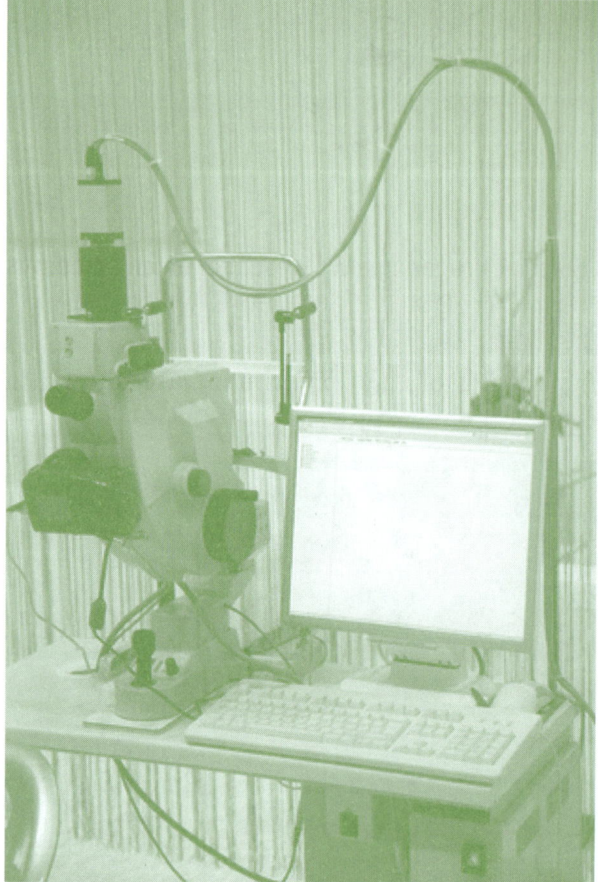

Figure 1.6 A digital camera is attached to a microscope so images can be viewed and saved on a computer.

1.2 Cells

Types of cells

Today scientists identify the cell as the basic unit of living matter from which all plants and animals are built. A living cell can carry on all the functions necessary for life. About 150 years ago scientists came to realise that:

- all living things are made up of one or more cells
- cells are the basic units of structure and function in living things
- all cells are produced by pre-existing cells.

These ideas are now known as the cell theory. Some plants and animals consist of only one cell. Bacteria and protozoa are organisms consisting of one cell. We say they are unicellular organisms. But the plants and animals we can ordinarily see are made up of millions of cells. We say they are multicellular organisms. Humans are multicellular organisms, each consisting of about a trillion cells.

All cells have certain basic similarities.

- Most cells are too small to be seen with the unaided eye. When you look at single-celled organisms, you need to use a microscope or a hand lens. The same is true if you look at skin cells or blood cells. You could fit about 50 000 red blood cells in this letter O. On the other hand, a chicken's egg is one cell.
- A cell is surrounded by a thin and flexible cell membrane. Inside is a chemically controlled watery environment composed of a nucleus and the cytoplasm (see Figure 1.7). Red blood cells do not contain a nucleus.

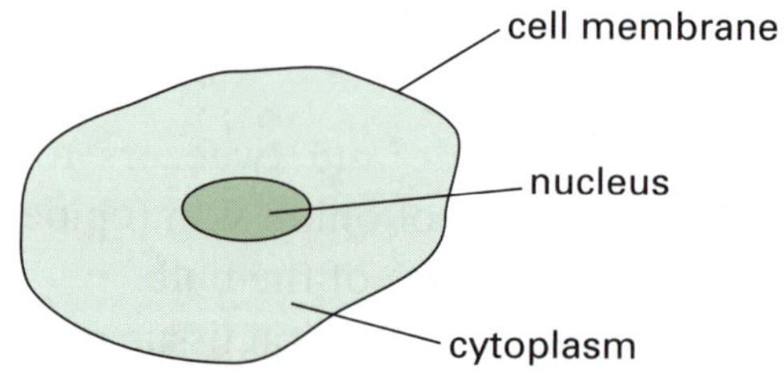

Figure 1.7 A simple cell

- A cell takes in raw materials and energy and uses them to carry out all the necessary functions required to grow and reproduce.
- All waste products are passed out of the cell through the membrane.
- Cells are alive (take in oxygen and food, eliminate wastes, grow and reproduce). In time all cells die.
- Different cells have different functions in a multicellular organism. For example, muscle cells contract and relax to allow muscles to

move, and nerve cells conduct messages to various organs.

Figure 1.8 shows the wide variety of cells in animals and plants.

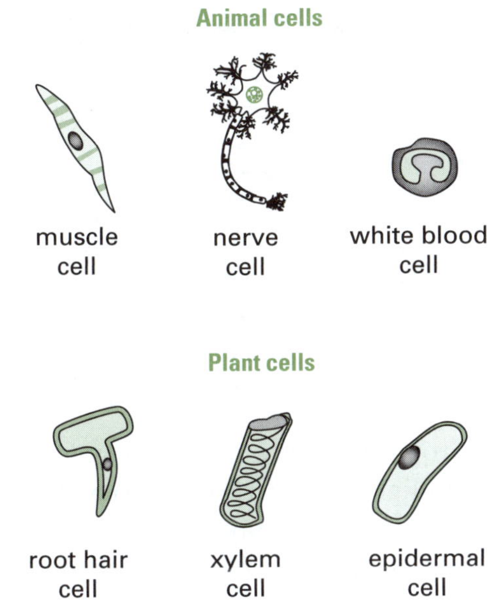

Figure 1.8 Animals and plants have a range of different cell types. For example, in plants xylem cells carry water through the roots, stems and leaves; epidermal cells are the cells on the outside surface of a plant and they vary in shape. In animals the nervous system carries electrical signals, while the bloodstream carries hormones from glands to other sites in the body.

Diffusion

Diffusion is the movement of materials across the cell membrane. Nutrients and water move into cells through pores in the cell membrane. In a similar way waste materials move out of the cells and are then transported away. The movement of materials across the cell membrane is an example of diffusion.

Diffusion is driven by differences in concentration. These differences in concentration happen all the time in our bodies and our cells because substances are separated by cell membranes. If the concentration of a chemical is higher on one side of the permeable membrane, the chemical will diffuse from the side of higher concentration to the side of lower concentration.

Osmosis is the process of diffusion of water across a membrane from a region of high water concentration to one of low water concentration. A wilted plant recovers when watered because there is a higher concentration of water in the soil outside the root hairs than inside. Therefore, water diffuses across the cell membranes into the root cells.

Cellular structure

Cells vary in structure between animals, plants and fungi.

Animal cells

All animal cells are surrounded by a thin cell membrane. Inside the cell are the nucleus and cytoplasm as well as other much smaller structures called organelles. The intracellular fluid (or cytosol) together with the organelles comprise the cytoplasm. Mitochondria are important organelles. Table 1.1 summarises the function of these components.

Table 1.1 Function of cell structures and organelles

Cell structure	Description
cell membrane	thin layer surrounding the cell; controls entry of nutrients (such as food and oxygen) and exit of materials (such as waste) from the cell; protects cell contents
nucleus	control centre of the cell; contains inherited (genetic) codes in structures called chromosomes; surrounded by a nuclear membrane
cytoplasm	clear, jelly-like liquid; many chemical reactions occur within the cytoplasm
mitochondrion	site where energy is generated by a process called respiration

Figure 1.9 shows the structure of a typical animal cell. Animal cells vary in size. Most are between 10 micrometres (μm) and 50 μm. (Note: a micrometre, or micron, is 1 thousandth of a millimetre—1 000 000 μm = 1000 mm = 1 m; μ is the Greek letter *mu* and in measurement means 'one-millionth of'.)

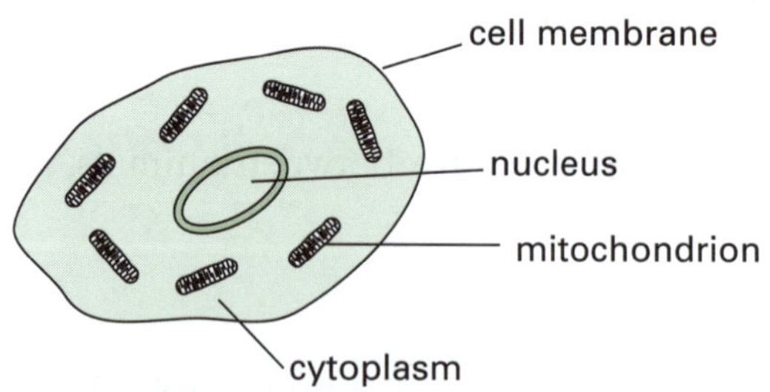

Figure 1.9 Structures in a typical animal cell

Plant cells

All plant cells have a cell membrane which is surrounded by a rigid cell wall. The strongest components of the cell wall are cellulose molecules that are derived from sugar molecules made by the plant. Green plant cells also contain organelles called chloroplasts which contain chlorophyll. As well, plant cells contain a large vacuole that is filled with a watery solution. Table 1.2 summarises the functions of the additional organelles in plants. The function of the cell membrane, nucleus, cytoplasm and mitochondria are the same as in animal cells.

Table 1.2 Functions of some structures and organelles in plant cells

Cell structure	Description
vacuole	storage area for water, food and waste substances; usually large in plant cells, but tiny or non-existent in animal cells
cell wall (plants only)	provides strength and support for the plant; composed of woody cellulose
chloroplast (green plant cells only)	contains the green chemical called chlorophyll which absorbs sunlight; site of photosynthesis (conversion of solar energy into chemical energy)

Figure 1.10 shows the structure of a typical green plant cell. Plant cells vary in size. Typically they are between 10 μm and 150 μm.

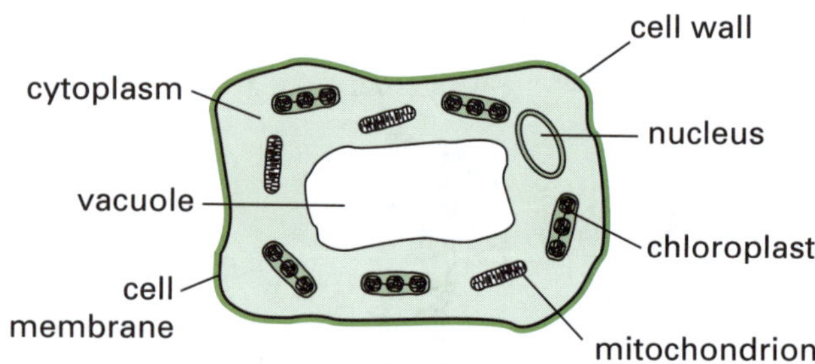

Figure 1.10 Structures in a typical green plant cell. Vacuoles full of water provide structural support. Chloroplasts are not present in non-green plant cells.

The chloroplasts of the leaves of green plants absorb sunlight and carbon dioxide from the air and use the water taken up by their roots to produce sugars and release oxygen. Sugars such as glucose store chemical energy which is later used by the plant. This process is called photosynthesis.

The word equation for photosynthesis is:

$$\text{carbon dioxide} + \text{water} \xrightarrow{\text{light}} \text{sugar (glucose)} + \text{oxygen}$$

Fungal cells

Fungal cells resemble plant cells. They have a cell wall but it is not made of woody cellulose. Instead it is made of another natural material called chitin that strengthens the cell. Chitin is also found in insect exoskeletons. Fungal cells have no chloroplasts as fungi do not photosynthesise. Fungal cells have spherical or tubular vacuoles which carry out the same functions as in plant cells.

1.3 Unicellular organisms

Many unicellular organisms exist. Some unicellular organisms have whip-like tails (flagella) or hairs (cilia) to help them move though the water (see Figure 1.11).

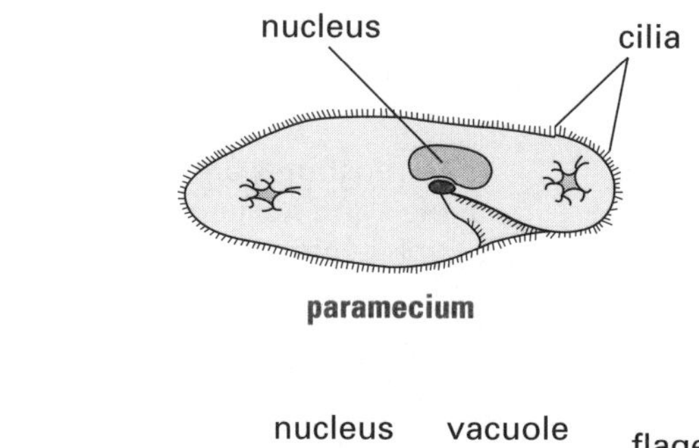

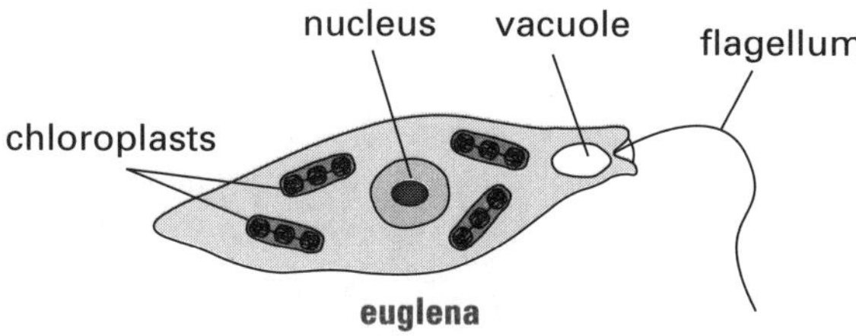

Figure 1.11 Some examples of unicellular organisms

Unicellular organisms with membrane-bound nuclei belong to the kingdom protista. Some protists catch their food using their cilia to sweep the food into a grooved opening in the side of the cell. Some protists feed through a process called phagocytosis. This involves engulfing their food by extending and surrounding the food particle with their cell membrane. The amoeba feeds in this way (see Figure 1.12).

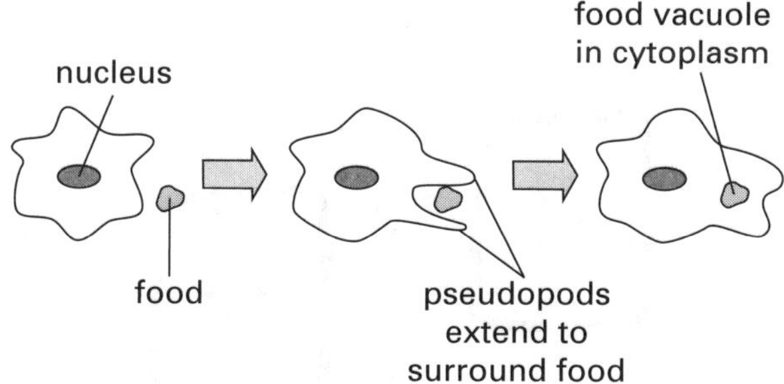

Figure 1.12 An amoeba feeds by phagocytosis

1.4 Cell division and reproduction

In multicellular organisms, cells divide to produce new cells. Cell division is necessary for:

- growth of the organism
- repair of tissues and organs
- reproduction by the formation of sex cells.

Figure 1.13 shows a simple model of normal cell division, which is called mitosis. In normal cell division, the information that is coded in the chromosomes inside the nucleus must be copied before the cell divides. In this way each new cell (daughter cell) has an exact copy of the information that controls the cell.

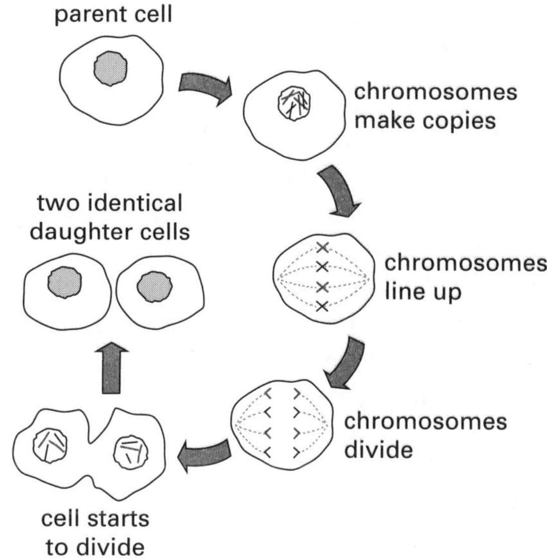

Figure 1.13 Normal cell division (mitosis)

Growth

Living things begin their life as one cell. In multicellular organisms, this first cell (the fertilised egg in sexually reproducing organisms) divides many times to produce all the cells that make up the adult organism. As an embryo grows, cells become specialised for different functions. Thus some cells will become muscle cells while others will become nerve cells.

An organism grows because:

- each new cell produced by the division of an old cell grows to its maximum size
- cells continue to divide to make sufficient cells for the adult organism.

Repair

Cell division becomes less frequent as an organism ages. However, it does not stop. Throughout the life of a living organism, cells become damaged or die. These cells need to be repaired and, in some cases, replaced by new cells. Our skin cells eventually die and are shed. They are replaced by newly dividing cells formed in a layer below the surface. New red blood cells are constantly being formed by cell division in the bone marrow.

Reproduction

In simple unicellular organisms, reproduction is achieved by simple cell division in a process called asexual reproduction. This type of cell division (mitosis) is the same as the method used to produce new body cells in multicellular organisms. All cells produced by mitosis are genetically identical to the original cell from which they were formed. They have the same number and types of chromosomes present as were present in the original cell. Asexual and sexual reproduction will be discussed in more detail later in this chapter.

Mitosis in animal cells

When an animal cell is about to divide, the information encoded on the chromosomes is copied before the nucleus starts to split. The cytoplasm then starts to pinch inwards as the two separate daughter nuclei separate. Eventually two separate and smaller cells are formed. This process is shown in Figure 1.14. In the cytoplasm, structures such as mitochondria also divide and are shared about equally between each daughter cell.

The new daughter cells absorb nutrients and grow. Often one of the daughter cells will keep its ability to divide, while the other may become specialised to form certain tissues and organs of the animal. For example, some cells may become muscle cells while others may become cells of the intestines.

Mitosis in plant cells

When a plant cell is about to divide, the nucleus becomes larger and the vacuole disappears. The nucleus then divides. New cell walls grow to separate the two daughter cells. Small vacuoles (which join to produce a large vacuole) form in one daughter cell. Chloroplasts also divide and are shared between each daughter cell. Water is then absorbed to make this cell larger. The second daughter cell prepares to divide again. This process is shown in Figure 1.15.

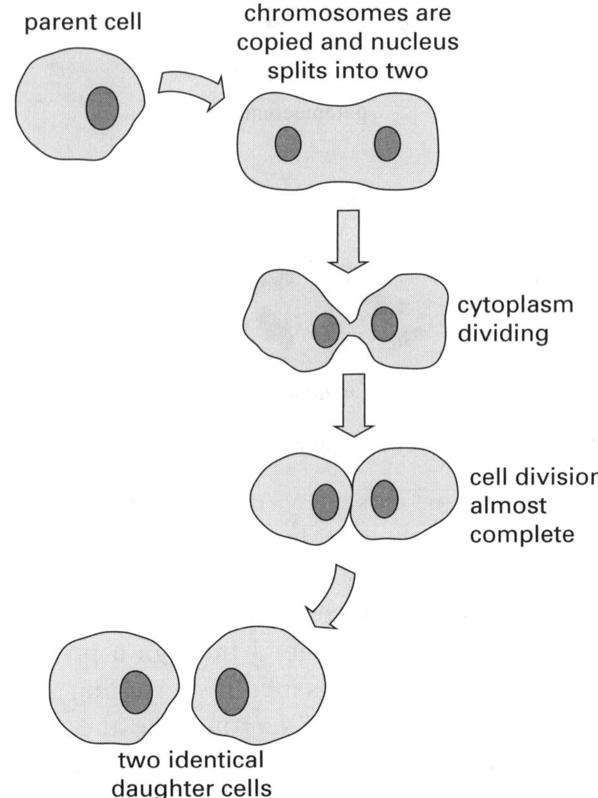

Figure 1.14 Normal cell division in animal cells

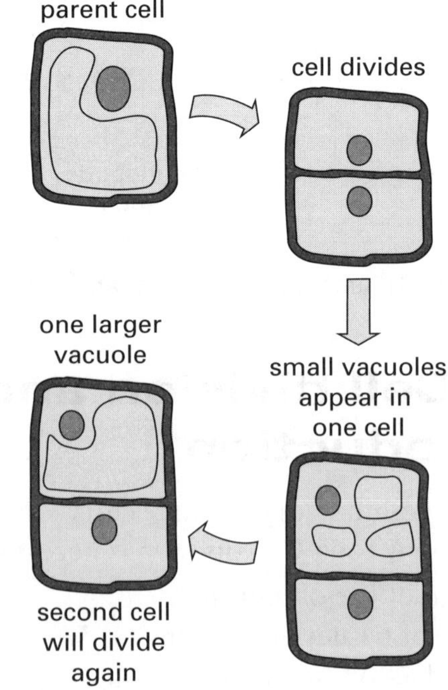

Figure 1.15 Cell division in a plant cell

1.5 Disease

Disease in a multicellular organism is the result of abnormal cell function and the breakdown of one or more organs or body systems.

Diseases of the human body can be classified as infectious or non-infectious.

- Infectious diseases are caused by pathogens. A pathogen is a disease-producing microorganism (microbe). Examples are bacteria, viruses, protozoa and some fungi. These pathogens enter the body in a variety of ways and their activities disrupt the proper functioning of the body cells. Infectious diseases (e.g. measles, HIV and diphtheria) can be passed from one person to another.
- Non-infectious diseases have a variety of causes. Some chromosomes have faulty information codes which lead to faulty genes and can cause disease. Toxic chemicals can cause organ and tissue disease. Some diseases are also classified as lifestyle diseases. Non-infectious diseases cannot be passed onto another person.

Figure 1.16 shows some examples of viruses and bacteria that cause disease in humans. Not all microbes cause disease. Some have a beneficial role in the body (e.g. in the production of vitamins) and in the environment (e.g. as decomposers).

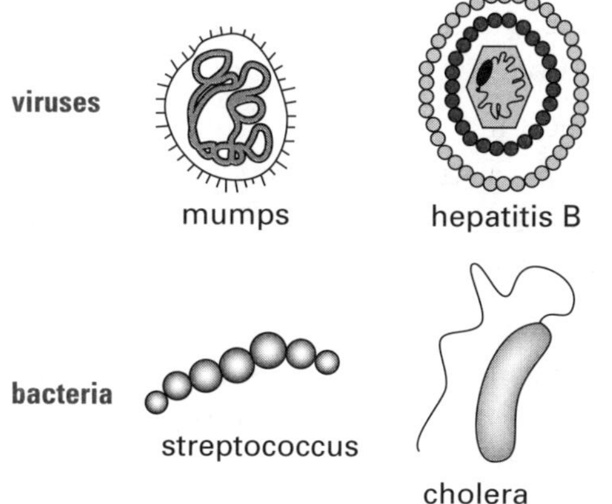

Figure 1.16 Microbes invade cells and cause disease.

- Bacteria belong to the kingdom monera. Their cells are very simple. Bacteria may have many shapes such as spheres, rods and spirals as well as chains of spheres and rods. Pathogenic bacteria can damage the cells of the host by producing poisons (toxins). Bacteria multiply very rapidly inside cells at body temperature. Every 20 to 30 minutes each bacterium cell divides by mitosis and produces two new bacteria. Over the course of a day the number of bacteria becomes quite large and their toxins start to disrupt the body.
- Viruses are not classified into the five major kingdoms of living things because they are not cellular life forms. They consist of particles that are usually made up of bundles of genetic material inside a protein coat. They invade a host and use the nucleus of the host to complete their reproduction. Eventually the host cell dies and splits open to release thousands of new viral particles which then infect other cells. Mumps is a common viral disease.
- Protozoa belong to the kingdom protista. They are single-celled microbes. They vary widely in shape and mainly live in water or moist soil. They can reproduce asexually or sexually. Not all protozoans are pathogenic.
- Fungi are mostly multicellular organisms. The fungal kingdom is distinct from plants as fungi do not produce their own food by photosynthesis. The fungi that cause disease in humans tend to attack the skin, hair, nails and moist membranes.

Table 1.3 classifies some common infectious diseases in humans according to the type of microbe that causes the disease.

Resisting the attack of microbes

The human body resists attack by pathogenic microbes in a variety of ways. The body provides barriers to microbe entry. If microbes enter, then the white blood cells of our immune system attack the invaders. Time is very important. Our immune system must defeat the pathogens before they overwhelm the body. Disease and even death may result if the body's defences cannot overcome the invasion.

Table 1.3 Diseases caused by microbes

Disease	Pathogen	Transmission	Symptoms of the disease
tuberculosis	bacterium	inhaling infected droplets in the air; drinking contaminated milk	lesions form in lungs producing fever and coughing
malaria	protozoan	protozoan enters human during a mosquito bite	severe fevers; shivering; sweating; nausea; reoccurs periodically
measles	virus	contact with infected people; inhalation of infected droplets	red spots (rash) on face and body; fever
tinea	fungus	contact of the foot with infected areas (e.g. communal bathing)	itching and blistering of skin between toes; cracking of skin

Barriers to microbe entry

The following are the body's natural barriers to microbe entry.

- Acidic environments. Acids that are secreted onto our skin and that are produced in the stomach kill many of the microbes or slow down their reproduction. Urine is also acidic and this prevents microbes entering the body via the urethra. The highly acidic gastric juices in the stomach kill many microbes.
- Mucus linings. Our airways are lined with a sticky mucus that traps microbes that may be inhaled.
- Hairs and cilia. The respiratory tract is lined with cilia (microscopic hairs) that trap microbes.
- Skin. The skin is a physical barrier to the entry of microbes. Hair follicles and sweat glands release chemicals that can destroy bacteria.
- Tears. Tears contain enzymes that attack microbes. Tears also flush foreign materials away.

Phagocytes

Phagocytes are cells that engulf and destroy the microbes. Pus is the remains of dead phagocytes and microbes.

A fever is often associated with the action of white blood cells. The increased temperature of the body helps to destroy some microbes.

Immunity

The body's third line of defence involves specialised white blood cells called lymphocytes. These are produced by the immune system as a response to infection. This process is called naturally induced active immunity. These white blood cells (which are made in lymph nodes that drain into veins) circulate in body fluids, attacking and immobilising the pathogen. Some lymphocytes produce antibodies and others produce special 'memory' cells that remain in the body for a long time and can react rapidly if re-infection occurs.

Antibodies are special proteins produced by our immune system. They bind to pathogens and immobilise or destroy them. Antibodies are specific to the antigen (a specific foreign substance) that triggered their formation.

Immunisation is the process of introducing into the body a serum or vaccine that will stimulate the body's immune system to produce specific antibodies. Vaccination is an example of artificially induced active immunity. Vaccines contain one of the following:

- killed microbes
- harmless strains of living pathogens
- modified toxins that stimulate the immune system without destroying it.

Babies are born with antibodies transferred from the mother's own immune system. More are received through the mother's milk while breastfeeding. This is an example of naturally induced passive immunity.

Antibodies for specific diseases can also be cultured in other animals (e.g. horses) and then a vaccine is developed containing these specific antibodies. This is normally done for travellers who may come in contact with specific diseases such as cholera and yellow fever. This type of vaccination is an example of artificially induced passive immunity.

The British doctor Edward Jenner produced the first vaccine in 1796. At the time smallpox was a very contagious and serious viral disease that regularly reached epidemic levels with extremely high death rates. He observed that milkmaids who had contracted cowpox, a relatively mild infection, from the cows they were tending were somehow resistant to smallpox.

Jenner assumed correctly that cowpox is similar to smallpox. He concluded that by giving humans an injection of the cowpox virus it would somehow cause the human body to respond to both viruses, without causing major illness or death. Jenner first used the word *vaccine*, which derives from the Latin word *vacca*, meaning 'cow'.

Mass immunisation is an important program for various diseases such as whooping cough, diphtheria, polio, tetanus, measles and mumps. It ensures the eradication of dangerous pathogens from the general population by ensuring that those that are more susceptible are protected.

Quarantine is used when a disease is likely to spread rapidly to the wider population and create an epidemic or pandemic. The government's medical officers monitor the threat of a disease to the general population and can impose quarantine on disease carriers. Quarantine stations were very commonly used in the past in Australia whenever immigrants arrived from overseas. All animals from overseas are still quarantined to ensure dangerous diseases do not enter the country. (In earlier centuries quarantine was an enforced isolation for 40 days; hence the derivation of the word from the Italian *quarantena*, meaning 'forty'.)

Antibiotics

When our body cannot resist the attack of pathogenic bacteria, treatment with antibiotics may lead to a cure. An antibiotic can be useful in many bacterial infections but cannot be used to cure viral infections. Penicillin was one of the earliest antibiotics, and was developed in the late 1940s. It was extracted from a green mould growing on rockmelons. Since then many antibiotics have been produced from other moulds and bacteria as well as being chemically synthesised, and millions of lives are saved annually.

Antibiotics work in a variety of ways including:

- damaging the microbe's cell membrane
- interfering with the metabolism of the microbe
- preventing replication of the genetic information in the nucleus of the cell
- preventing a cell wall forming around new microbes.

Unfortunately, bacteria can become resistant to antibiotics. This resistance is an example of the evolutionary principles in action. Other anti-microbial drugs continue to be developed to cure fungal and protozoan diseases.

Antibiotics need to have certain properties. They should:

- kill (or inhibit) the pathogen
- not harm the person receiving the antibiotic
- cause minimal side-effects
- not disturb the balance of useful microbes in the body.

It is important that antibiotics are not prescribed for viral infections. Antibiotics only act on cellular life forms, not viruses. Overuse of antibiotics can lead to antibiotic resistance.

Experiment 2

Culturing bacteria and moulds

Nutrient agar is a special jelly which can be used to grow microbes such as bacteria and moulds. Bacterial colonies tend to be rounded and pearly in appearance. They are often white or cream in colour. Moulds form irregular-shaped colonies that are furry in appearance.

Aim

- To use nutrient agar to culture bacteria and moulds

Method

1. The nutrient agar is prepared in Petri dishes *with lids*. The dishes are pre-sterilised and provided for you to use.
2. In order to grow microbe colonies, remove the lids and expose the agar to various conditions before resealing and taping down. (For example, open it to the air for 20 minutes and then cough on the exposed plate, gently place your fingers on the agar surface or add a few drops of water from an aquarium.)
3. The labelled, sealed dishes are placed in an incubator at 35 °C for several days. The microbial colonies should grow in that time. It is important to keep the dishes sealed in case pathogens have been cultured.

Results

Figure 1.17 shows a typical result when the dish is exposed to the air for about 20 minutes.

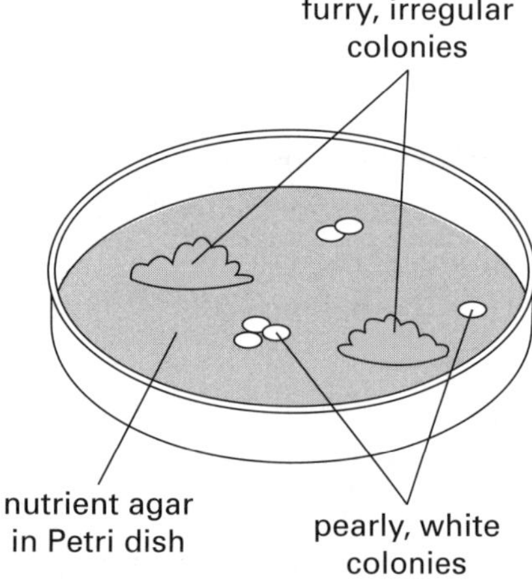

Figure 1.17 Experiment results: culturing microbes

Analysis

Describe what you see in the Petri dish and classify the colonies as bacterial or fungal.
Go to p. 184 to check your answer.

Conclusion

Write a suitable conclusion.
Go to p. 184 to check your answer.

Non-infectious diseases

During the life of an organism, the process of cell reproduction and cell differentiation may go wrong. There are a number of different ways in which genetic disease may occur:

- gene mutations in the genetic codes in the nucleus
- deletion of one or more genes from a chromosome
- whole chromosome(s) missing or extra chromosomes present.

Disease can also be caused by environmental agents and lifestyle.

- Environmental agents (e.g. toxic chemicals, high energy radiation, viruses) may damage the chromosomes, leading to changes or mutations in genes.
- Lifestyle diseases are non-infectious. They are caused by organs or other parts of the body failing to work properly. Heart disease and obesity can be caused by lifestyle choices. The chemicals in tobacco smoke, for example, can cause emphysema, coronary heart disease, strokes and lung cancer.

Health, diet and exercise

Many scientific studies have shown the medical benefits of a healthy lifestyle. Lifestyle diseases such as obesity, diabetes and heart disease can be reduced in frequency by changes in what we eat and how we exercise, and by eliminating high-risk behaviours such as drug taking and binge drinking.

Healthy diet

The recommendations for a healthy diet include:

- eat more vegetables, whole grains and fruit as these contain more complex carbohydrates, vitamins, minerals and fibre
- reduce saturated fats and total fat intake and replace with smaller amounts of unsaturated fats to control blood cholesterol
- reduce simple sugar intake and fast foods containing them.

Exercise

Regular daily exercise (e.g. walking, bike riding, swimming, running) helps to control weight and strengthen muscles and the cardiovascular system.

Test yourself 1

Part A: Knowledge

1. A plant cell is distinguished from an animal cell because of the presence of *(1 mark)*
   A a nucleus.
   B a cell membrane.
   C a cell wall.
   D mitochondria.

2. Which one of the following statements about cells is true? *(1 mark)*
   A Plant cell walls are largely composed of cellulose.
   B All cells contain cell walls.
   C Mitochondria are the control centres of a cell.
   D Vacuoles are quite small in most plant cells.

3. The following ideas form part of the cell theory, except one. Which one is incorrect? *(1 mark)*
   A All living things are made up of one or more cells.
   B Cells are the smallest units of living things.
   C Cells are too small to be seen without a microscope.
   D All cells come from other cells.

4. The control centre of all cells is the *(1 mark)*
   A cytoplasm.
   B nucleus.
   C mitochondrion.
   D vacuole.

5. Which statement is true of the unicellular organisms shown in Figure 1.18? *(1 mark)*
   A All unicellular organisms cause disease.
   B Unicellular organisms are photosynthetic.
   C These unicellular organisms belong to the kingdom protista.
   D They all have flagella that help them move.

colpoda paramecium vorticella

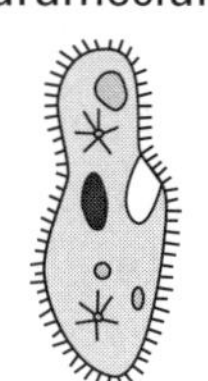

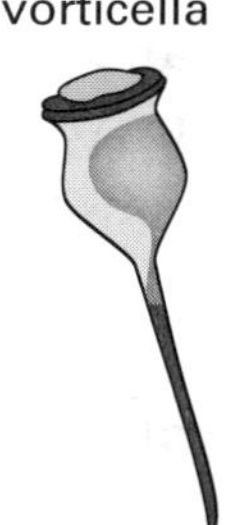

Figure 1.18 Unicellular organisms

6. Complete the following restricted-response questions using the appropriate word. *(1 mark for each part)*
   a) Antoine van Leeuwenhoek developed a single-lens .... in 1676 that could magnify 270 times.
   b) A cell is surrounded by a thin and flexible cell ....
   c) Inside the cell is the nucleus and the cytoplasm as well as other much smaller structures called ....
   d) Green plant cells have .... which contain chlorophyll.
   e) Phagocytes are white .... cells that attack foreign substances including microbes.

7. Use the code letters to match the terms or phrases in each column. *(1 mark for each part)*

Column 1	Column 2
A eyepiece lens	F immunity
B nucleus	G penicillin
C antibodies	H objective lens
D antibiotic	I mutation
E toxic chemicals	J chromosomes

Part B: Skills

8. A student uses a monocular microscope with the following lens magnifications. Calculate the total magnification of the lens system used. *(1 mark)*
   - eyepiece lens: 15×
   - objective lens: 40×

9. The following steps refer to the correct procedure for setting up a monocular microscope to view a specimen of a plant root. The steps are jumbled. Arrange them in the correct order. *(3 marks)*
   a) Place the slide on the stage.
   b) While looking through the eyepiece, adjust the lamp and the angle of the mirror to get a bright circle of light in the field of view.
   c) While looking through the eyepiece, use the coarse focus knob to wind the tube up until the specimen is in focus.
   d) Look from the side and lower the tube until the objective lens is just above the slide.
   e) Rotate the nosepiece to select the lowest power objective lens.
   f) Adjust the position of the slide and use the fine focus knob to complete the focusing.
   g) Adjust the light intensity with the iris lever.

10. Bacterial cells divide by fission. Read the following description and draw a labelled diagram to show the appearance of the bacterial cells after 1½ hours. *(2 marks)*

    A bacillus is a bacterium that appears rod-shaped in a two-dimensional view, but like a long breakfast sausage in a three-dimensional view. No membrane-bound nucleus can be seen in the cell, although the nuclear material is present in the central zone of the cytoplasm. The bacillus undergoes fission to produce two new daughter cells which enlarge and grow to maturity each 30 minutes. After 1½ hours, a chain of bacilli has formed.

11. In plants, the growth of the shoot is restricted to the tips. Cells just behind the tip are dividing. Below this region the cells are expanding as the young cells absorb water into their vacuoles.

    Some of the cells in Figure 1.19 come from the actively dividing region and others come from the expanding region. Unfortunately the cells are all jumbled up and not in a logical sequence. Use the code letters to arrange the cells in a logical sequence from the top of the shoot down. *(3 marks)*

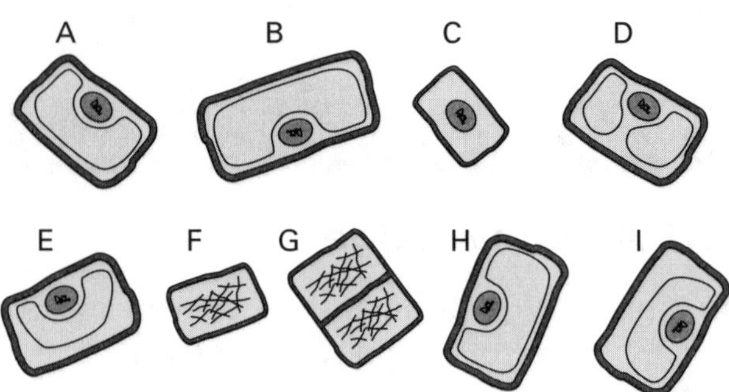

Figure 1.19 Dividing cells

12. A student used a microscope to examine some cells. The diameter of the field of view was 1000 μm (1000 μm = 1 mm). A drawing of the cells he observed is shown in Figure 1.20.

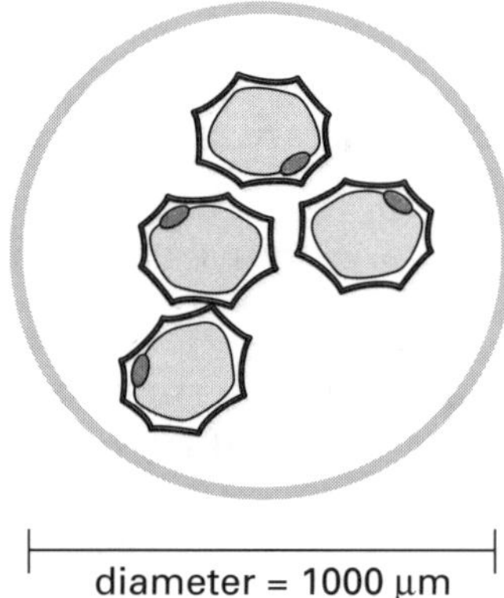

Figure 1.20 Cells

   a) Estimate the diameter of these cells. *(1 mark)*
   b) Are these cells from an animal or a plant? Explain. *(2 marks)*

13. The following is a description of what Robert Hooke (1635–1703) saw when he studied the bark of a cork oak tree.

    I took a good clear piece of Cork, and with a Pen-knife Sharpen'd as keen as a Razor, I cut a piece of it off, and thereby left the surface of it exceeding smooth, then examining it very diligently with a Microscope, me thought I could perceive it to appear a little porous … I could plainly perceive it to be all perforated and porous, much like a Honey-comb.

a) What had Hooke observed? *(1 mark)*

b) Suggest why cork is so light. *(1 mark)*

c) Why doesn't cork soak up water? *(1 mark)*

d) When cork is compressed it springs back to its original shape. Suggest a reason for this. *(1 mark)*

e) How has the use of the microscope assisted biologists in their studies of cells? *(1 mark)*

14. Until the middle of the 1800s it was commonly believed that certain low forms of living matter came into existence from non-living material. This theory, called spontaneous generation, explained that worms came from cheese and wood, and maggots from the juices of decaying meat. How does spontaneous generation contradict the cell theory? *(1 mark)*

Go to p. 184 to check your answers.

1.6 Multicellular organisms

Single-celled organisms, including bacteria and protozoa, comprise more than half of the total biomass on Earth. This is because they have been so successful in adapting to a variety of different environments. These unicellular organisms can synthesise all of the substances they require from a few simple nutrients.

Division of labour

Multicellular organisms consist of more than one cell. Most living things that can be seen with the unaided eye are multicellular, such as all animals and land plants. In multicellular organisms cells are specialised for different functions, carrying out extremely varied activities. Some cells are specialised for sensitivity to light, heat, sound or pressure; others for synthesising materials such as hair, bone, shell or bark. Some cells are associated with transport, such as xylem cells in plants, which conduct water and help support the plant, and phloem cells, which transport food materials (see Figure 1.21). Still other cells are specialised for producing such substances as milk, hormones and poisons. Cells become efficient in one process and depend on other cells for the necessities of life.

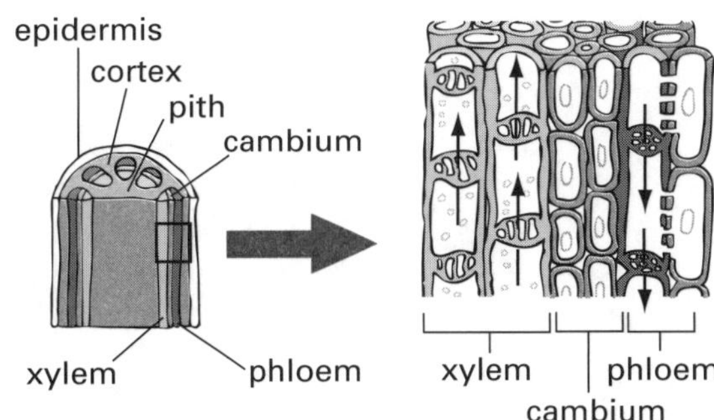

Figure 1.21 Specialised cells in the stem of a plant. Xylem cells carry water and dissolved minerals from the roots to the rest of the plant and may also give it mechanical support. Phloem cells convey organic nutrients to all parts of the plant where needed. Between these two tissues, a layer of cells called the cambium gives rise to new phloem and xylem cells.

In a vertebrate, more than 200 distinct cell types can be distinguished. By collaborating and dividing labour, it becomes possible to use resources far better than single-celled organisms can.

1.7 Tissues and organs

There are many parts to a computer: the RAM, the ROM, the hard drive, the transformer, the disk drive, the keyboard, the mouse and the monitor—all need to do their part so that you can effectively operate the machine. Each of the parts of a computer is different, indicating their special functions. The same is true of multicellular organisms.

Multicellular organisms can be thought of as complicated machines with each part needing to work with other parts. Living things are made up of cells. Multicellular organisms contain many similar cells working together. These cells form groups called tissues. The tissues are specialised working groups that form part of a larger organ. In animals, examples of tissues include muscle, skin and bone. A number of related organs form an organ system. The organ system carries out a specific function. For example, the stomach includes mucus membrane tissue, muscle tissue, a layer of tissue lining the abdomen, and so on.

Figure 1.22 shows the different levels within a multicellular organism.

Our skin is a very large and important organ made up of various tissues. The skin, hair and nails belong to the integumentary system. Table 1.4 outlines the other main systems of the human body.

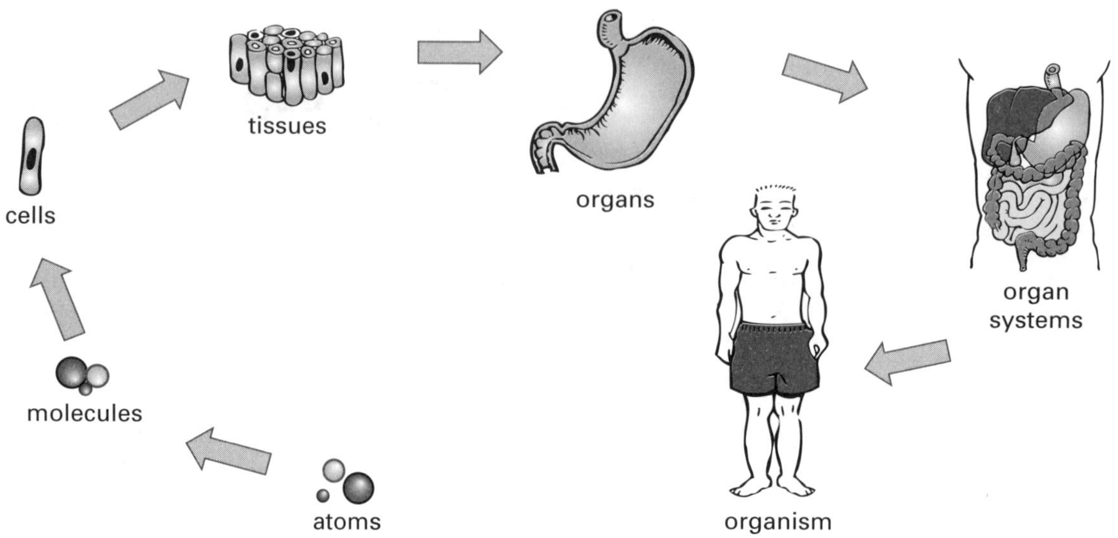

Figure 1.22 Levels of organisation in multicellular organisms

Table 1.4 The main systems in our bodies

Body system	Major organs	Roles
skeletal system	bones	supporting the body (legs, backbone); protecting organs (skull, ribcage)
muscular system	muscles	move the whole body (leg muscles); assist circulation (heart muscle); assist breathing (diaphragm muscle)
digestive system	mouth	chews and breaks down food; mixes with saliva
	oesophagus	tube carrying food to stomach from mouth
	stomach	churns food with acid
	small intestine	completes digestion and nutrients are absorbed into the bloodstream
	liver	makes bile (for fat digestion) and stores glycogen
	pancreas	produces digestive enzymes; regulates sugar balance
	large intestine	removes water from waste
circulatory system	heart	pumps blood (containing oxygen and nutrients) around the body
	blood vessels	transfer blood to organs, tissue and cells
respiratory system	lungs	take in oxygen and remove carbon dioxide waste (gas exchange)
	diaphragm	muscle that contracts/relaxes to change volume of chest cavity
excretory system	kidneys	filter the blood and remove liquid waste
	bladder	stores urine
	liver	detoxifies blood and removes nitrogenous waste
	lungs	remove carbon dioxide

Green plants are also multicellular organisms. In green plants various structures have specific functions. Table 1.5 provides information about the features of structures in a green plant.

Table 1.5 Features of green plants

Part of the plant	Features
roots	• help to anchor a plant in the soil and prevent soil erosion • absorb water and dissolved minerals from the soil through tiny root hairs • some roots (such as carrots) store food
stems	• support the branches and leaves so they can be exposed to the light • some store food (potatoes are swollen underground stems) • help transfer water, minerals and food around the plant • bark around the trunk helps insulate and protect the plant
leaves	• contain pigments such as chlorophyll to trap sunlight to make food (photosynthesis) • have pores (stomata) in their surface to allow carbon dioxide to enter and allow water vapour and oxygen to escape
flowers and fruits	• flowers are the sexual reproductive organs of flowering plants • fruits (fleshy or dry) develop from ovaries after pollination • seeds develop inside the fruits until they are ready to be shed (dispersed)

1.8 Digestive system

Digestion is the process of breaking down the complex molecules in food to simpler molecules. The broken down particles of food are then absorbed into the bloodstream and taken to where they are needed or stored.

Like most other mammals, digestion in humans begins in the mouth. As Figure 1.23 shows, we have four different kinds of teeth to cut, crush and grind food. In the mouth, food particles are mixed with saliva to lubricate the pieces. Saliva also contains an enzyme to begin the breakdown of starch.

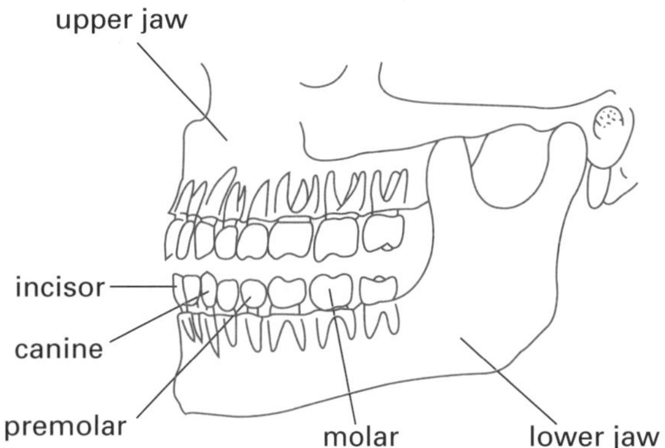

Figure 1.23 Human teeth

When food is swallowed, it passes into the food tube or oesophagus. The oesophagus contains muscles that squeeze the balls of food into our stomach. This is shown in Figure 1.24.

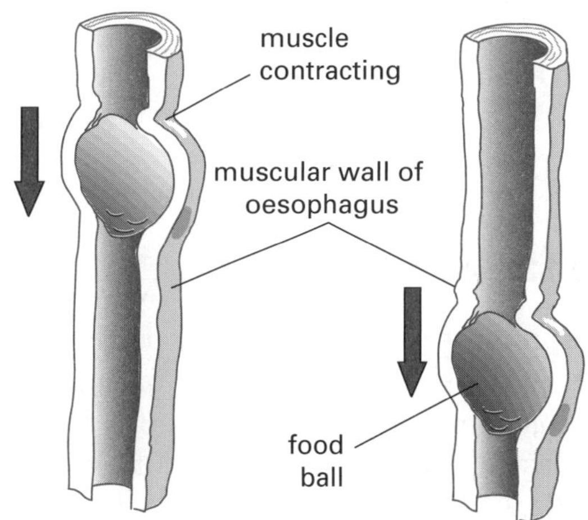

Figure 1.24 Muscle contraction in the oesophagus

The stomach is a bit like a cement mixer. It churns the food and mixes it with other juices such as acids and enzymes to begin the breakdown of proteins. In time the food is pushed into the small intestine.

In humans the small intestine is a coil about 7 m long. More enzymes are added here, with some coming from the pancreas and others from the wall of the intestine itself. The gall bladder releases bile to help break down fat particles. Digestion is completed in the small intestine. Figure 1.25 shows the parts of the human digestive system.

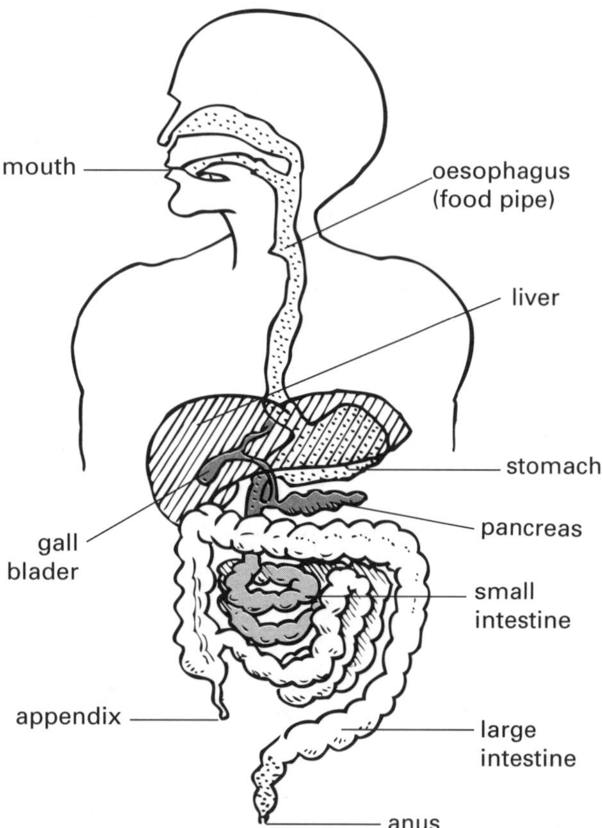

Figure 1.25 Human digestive system

Absorption of the digested food particles also occurs in the small intestine. Millions of finger-like projections called villi enormously increase the surface area of the inside of the small intestine. Absorption of food particles occurs by the slow process of diffusion into the villi and then into the blood.

Undigested food passes into the large intestine, and it is here that water is absorbed. It is also where millions of bacteria live that feed on the accumulating faeces. They produce certain vitamins, gases and other compounds. The vitamins are absorbed through the walls of the large intestine into our blood stream.

The partly dried out faeces are eliminated through the anus. So faeces are not totally composed of waste; it is food we did not use ourselves, some indigestible fibre and bacteria.

The absorbed fragments of digested food are transported by our circulatory system to the many cells of our body where they will be used.

1.9 Comparing digestive systems

The digestive systems of different organisms vary, depending on their diets.

Cattle

Cows are herbivores that feed on plants. They are also ruminants. This means their digestive system allows them to eat otherwise indigestible food by continually regurgitating and re-chewing it as 'cud'.

A cow's tongue helps it to gather grass and pinch it off between their incisors and dental pad. Saliva, which is secreted by the salivary glands, initiates starch digestion. This saliva contains sodium bicarbonate that helps to maintain a neutral pH (6.5 to 7.2) encouraging microbes to grow.

Feeding

When cattle feed, they grip a little grass with their tongues. Then using the upper jaw and the teeth of the lower jaw, they jerk their heads to pull off the grass.

As Figure 1.26 shows, the upper jaw has no front teeth while the lower jaw has eight front teeth. Towards the back of the mouth are the molars, which cows use to chew grass. Being herbivores, cattle do not need canines. These are more prominent in carnivores. Humans, being omnivores, have a variety of teeth.

The older the animal is, the more the teeth are worn. You can tell the age of a cow by looking at its front teeth.

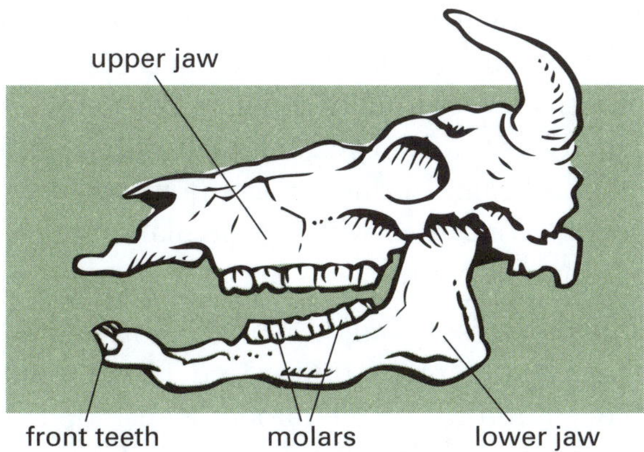

Figure 1.26 Teeth in the jaws of a cow

Digestive tract

Humans, dogs, poultry and pigs have single stomachs. They do not produce enzymes capable of breaking down cellulose. But the stomachs of cattle and sheep contain microbes that produce these enzymes.

A cow's stomach is divided into four compartments: rumen, reticulum, omasum and abomasum (see Figure 1.27). Food is first subjected to fermentation by microbes in the reticulum and rumen. Food passes readily between these two compartments, which together occupy over 50% of the cow's stomach and are where most of the microbial activity occurs. The food (cud) is often sent back up to the mouth for re-chewing. The cud is again swallowed and further digested by microorganisms present in the rumen.

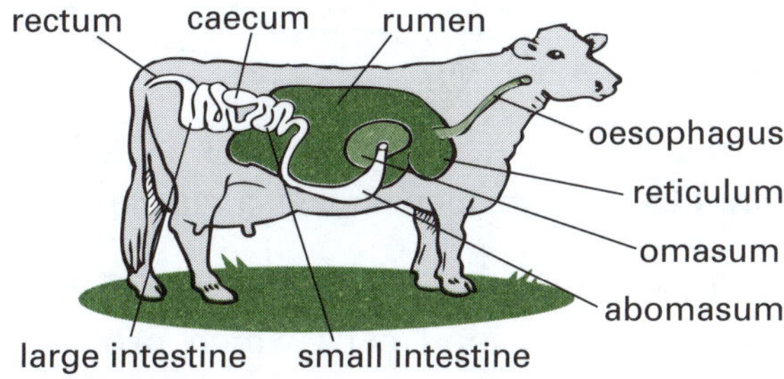

Figure 1.27 The digestive system of a cow

The action of microorganisms (bacteria, protozoans, moulds and yeasts) allow cattle to obtain nutrition from the fibrous material in plant matter. Fermentation in the rumen is possible because the environment is stable enough for microbial growth. Temperature ranges from 37 to 40 °C, while pH ranges from 5.5 to 7 (slightly acidic to neutral).

After sufficient time the partly digested food passes to the omasum, which contains many folds of tissues thus increasing its surface area. Water is absorbed here and the particle size of food is reduced. Food then passes to the abomasum (also called the 'true stomach'). It is here that digestive enzymes prepare the food for absorption in the small intestine.

Food broken down through the ruminant's stomach is largely absorbed in the small intestine. It is here that absorption of nutrients including vitamins takes place. Undigested material passes to the large intestine where excess moisture is absorbed. Microbes also pass into the cow's lower digestive tract where they are digested and their proteins, vitamins and other nutrients beneficial to the cow are absorbed.

The relationship between the cow and the microbes is mutually beneficial. Through this microbial action, high-fibre food sources become assets and nitrogen compounds are turned into protein. Animals with simple stomachs cannot use cellulose or non-protein nitrogen.

Dogs

Although they are carnivores, dogs are not as exclusively dependent on meat as, for instance, cats are. Dogs can best be classified somewhere in-between carnivores and omnivores, with a strong leaning towards carnivore preferences.

The digestive system of carnivores is the simplest among mammals. Other species (including humans) have a more extensive large intestine, and as we have seen, ruminants like cattle and sheep have a large set of stomach compartments through which food passes before it reaches the intestine.

Figure 1.28 shows the teeth and jaws of a dog. The large canine (fang) teeth of all carnivores are there to grab, hold, tear and cut. The front incisor teeth are used to grab, hold and scrape meat off bones. The premolars and molars behind are for crushing and some grinding.

The jaws are fairly long, compared to the face, allowing the fangs to grab fairly large objects.

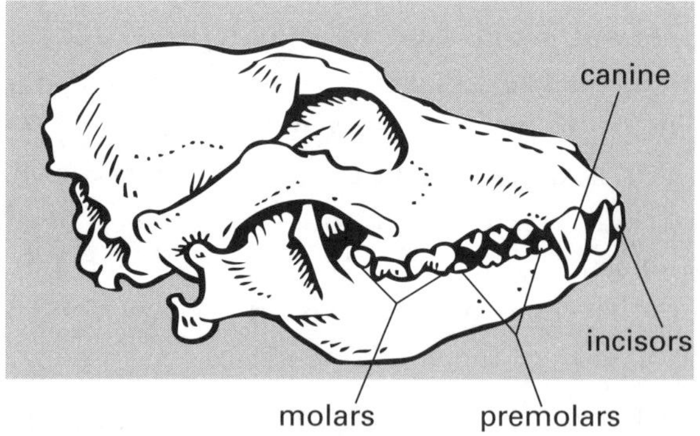

Figure 1.28 Dog's teeth; notice the prominent canines

The muscles that control the jaws are some of the most powerful muscles in the dog's entire body. The jaw joint of dogs is a stiff hinge allowing for only up and down movement. Dogs 'wolf' their food down in big chunks. They don't chew their food and so saliva is only a lubricant for swallowing. The dog's stomach takes care of all the digestion.

As raw food is not easy to digest, a dog's stomach produces large quantities of strong acid thus making the acid levels in a dog's stomach higher than in a human's. Unlike human stomachs, the stomach juices of canines are capable of dissolving large chunks of food and bones. So by the time food reaches the small intestine, the digestive juices have broken it down to a state where it is easy to process and assimilate.

1.10 The circulatory system

The many cells of our bodies need to be nourished with food and oxygen, and their waste products eliminated. This job is done by the circulatory system that carries materials to and from the cells. As well, the blood transfers heat generated from our muscles and other organs to all our tissues to keep us warm.

The heart is the organ that pumps blood around our body. The many blood vessels form the highways along which nutritional materials, water, heat and wastes move. Some fluid can also move through the walls of our blood vessels and into the surrounding tissue. Fluid that bathes this tissue eventually collects in small tubes called lymphatic vessels and is returned to the blood stream. In this section, however, we will only be concerned with the circulation of blood around our bodies.

Our blood is a complex mixture of chemicals. There are some 5 billion red blood cells in every millilitre of our blood. Red blood cells are continuously made in our bone marrow and do not have a nucleus (see Figure 1.29). They only survive for a few months of coursing around and being battered by our blood vessels, so they need to be replaced. About a million are replaced each second.

Red blood cells contain a red iron pigment called haemoglobin, which carries oxygen to all cells. In our lungs, haemoglobin combines easily with oxygen to form oxyhaemoglobin. Around our cells where the oxygen concentration is low, oxyhaemoglobin breaks down releasing oxygen.

$$\text{haemoglobin} \underset{-\text{ oxygen}}{\overset{+\text{ oxygen}}{\rightleftharpoons}} \text{oxyhaemoglobin}$$

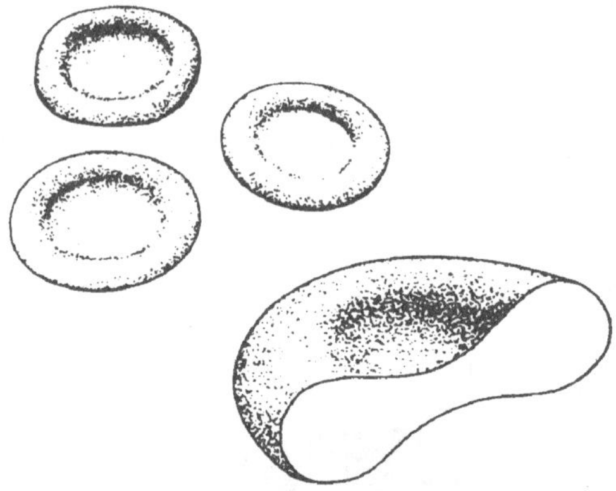

Figure 1.29 Cross-section through a red blood cell

Our blood stream also contains several different types of white blood cells which protect us from diseases by engulfing bacteria, dead cells and foreign bodies. Some are responsible for producing antibodies to help us attack infection. There are also platelets to help blood to clot. All this is contained within a straw-coloured liquid (plasma) consisting of water, salts and many other dissolved substances such as food and wastes.

The heart is a pump which pushes the blood around our bodies. The human heart is a muscular organ consisting of four chambers, as shown in Figure 1.30. It is vertically divided into two halves so that the blood on the left side does not mix with the blood on the right side.

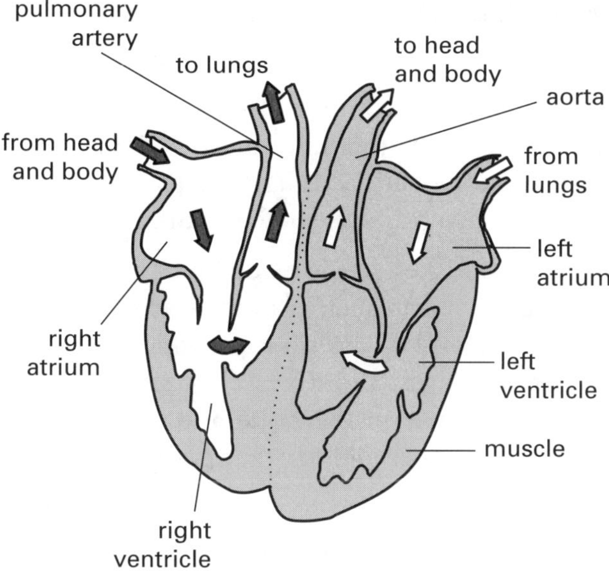

Figure 1.30 The human heart

Blood, having circulated around certain body parts, enters the right atrium of the heart. This blood is deoxygenated. That is, it has given up its oxygen load to the various cells of the body. From here it is pumped to the right ventricle. When the right ventricle contracts, it forces blood out through the pulmonary artery towards the lungs where it will be oxygenated (replenished with oxygen). This oxygenated blood returns to the left atrium through the pulmonary vein. It then passes to the left ventricle from where it is pumped out into the aorta to carry oxygenated blood once again to body cells. Deoxygenated blood returns to the heart in veins.

Human blood circulation is sometimes referred to as double circulation. The blood passes through the heart twice in going once around the circuit. The right side is concerned with pumping deoxygenated blood from the body to the lungs. The left side is concerned with pumping oxygenated blood from the lungs to the body.

Blood moves around our bodies in tubes of different sizes.

- Blood moving away from the heart is carried in tubes called arteries. Arteries have thick muscular walls.
- Blood moving towards the heart is carried in tubes called veins. The walls of veins are thinner and less muscular than arteries.

Around our cells are very delicate networks of capillaries (see Figure 1.31). These are very fine tubes which present a large surface area for the exchange of materials by diffusion. Some of the capillaries are so narrow that red blood cells can only pass through them in single file. At either end of capillary beds are larger tubes, and behind them even larger ones, eventually joining into the arteries and veins.

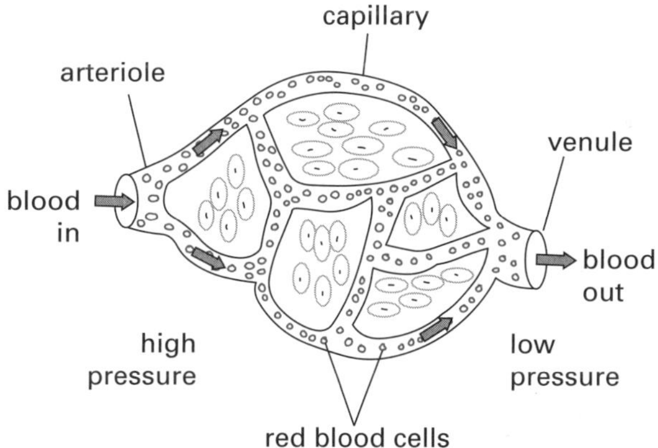

Figure 1.31 A capillary bed

1.11 The excretory system

Through normal functioning, wastes are constantly produced in our cells. These wastes need to be eliminated, otherwise the build-up would slow us down and eventually poison us. The removal of waste products produced by cell metabolism (reactions in the cells) is called excretion. Leftovers from the digestive system are not excreted products since they were never absorbed into the cells of the body.

We excrete carbon dioxide from our lungs. This is a waste product from cellular metabolism. Carbon dioxide diffuses out from our blood stream and into the air sacs of our lungs before being breathed out.

We also excrete some waste products through our skin. Our sweat glands secrete water, minerals salts and some glucose. Elimination of wastes this way is not the skin's primary function.

The major excretory organs in the human body are our two kidneys (see Figure 1.32). The kidneys continually filter blood plasma removing excess water, ions and urea. Urea is a waste product that is formed when the liver has broken down the excess amino acids (the building blocks of proteins) our bodies are unable to store. This waste product coming from our kidneys is called urine.

The kidneys are located just to the sides of the small of the back and look like beans. The urine filtered from the blood stream flows down the ureter into the urinary bladder. Here it is temporarily stored until the bladder is full. Then a message is sent to your brain for you to relax a sphincter muscle allowing urine to pass out through the urethra.

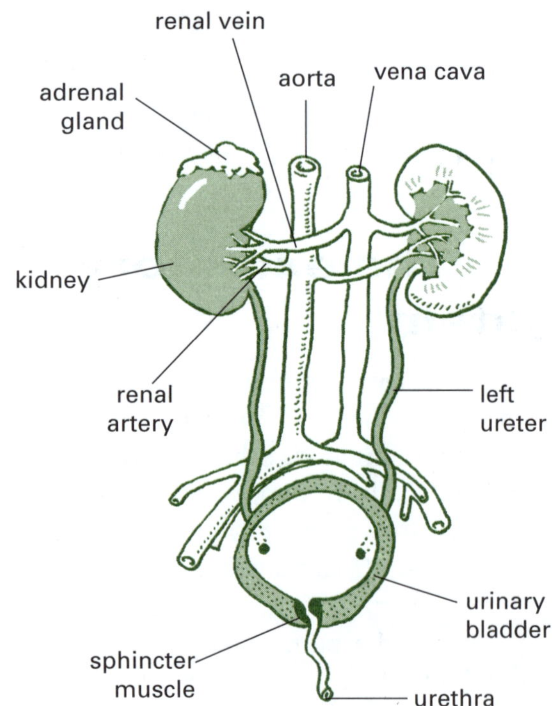

Figure 1.32 The human urinary system

Because our kidneys are continuously functioning, they help maintain a fairly constant composition of the concentrations of various substances in our blood stream.

1.12 The skeletal system

In humans, movement is usually achieved through the combined effects of specialised tissues: the skeleton and muscles. The skeleton also supports our body and helps give us our shape. It also protects some organs.

Most of our skeleton is made up of bone. There are smaller parts, such as at the end of the nose and at the end of the rib cage, which are made up of cartilage. Blood vessels pass through a series of canals in the bone and connect cells with one another and to other parts of the body.

Muscles are attached to our bones by tendons. When muscles contract they cause part of the body to move. When the biceps muscle contracts, it raises our arm. At the same time the triceps muscle relaxes, as shown in Figure 1.33. In order to lower your arm, you contract the triceps and relax the biceps.

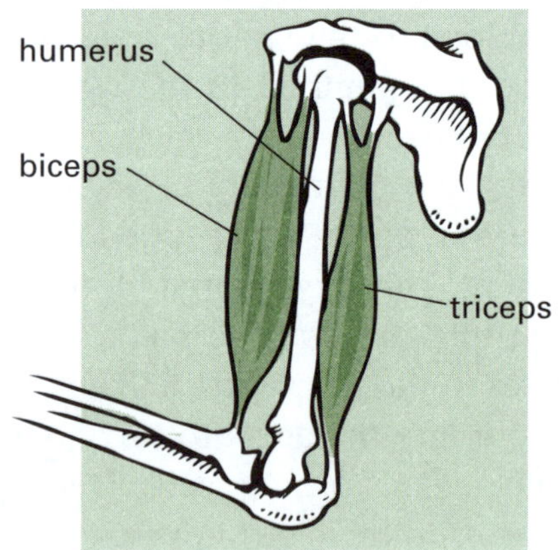

Figure 1.33 Muscles in the upper arm

The various parts of our skeleton also have other functions. (See also Figure 1.34.)

- The vertebral column consists of a number of bones separated from each other by a disc of cartilage forming a protective hollow tube through which the spinal cord passes.

- The rib cage of curved bones protects the delicate lungs and heart, and allows breathing.
- The skull consists of a number of bones fused together and protects the delicate organ of the brain.
- The limbs of mammals are four in number: two forelimbs and two hindlimbs. Humans walk upright on hindlimbs. This frees up the forelimbs for other functions.

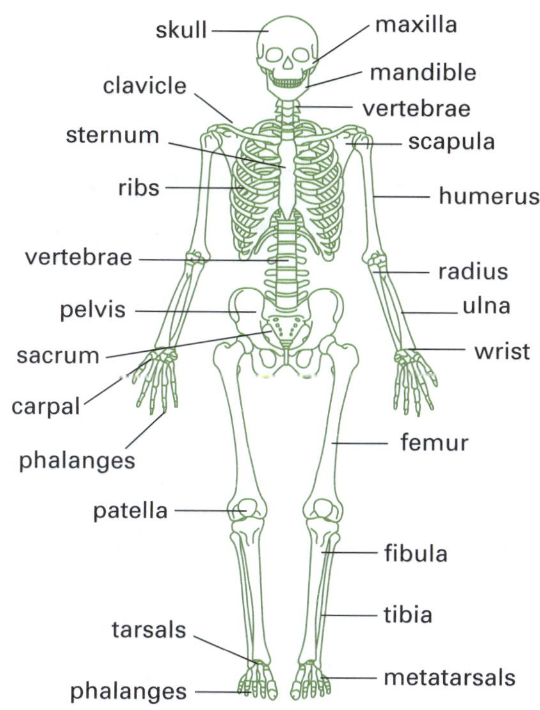

Figure 1.34 The human skeleton

1.13 The human respiratory system

The main function of the respiratory system is to exchange waste carbon dioxide produced by the cells, for oxygen. Carbon dioxide diffuses out of the blood stream and through the lining of the lungs to the outside, while oxygen diffuses in. Since diffusion is a relatively slow process, having a large surface area speeds up gaseous exchange.

Each lung contains several hundred million grape-like structures called alveoli. These are basically very small air pockets in contact with numerous fine blood capillaries to allow the diffusion of gases (see Figure 1.35).

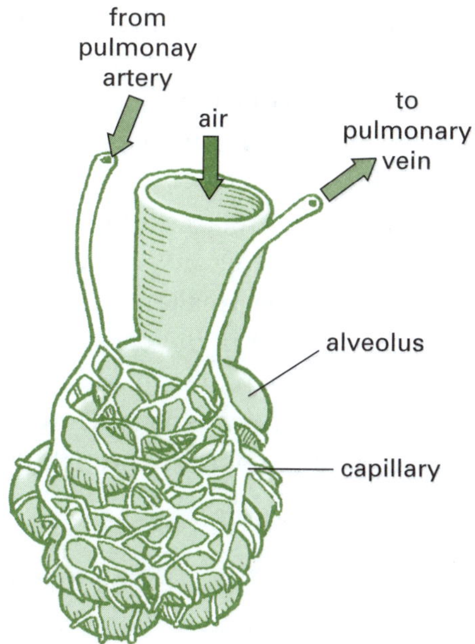

Figure 1.35 Blood vessels and alveoli

A thin sheet of muscle called the diaphragm and other muscles attached to the rib cage allow breathing (see Figure 1.36). These contract when we breathe in and relax when we breathe out. The sudden change in air pressure forces air into and out of our chests.

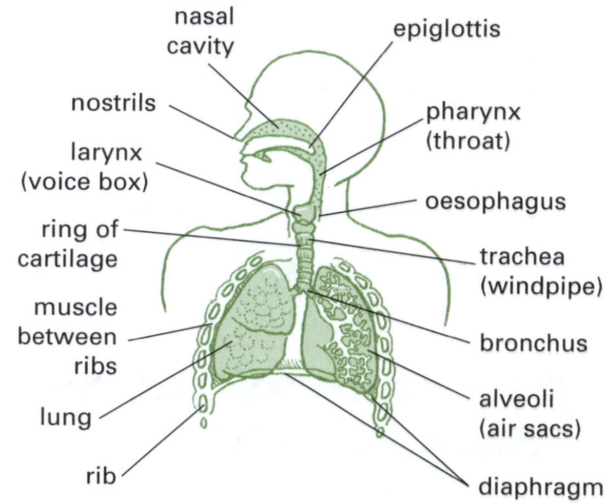

Figure 1.36 The human respiratory system

The sequence for inhaling and exhaling is as follows.

- Inhaling
  - Diaphragm contracts down and rib cage moves up and outwards—the chest (thoracic) cavity is now larger.

- Air pressure in the chest cavity drops and is lower than outside air pressure.
- Lungs expand and air is drawn in from the outside through the nose.

Exhaling

- Diaphragm relaxes upwards and rib cage moves down and inwards—the chest cavity is now smaller.
- Air pressure in the chest cavity rises and is higher than the outside air pressure.
- Lungs contract and air is forced out of the lungs to the outside through the nose.

Respiration at the cellular level

Animal and plant cells use molecules like glucose to supply their energy needs. In the cytoplasm of the plant or animal cell, some energy can be obtained by splitting the glucose in two. Oxygen is not required for this first step. Most energy is generated in the cell by aerobic respiration which requires oxygen. Where does this process happen in the cells of a plant or animal? Scientists have discovered another important structure in cells called a mitochondrion. It is a miniature powerhouse where chemical energy is converted into other useful forms. Muscle cells (especially heart muscle cells) are packed with thousands of mitochondria. In sperm cells, the mitochondria are clustered near the tail of the sperm where the energy is required. Cellular respiration can be summarised using a word equation.

glucose + oxygen → carbon dioxide + water + usable energy

1.14 The respiratory system in fish

Fish live in fresh water, salt water and estuaries (where fresh and salt water meet), and some even on land (fish such as the mudskipper can survive on land for a period). As the majority of fish live in water, they breathe through very efficient gills which are slit-like structures on the front side region. A fish draws in water, containing dissolved oxygen, by closing the lid over its gills and opening its mouth. When the fish closes its mouth and opens the gill lid, the water is forced out and over the respiratory surfaces of the gill filaments (see Figure 1.37). Through diffusion, gases in the water enter the blood stream of the fish, with the oxygen spreading throughout its body. Carbon dioxide diffuses out.

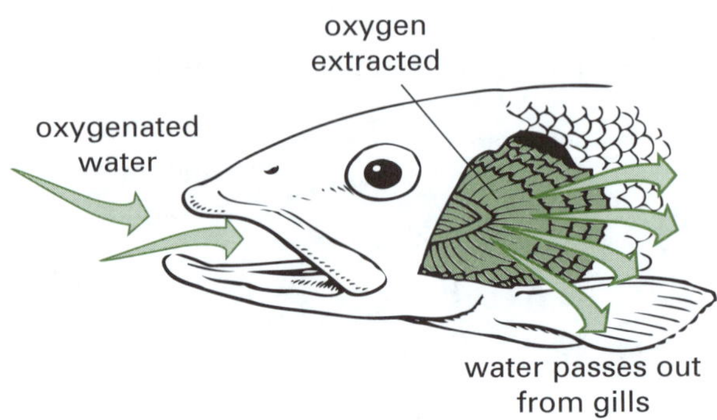

Figure 1.37 Water flows in through the mouth and out over the gills.

Gills have a large surface area for gaseous exchange. Since water contains much less oxygen than air, a fish must have an organ with a large surface area in order to absorb enough oxygen from the water to survive. Between the water and the bloodstream in a gill is a very thin membrane between 1 and 3 microns thick (1 micron = 1 μm = 1-thousandth of a millimetre). This short diffusion distance increases the rate of oxygen entry into the blood.

A single gill consists of a curved arch bearing a double row of gill filaments as shown in Figure 1.38. Each filament has many tiny folds in its surface, increasing the amount of surface area along a given length. The folds of the filament are packed so closely that most of the water passing between them is involved in the gas exchange process. Also, the blood in the filament folds travels forwards, in the opposite direction to the water flow. This constant imbalance allows optimal passage of oxygen to the blood.

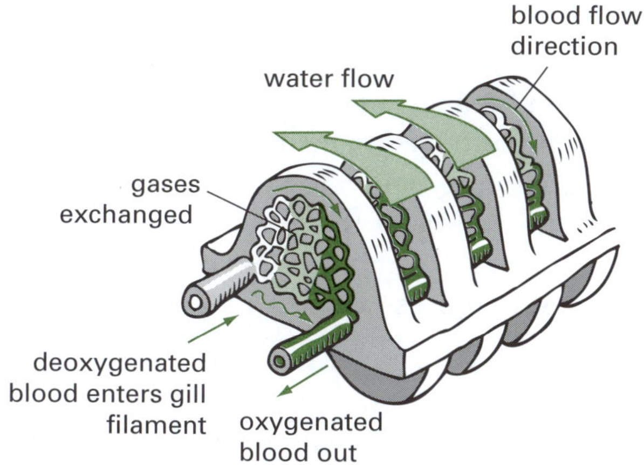

Figure 1.38 Gaseous exchange in gill filament

Water flows continuously in only one direction over the gills. Contrast this with the interrupted, two-way flow of air in and out of the lungs of mammals.

1.15 Asexual and sexual reproduction

Reproduction is the biological process by which new individuals are produced from their parents. It is an important feature of all known life, as each individual organism exists as the result of reproduction.

Asexual reproduction

Asexual reproduction is a method by which offspring arise from a single parent, and inherit the genes of that parent only. It is the main form of reproduction for single-celled organisms such as bacteria and some protists. Many plants and fungi reproduce asexually as well.

There are several ways these organisms can do this.

- Binary fission. The parent organism literally divides in two, producing two daughter organisms (see Figure 1.39).
- Budding. Some cells split off, resulting in a 'mother' and 'daughter' cell. The offspring is smaller than the parent (see Figure 1.40).
- Spore formation. Some plants and many algae form spores that grow into multicellular individuals without fertilisation (see Figure 1.41).
- Vegetative reproduction. This is usually found in plants where new individuals are formed without producing seeds or spores. Some plants reproduce by forming bulbs or tubers, or forming adventitious shoots and suckers along their lateral roots (see Figure 1.42).
- Fragmentation. A new organism grows from a fragment of the parent. This process is used to artificially propagate many plants using cuttings and grafting (see Figure 1.43).
- Parthenogenesis. An unfertilised egg develops into a new individual.

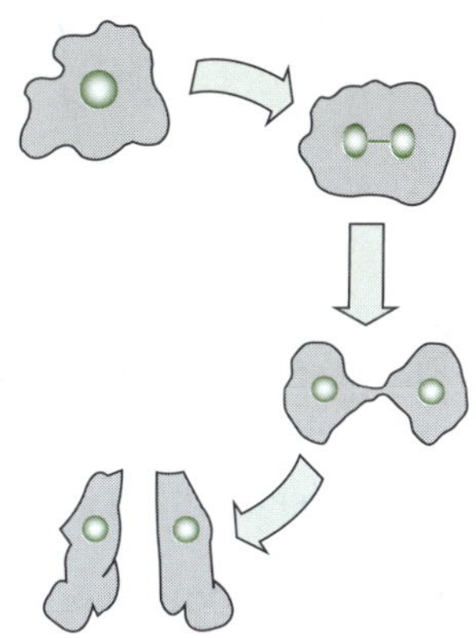

Figure 1.39 Binary fission in the amoeba, a single-celled organism, forming two daughter cells

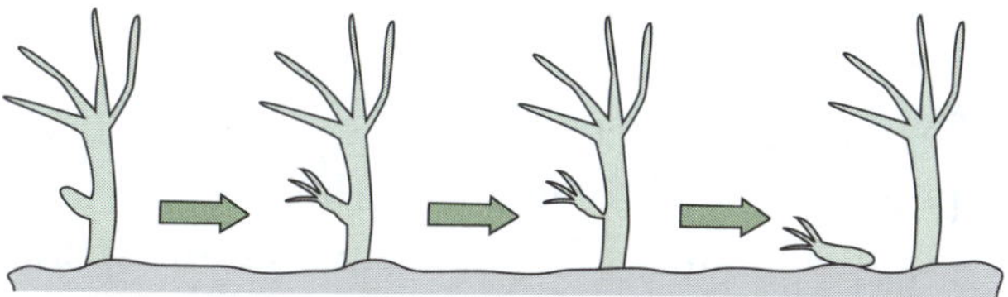

Figure 1.40 Hydra is a simple animal that lives in freshwater ponds and streams. It divides by budding.

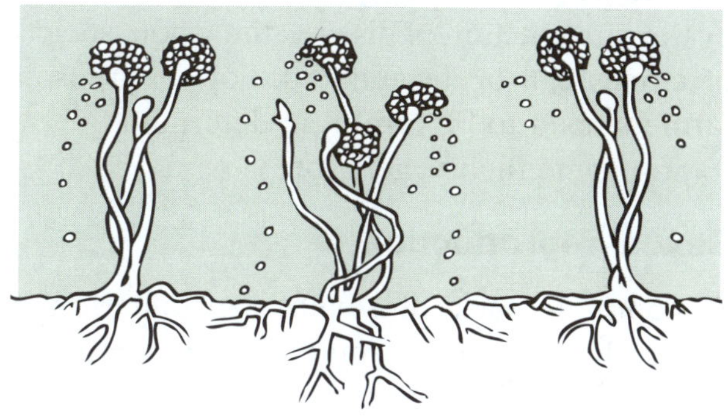

Figure 1.41 Spore formation in mould

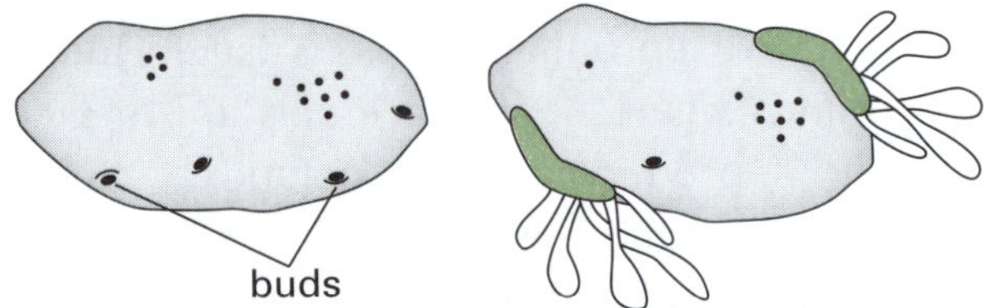

Figure 1.42 The many buds on a potato tuber act as organs for vegetative reproduction.

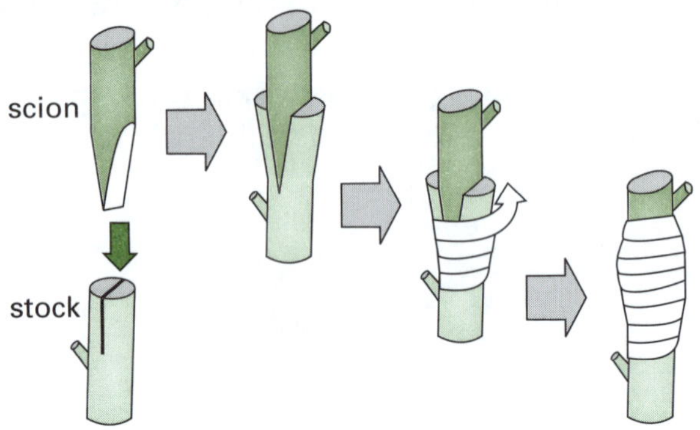

Figure 1.43 Reworking established citrus trees to a different variety by grafting is sometimes necessary. The new tree grafts should make good growth because of the established root system.

The 'offspring' from asexual reproduction are direct copies of the original, with no input from another individual as in sexual reproduction. This is because the new organisms are produced by ordinary cell division so no variation is introduced, unlike with sexual reproduction.

Asexual reproduction can create individuals rapidly and in large numbers. This is important, especially in times of dryness, since motile sperm require water to fertilise the egg. Further, plants with the desired characteristics can be cloned for economic reasons in agriculture. However, if something goes wrong, such as a fatal mutation or the introduction of disease, the whole society of clones can be terminated. For this reason, farmers need to be careful in determining how to propagate their vegetation.

Sexual reproduction

A complete lack of sexual reproduction is relatively rare among multicellular organisms, particularly animals. Sexual reproduction is the primary method of reproduction for the vast majority of multicellular organisms, including almost all animals and plants.

Animals

Animals typically produce male gametes called sperm, and female gametes called eggs or ova. In sexual reproduction the two distinct gametes fuse during fertilisation to form a fertilised egg. The fertilised egg is the zygote which divides many times to form an embryo. The embryo becomes a foetus when all the organs are formed. The foetus then grows until the end of the pregnancy.

Through sexual reproduction there is:

- greater variation within a species.
- greater chance that genetically stronger individuals will be produced
- less chance of harmful mutations in the species.

Humans

In humans, like other animals, the sexes are separate. The two sexes need to copulate so that sperm is transferred from the male to the female. Figure 1.44 shows the structure of the male and female reproductive systems in humans. The role of each component is described in Tables 1.6 and 1.7.

Fertilisation and pregnancy

The major events in the fertilisation of an egg and the development of a baby are as follows (see also Figure 1.45).

- Ovulation is the process of an egg being released from an ovary. This occurs about every 28 days.
- The egg is transported through a fallopian tube (also known as the oviduct) towards the uterus. If the egg is not fertilised by a sperm, it will pass out of the body in the menstrual blood flow produced by the breakdown of the thickened lining of the uterus. This blood flow lasts about 4 to 5 days. The uterine lining will regrow in the next cycle.
- If the egg is fertilised by one of the millions of sperm present in the fallopian tube, it forms the zygote. The zygote will begin to divide into a ball of cells. This ball of cells takes 8 days to travel into the uterus where

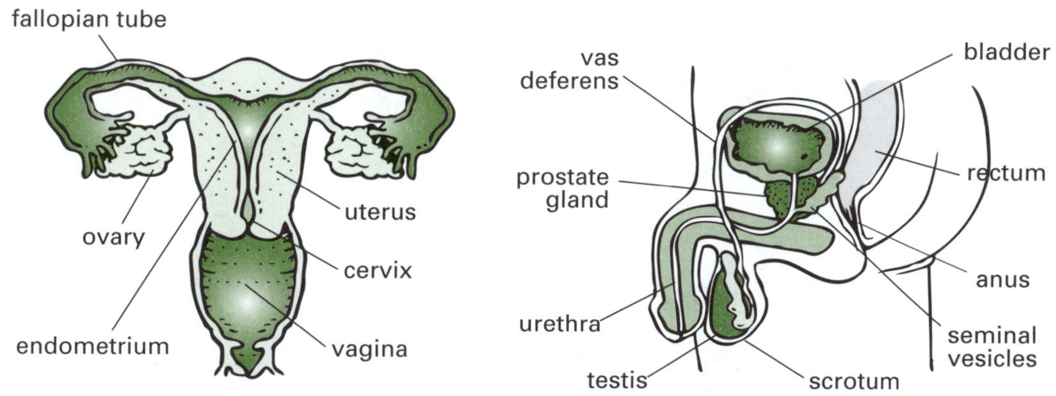

Figure 1.44 Front view of the female reproductive system, and side view of the male reproductive system

Table 1.6 Male reproductive system

Structures	Function
scrotum	external sac holding the testes; helps to regulate the temperature of the testes
testes	site of sperm production; glands that produce male hormones such as testosterone
vas deferens (sperm duct)	tube that carries sperm away from the testes
glands: prostate, seminal vesicles, Cowper's gland	produce protective and nutrient fluids for the sperm; the combination of sperm and these fluids is called semen, which is a slightly alkaline (basic) suspension of sperm cells
urethra	tube that carries semen during intercourse; also discharges urine from the bladder
penis	organ used to deposit semen in the vagina of the female during intercourse

Table 1.7 Female reproductive system

Structures	Function
ovaries	site of egg production; glands which produce female hormones such as oestrogen and progesterone
fallopian tubes (oviducts)	tubes through which an egg moves following ovulation; site of egg fertilisation if sperm is present in the reproductive tract
uterus (womb)	muscular organ in which a fertilised egg will implant and grow to produce a baby
vagina	canal in which semen is deposited during intercourse; birth canal through which a baby is born
cervix	narrow entryway between the vagina and the uterus; its muscles are flexible so that it can expand to let a baby pass through during birth
endometrium	the mucous membrane that lines the uterus; thickens under hormonal control and, if pregnancy does not occur, is shed in menstruation; the site where fertilised eggs are implanted

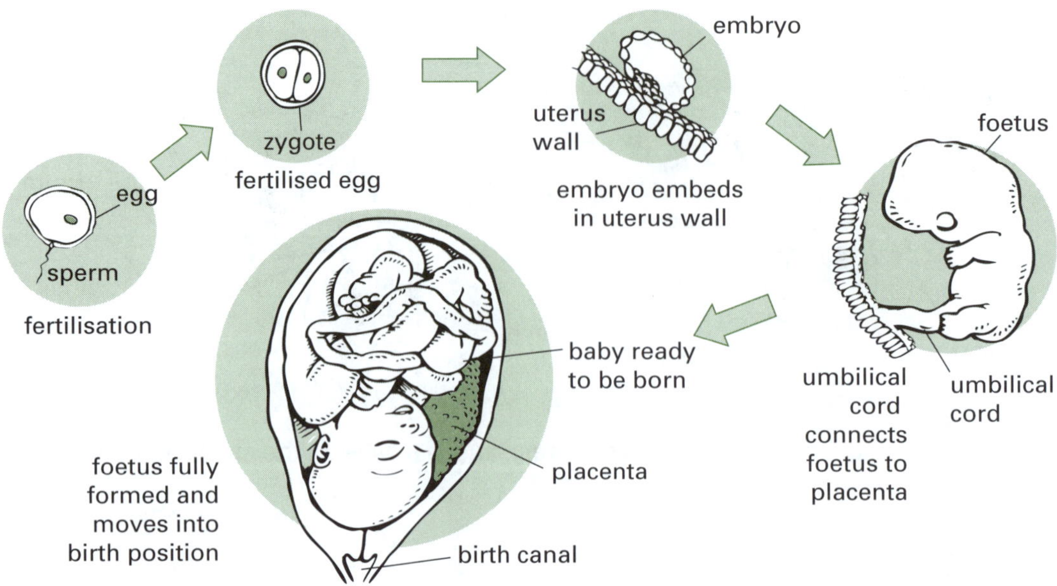

Figure 1.45 Stages during pregnancy in a human female

it becomes implanted in the uterine wall. The pregnancy or gestation period has begun.

- The developing embryo is connected to its mother by a special tissue called the placenta. The umbilical cord is an extension of the placenta. Nutrients and oxygen pass across the placenta to the developing embryo. Wastes pass the other way and are eliminated by the mother. The blood of the baby and of the mother never mix.
- For the first 8 weeks the embryo's organs grow.
- From 8 weeks until its birth the developing baby (or foetus) grows and matures.
- When the baby is mature it is expelled from the uterus by strong muscular contractions. It passes through the vagina (birth canal) and is born. This normally occurs 270 days after fertilisation.

Flowering plants

The flower is the reproductive organ of a flowering plant. Throughout the centuries flower breeders have produced plants with large and fragrant flowers, different colours and exotic shapes. There is a very large range of different kinds of flowers. But there are many plants, especially grasses and shrubs, which have inconspicuous flowers.

Plants have colourful petals and fragrances to attract insects, especially bees and flies, and birds. While bees are foraging for nectar inside the flower, they rub themselves along the anther of the male organ, which is called the stamen. The anther contains pollen. The bee may then fly to another flower of the same species in its search for more food and transfer this pollen to the sticky tip of the stigma. As Figure 1.46 shows, the flower is the reproductive organ of a flowering plant.

Other plants, without showy flowers, use wind to transfer pollen from one plant to another.

Pollination occurs when pollen grains are transferred to the sticky stigma of the pistil (female reproductive structure). Here it germinates and grows a long pollen tube down the style. While this pollen tube is growing, sex cells travel down the tube behind the tube nucleus. Eventually the tube reaches the ovary and the sperm cells are released into the ovule to fertilise it. The fusion of the two gametes results in the formation of a zygote. As the zygote develops into an embryo, another cell develops into the endosperm, which serves as the embryo's food supply. The ovary now becomes the fruit and the ovule becomes the seed. When the seed germinates, it grows into a new plant.

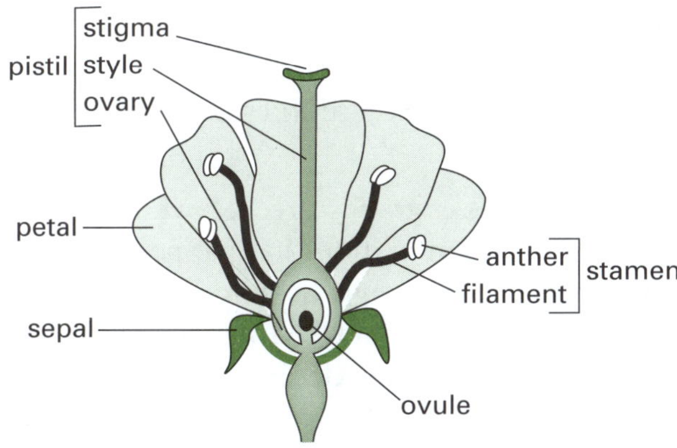

Male
Stamens: anther/filament
Anther: contains pollen

Female
Ovary: contains ovules
Stigma: pollen sticks to its surface
Style: supports the stigma

Sepals: protect flower before it opens
Petals: colourful to attract insects/birds to aid in pollination

Figure 1.46 The flower is the reproductive organ of a flowering plant.

Most plants (over 85%) have complete flowers that have both male and female parts. Others have only male or female parts, and still other plants have flowers on the same plant that are a mix of male and female flowers.

Experiment 3

Seed germination

Aim

- To investigate how the presence or absence of water affects seed germination

Method

1. Choose 10 seeds (such as wheat, mustard or cress) to germinate.
2. Investigate whether the seeds will germinate in the presence or absence of water.
3. Conduct your germination experiments in wide-mouthed test tubes or in Petri dishes with lids.
4. A bed of cotton wool is a good base for the seeds.
5. You will need to control the temperature and the amount of sunlight.
6. Record your observations over a number of days.

Results

The following observations were made.

1. In the presence of water, the seed coat cracks open and the root emerges. The root begins to grow in length and branching rootlets appear. The shoot then develops and the first leaves unfurl.
2. In the absence of water, the seeds do not germinate.

Conclusion

Write a suitable conclusion.

Go to p. 184 to check your answer.

1.16 Cloning in agriculture

While cloning sounds like an artificial process devised by humans, many plants clone themselves naturally to reproduce. They send out a small shoot-like structure, called a runner, along or just under the soil surface. This grows into a new separate plant that is genetically identical (a clone) to the original plant (see Figure 1.47). Mondo grass is a versatile plant used as a border or as a lawn in areas where there is no traffic. It spreads by producing runners.

Figure 1.47 Most strawberry plants spread by runners from a mother plant.

Vegetative propagation is just cloning where cuttings of the plant, such as a twig or stem, are taken and planted. In horticulture (the industry and science of plant cultivation) cloning is used to grow plants with specific qualities such as height, flower colour, crop yield, taste, texture or resistance to pathogens. It can also be used to propagate rare plants.

Tissue culture

In order to do this, horticulturalists use a more complex method called tissue culture (micropropagation). A small piece of the desired plant such as a bud, node, leaf segment or root segment is grown in a test tube or Petri dish on a nutrient culture medium. [This culture medium contains a gel (agar) with the proper mixture of nutrients, sugars, vitamins and hormones, causing the plant to grow rapidly.] It is then chemically treated to produce shoots. Separating the buds allows more shoots to grow. The shoots can be coaxed to grow roots so that whole plants develop. Plants with the desired characteristics can then be placed in soil and grown on farms.

Tissue culture is used for rapid multiplication of plants, and can produce up to 1 million new plantlets of some varieties in one year. The steps in tissue culture are shown in Figure 1.48.

All the plants produced in this way are genetically identical because they have all come from the same plant initially and so share that plant's genetic make-up.

Example 1: Seedless grapes

Nearly all grapevines in production today produce seedless grapes. Most fruits today do not come from seeds; they come from cuttings instead. This is true of grapes, blueberries, apples and cherries. Scientists are still working on citrus fruits.

The first seedless grapes are believed to have occurred from a mutation that happened some thousands of years ago around Afghanistan. The grapevine that mutated produced grapes without any seeds. Seedless grapes are grown from buds, canes or cuttings, or by tissue culture. Genetic engineers have come up with several ways of incorporating genes into plants to make the fruits seedless. Therefore, it may soon be possible to have seedless fruit of almost any species.

In the past, most of the seedless varieties of grapes weren't cold hardy enough to withstand harsh winter conditions but, with a lot of scientific experimentation and effective breeding programs, new seedless varieties have been specially bred for these conditions. There are currently over a dozen varieties of seedless grapes.

Eliminating plant disease

Tissue culture can be used to eliminate plant diseases. A new plant can be produced from a plant with partially infected tissue. Particular plants can be cleaned from viral or other infections.

Tissue culture allows the production of plants in sterile containers that can be moved with greatly reduced chances of transmitting diseases, pests and pathogens. It also enables cold storage of large numbers of viable plants in a small space.

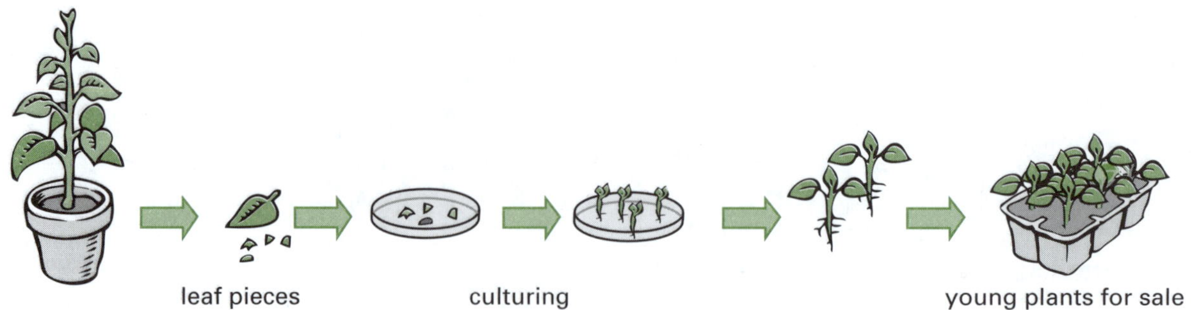

Figure 1.48 Steps in tissue culture

Example 2: Fungus-resistant bananas

There are a number of diseases that affect banana plants. One disease becoming more and more widespread is black sigatoka (a leaf spot disease of bananas caused by the fungus *Mycosphaerella fijiensis*). Applying massive doses of fungicides is losing its effectiveness as the fungus is becoming more resistant. Many hope that genetic engineering can offer a solution. Tissue culture can also be used to produce bananas that are free from disease.

Table 1.8 lists various advantages and disadvantages of using tissue culture.

1.17 Reproductive technologies

Reproductive technology covers all current and anticipated uses of technology in human and animal reproduction, including assisted reproductive technology and contraception.

- Contraception. This is a form of reproductive technology that enables people to control their fertility. It includes barrier methods, such as condoms or diaphragms, and hormonal contraception (oral contraception). The most common method of hormonal contraception is the 'pill'.
- Artificial insemination. Sperm is placed into the reproductive tract of a female to impregnate her. In humans, it often involves a sperm donor in cases where the male partner produces no sperm, the woman has no male partner or in instances of surrogacy. Artificial insemination is common in livestock breeding, especially for dairy cattle and pigs. Techniques developed for livestock have been adapted for use in humans.
- In-vitro fertilisation (IVF). A process where egg cells are fertilised by sperm outside the body. It is a major treatment for infertility when other methods of assisted reproductive technology have failed. The process involves hormonally controlling the ovulation, removing ova (eggs) from the woman's ovaries and fertilising them with sperm in a fluid medium. The fertilised egg (zygote) is then transferred to the patient's uterus. The first successful test-tube baby birth occurred in 1978.

Table 1.8 Advantages and disadvantages of tissue culture

Advantages	Disadvantages
• produce multiple copies of identical mature plants in a short period of time • produce plants with desired properties such as better flowers, improved odours, larger or seedless fruits, disease resistance, fewer environmental stresses or other favourable traits; greater yield can increase profits and food supply • produce plants at any time regardless of climate variations • produce plants without requiring seeds or the necessary pollinators to produce seeds (especially useful for plants having low levels of seed production, or seeds that do not readily germinate) • useful in the genetically modified organism studies; regeneration of whole plants from plant cells that have been genetically modified • process highly efficient, and a possible solution for the prevention of starvation in developing countries	• equipment can be expensive for large-scale production • the procedure can be variable depending on the species; trial and error may be required for untried species • infection may continue through a number of generations if precautions are not taken • sterile conditions need to be maintained throughout • decreased genetic variability • cloned plants are all equally susceptible to the same diseases and insects

- Somatic cell nuclear transfer. Embryos are created for research or therapeutic purposes, such as in stem cell research. It is not to create cloned human beings, but rather to harvest stem cells that can be used to study human development and to potentially treat disease.
- Reproductive cloning. This process involves transferring a nucleus from a donor adult cell to an egg that has no nucleus. Dolly, a Finn-Dorset ewe, was the first mammal to have been successfully cloned from an adult cell in 1996 (see Figure 1.49). Since then several other cloned animals were announced.
- Cryopreservation. A process where cells or whole tissues are preserved by cooling to low subzero temperatures, such as –196 °C. At these low temperatures any biological activity is effectively stopped. This is used for men who would become sterile after treatment for testicular cancer, or for men undergoing a vasectomy to still have the option of having children. There is a new technology in which a woman's eggs (oocytes) are extracted, frozen and stored. The eggs can be thawed, fertilised and transferred to the uterus as embryos, when the woman is ready to become pregnant.
- Fertility medications. Drugs to enhance reproductive fertility. In women this stimulates the release of an egg from the ovary. There are presently not many fertility medication options available for men.

Ethical considerations

Reproductive technology has raised bioethical issues, religious issues and questioned sexual and reproductive morality. Some people question whether we should be allowed to play 'god' in this manner; others may argue this is going against nature's laws. Most countries have laws in place that outline what is and isn't allowed.

A number of advanced nations today are experiencing declining populations. With more women in the workforce and greater numbers of women furthering their education, this has led to many of them delaying or deciding against having children, or to not have as many. However, in other places, such as in parts of Africa, reproduction is a cultural issue where large families are seen as a source of free labour, status and wealth.

1.18 Disease and organ transplantation

Organ transplantation is the process of transferring an organ from one body to another, or from a donor site on the patient's own body. This is done to replace the patient's damaged or absent organ. Regenerative medicine is able to create and grow organs from the patient's own cells (stem cells, or cells extracted from the failing organs).

Organs that can be transplanted include the heart, kidneys, liver, lungs, pancreas, intestine

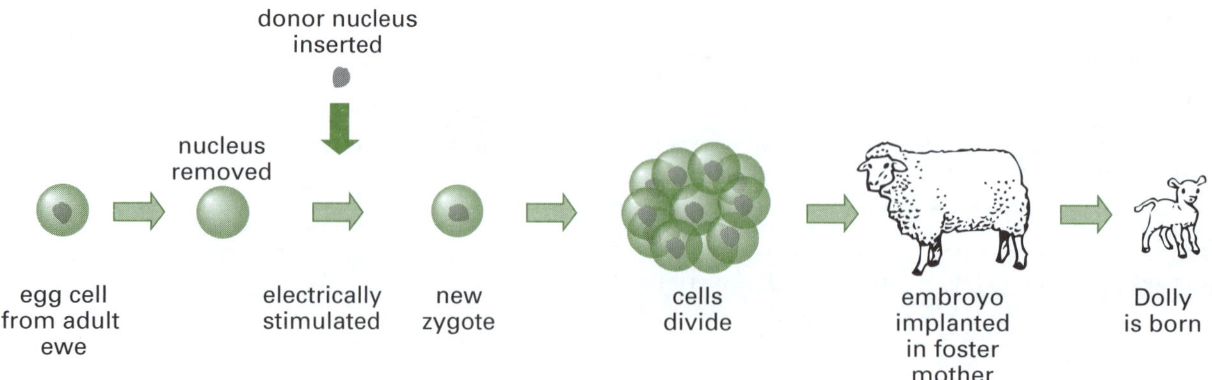

Figure 1.49 Dolly the sheep was the first mammal to be cloned.

and thymus. Tissues include bones, tendons, cornea, skin, heart valves and veins. The kidneys are the most commonly transplanted organs (see Figure 1.50), then the liver and the heart. But transplanted tissues, such as cornea, muscles and bone grafts, are far more common.

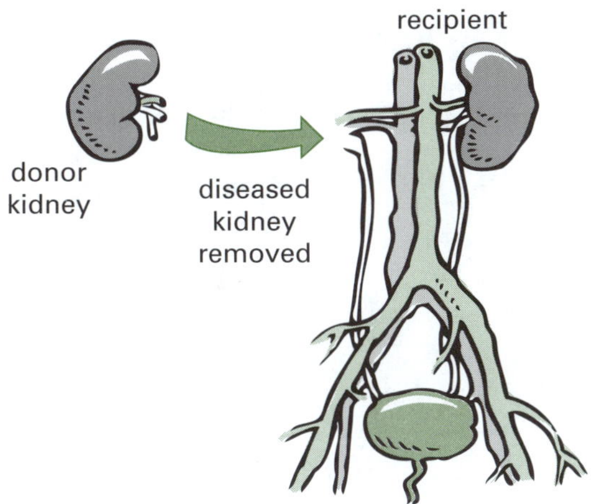

Figure 1.50 Transplanting a healthy kidney from a donor to replace the non-functioning kidney of the recipient

A problem of transferring organic matter from one person to another is that of transplant rejection. This is where the body has an immune response to the transplanted organ. This may lead to transplant failure and possibly the need to remove the organ from the recipient. Transplant rejection can be reduced by matching the donor and recipient as closely as possible before transplant and by using immunosuppressant drugs (drugs that lower the body's normal immune response).

Table 1.9 lists the year when the first successful transplant of that organ or tissue occurred.

Table 1.9 Some successful transplants

Year	First successful transplant
1905	cornea
1954	kidney
1966	pancreas
1967	liver
1967	heart
1981	heart/lung
1998	hand
2005	partial face
2008	full double arm
2010	full face

Transplanting from one part of the body to the other does not pose the problem of transplant rejection. Transplanting skin is one example, as outlined in Figure 1.51.

- The surgeon harvests the graft from a healthy part of the donor site.
- A mesher puts tiny holes evenly throughout the graft, allowing the graft to stretch and cover a larger area. The holes also allow fluid to drain from the site.
- The graft is carefully fitted to the site where it is to grow.
- After some 36 hours, the graft begins to develop new blood vessels that tether it to the underlying tissue and provide a source of nourishment.

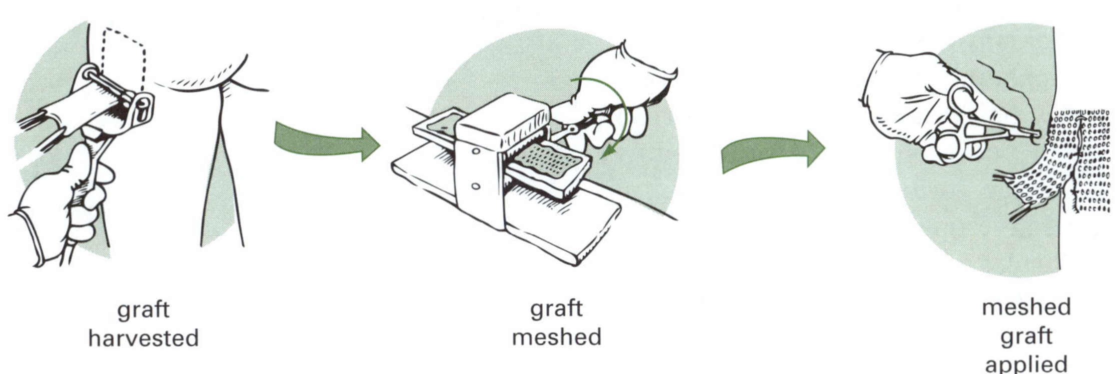

Figure 1.51 Steps in a skin graft

The main risk of skin replacement is graft failure. This may occur because:

- the graft fails to develop an adequate blood supply
- there is a poor match between donor and recipient sites
- excessive bleeding occurs during or after surgery
- an infection develops.

Another procedure is the incubation of bio-artificial organs in the laboratory using the patient's own cells. One strategy is outlined below and shown in Figure 1.52.

- Take healthy cells from a patient's diseased organ.
- Cause them to multiply in a Petri dish.
- Apply them to a collagen scaffold.
- Incubate at 37 °C (body temperature) in a bioreactor delivering oxygen and nutrients to the growing tissue.
- Implant the new functioning organ into the patient.

This new artificial organ will share the same DNA, proteins and other chemicals as the patient, thereby creating a viable donor organ that the patient's body shouldn't reject.

Ethical considerations

Transplantation raises a number of issues related to bioethics, especially given there are hundreds of thousands of people in the developed world waiting for organ transplants. This includes the definition of death, when and how consent should be given for an organ to be transplanted and whether payment should be made for transplanted organs. Other ethical issues include transplantation tourism where patients from rich countries pay large sums to have organ transplants in poor countries. More broadly concerning is the socio-economic context in which organ harvesting or transplantation may occur. A particular problem is organ trafficking, as there is a worldwide shortage of organs available for transplantation. Some people in poor countries might be tempted to sell, say, one of their kidneys.

In most countries there are strict regulations on the safety of transplants, mainly aimed at preventing the spread of contagious diseases. Regulations include criteria for donor screening and testing as well as strict regulations on the processing and distribution of tissue grafts.

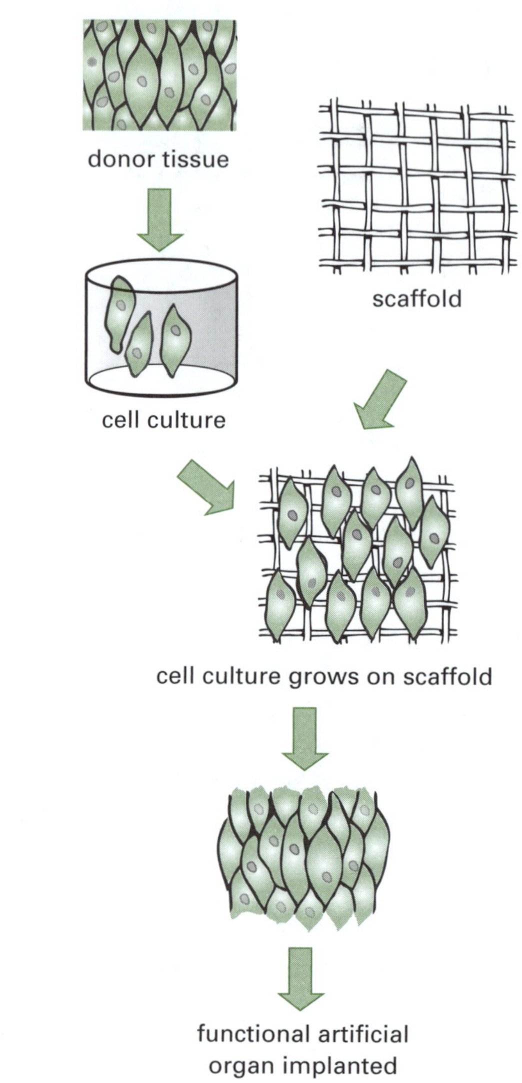

Figure 1.52 Steps in incubating 'bio-artificial' organs

Test yourself 2

Part A: Knowledge

1. Digestion in humans begins in the *(1 mark)*
   A mouth.
   B oesophagus.
   C stomach.
   D intestines.

2. Which is *not* a method used by unicellular organisms to reproduce? *(1 mark)*
   A binary fission
   B sexual reproduction
   C spores
   D budding

3. Red blood cells are made in *(1 mark)*
   A the heart.
   B the plasma.
   C bone marrow.
   D haemoglobin.

4. Each of these is a type of tooth, except *(1 mark)*
   A incisors.
   B canines.
   C molars.
   D dentines.

5. Which of the following is the correct sequence of organisation in multicellular organisms? *(1 mark)*
   A cells → tissues → organs → organ systems
   B cells → organ systems →organs → tissues
   C organ systems → organs → tissues → cells
   D organs → tissues → cells → organ systems

6. Complete the following restricted-response questions using the appropriate word. *(1 mark for each part)*
   a) The .... consists of a number of bones fused together to protect the brain.
   b) The removal of waste products produced by cell metabolism is called ....
   c) Oxygenated blood leaves the heart for the rest of the body through an artery called the ....
   d) .... occurs when pollen grains are transferred from the anther to the stigma.
   e) The .... system is responsible for taking in and processing food.

7. Use the code letters to match the terms or phrases in each column. *(1 mark for each part)*

Column 1	Column 2
A four stomach compartments	F cold temperatures
B cryopreservation	G male part
C asexual	H asexual
D stamen	I cow
E spore	J vegetative propagation

Part B: Skills

8. Figure 1.53 shows how a cutting (scion) can be taken and inserted into the stock of an apple tree to grow a new one.

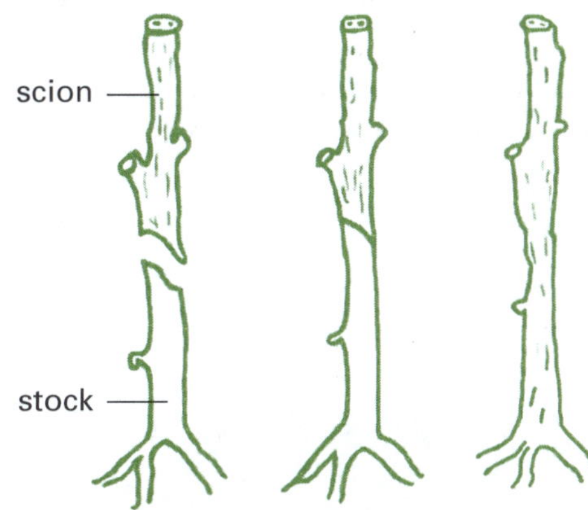

Figure 1.53 Scion

   a) Is this kind of reproduction sexual or asexual? How do you know? *(2 marks)*
   b) What is the difference between sexual and asexual reproduction? *(2 marks)*
   c) The following statements describe grafting in apple trees, but the sentences are not in order. What is the correct sequence? *(7 marks)*
   P. After the year's growth, most of the stem is removed, leaving behind the stock.

Q. As long as the scion and stock are united, and disease and drying out is prevented, the scion will grow.
R. Grafting is widely used to propagate a desired variety of shrub or tree, such as apples.
S. A twig (scion) is taken from a mature plant and inserted into a notch in the stock.
T. Apple trees are planted only for the root and stem system that grows from them.
U. The fruit the scion eventually produces will be identical to the fruit of the tree from which the scion came.
V. The scion will get all its water and minerals from the root system and stock.

9. Read the following description of a tooth, and then use it to insert the missing labels in Figure 1.54. *(6 marks)*

Low-crowned teeth are the types of teeth found in humans. This type of tooth consists of a crown above the gum line and a root embedded in the jawbone. The crown is encased in enamel and the root in cementum. Enamel is the hardest substance in the body, while cementum is calcified connective tissue. Dentin, a bone-like material, lies under the enamel and makes up most of the tooth. The pulp cavity includes blood vessels, lymphatic tissue and nerves.

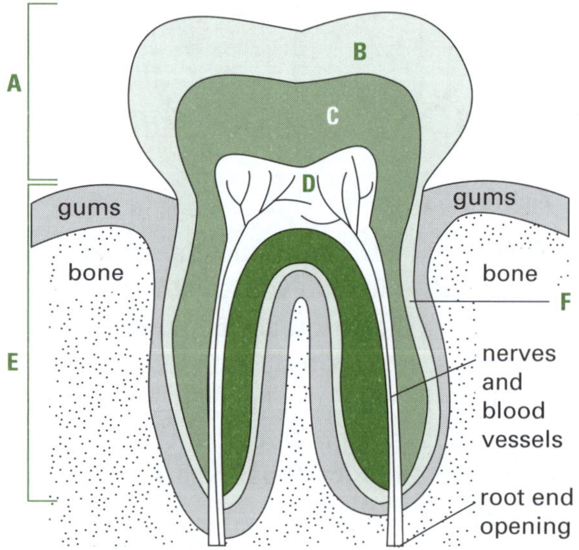

Figure 1.54 Tooth

10. Figure 1.55 shows the average volume of air that can forcibly be blown out, for both men and women, after full inspiration (filling you lungs as far as they will go). This is called the forced vital capacity (FVC).

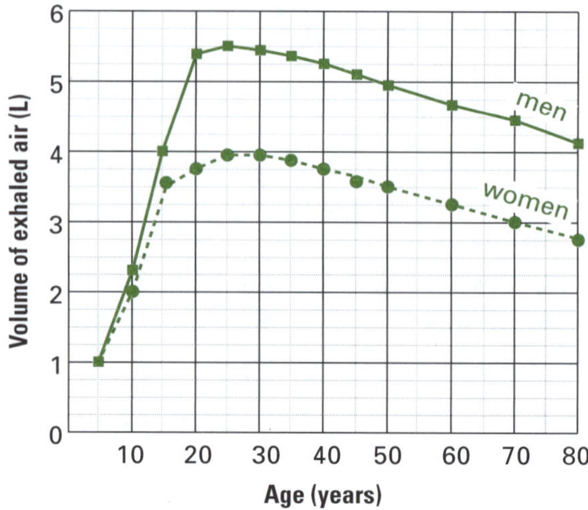

Figure 1.55 Volume of air expelled

a) What is the FVC for a 50-year-old woman? *(1 mark)*

b) At what age can a male first blow out 4.5 L of air? *(1 mark)*

c) At what two ages does a woman have a FVC of 3.6 L? *(1 mark)*

d) What is the maximum FVC for a male? At what age does it occur? *(2 marks)*

e) Suggest a reason why FVC increases up to age 25 years. *(1 mark)*

11. Figure 1.56 shows a menstrual cycle for a 19-year-old woman.

a) On what day did ovulation occur? *(1 mark)*

b) How long was her menstrual period? *(1 mark)*

c) If she has intercourse on day 8, is it likely that she can become pregnant? Explain. *(1 mark)*

d) Once ovulation occurs, it takes about 7 days for the egg to reach the uterus. If sperm are present on day 22, will she be likely to become pregnant? *(1 mark)*

e) How long is this woman's menstrual cycle? *(1 mark)*

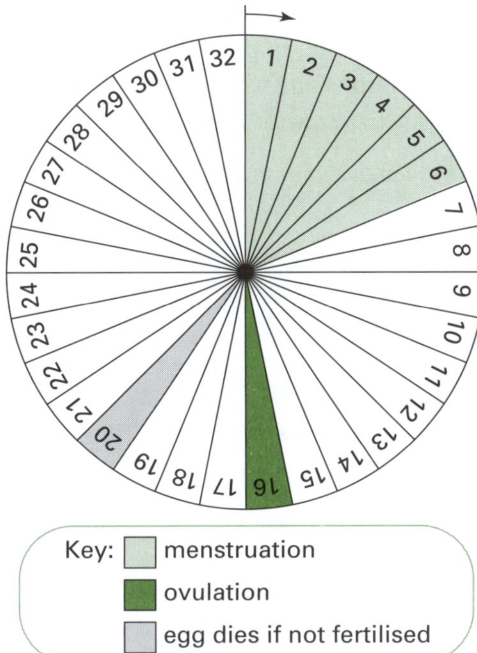

Figure 1.56 Menstrual cycle

12. Food spends different amounts of time in the various organs of the digestive system. Table 1.10 shows how long food spends in each of the organs.

Table 1.10 The length of time food spends in the organs of the digestive system

Organ	Time
mouth	2.5 minutes
stomach	4 hours
small intestine	6 hours
large intestine	20 hours

a) Draw a pie graph of the information in the table using 1 degree of the circle to represent 5 minutes of time. *(4 marks)*

b) The small intestine is much longer than the large intestine, and most nutrients are absorbed into the blood here. However, food passes through the small intestine more than three times faster than through the large intestine. Use the diagram of transverse sections of the digestive tract (Figure 1.57) to answer these questions.

i) Which diagram shows the section with the greatest inside surface area? *(1 mark)*

ii) Which of the diagrams could represent the inside surface of the large intestine? *(1 mark)*

iii) Which could be the small intestine? *(1 mark)*

iv) Which could be the mouth? *(1 mark)*

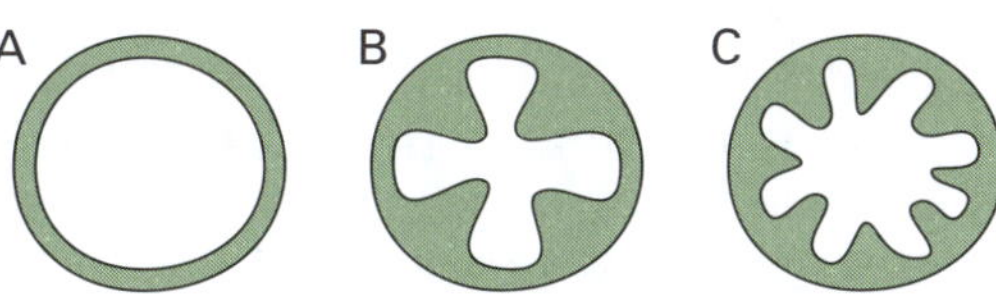

Figure 1.57 Blood vessels

13. Tim constructed a model chest from a glass bell jar, a glass T-piece and a rubber membrane as shown in Figure 1.58. He designed his model after reading some information about the pressure of gases. He read that the pressure of a gas can be increased by decreasing its volume.

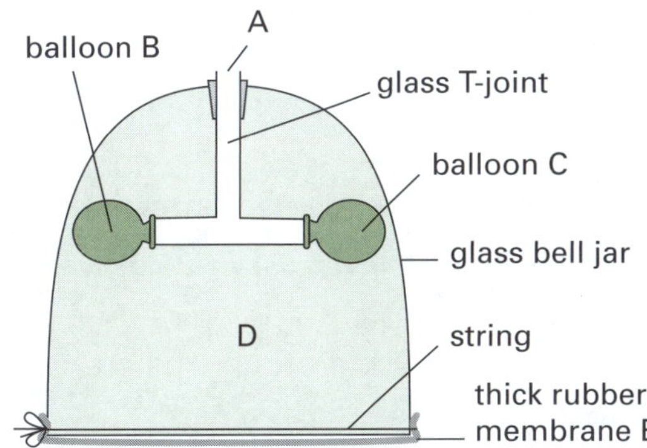

Figure 1.58 Model lung

a) When the rubber membrane is pulled downwards, the balloons inflate partially.

i) Explain what happens to the air pressure at A as the rubber membrane is pulled down. *(1 mark)*

ii) If the balloons partly inflate, is air entering or leaving the glass T-piece at A? *(1 mark)*

b) Predict what will happen if the rubber membrane is now pushed upwards into the jar. Explain your answer. *(2 marks)*

c) Which structures of the human respiratory system correspond to each of the following: the rubber membrane, the balloons, A and D? *(4 marks)*

d) What are the limitations of this model? *(3 marks)*

14. The apparatus in Figure 1.59 was constructed to compare the amounts of carbon dioxide in inhaled and exhaled air. Each flask contains a colourless solution of limewater that turns milky when it absorbs carbon dioxide. The more carbon dioxide there is, the milkier the solution becomes.

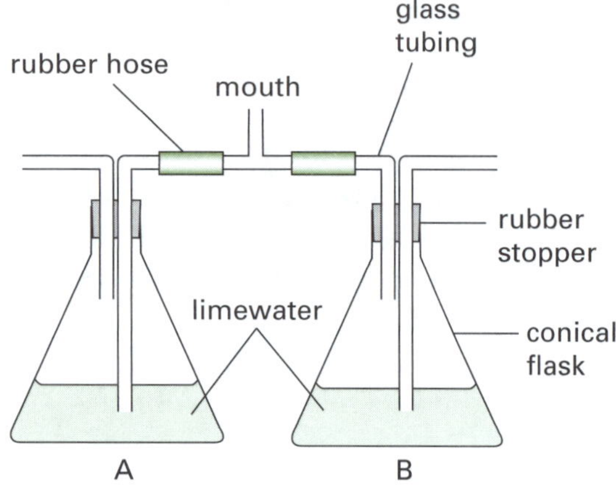

Figure 1.59 Inhaled and exhaled air

The student breathes in and out through her mouth, which is placed over the glass tube. Explain what would be observed in each flask. *(3 marks)*

Go to p. 185 to check your answers.

Summary

1. All living things are made up of one or more cells. Some organisms are unicellular.
2. Cells are the smallest units of living things. They vary in shape and have many different functions.
3. Cell membranes control the diffusion of nutrients into the cell and the removal of wastes to the external environment.
4. Light microscopes allow us to view the structures within a cell.
5. Organelles called mitochondria produce energy for the cell.
6. Photosynthesis occurs in organelles called chloroplasts which contain the pigment chlorophyll.
7. The nucleus is the control centre of the cell. The genes contain coded information and they are located on the chromosomes.
8. New cells are produced from old cells by the processes of cell division (mitosis) and cell enlargement.
9. Cell division in multicellular organisms is necessary for growth of the organism, repair of tissues and organs, and reproduction by the formation of sex cells.
10. Disease in a multicellular organism is the result of abnormal cell function.
11. Multicellular organisms consist of more than one cell where cells are specialised for different functions, carrying out extremely varied activities.
12. Multicellular organisms consist of many similar cells working together and forming groups called tissues. The tissues are specialised working groups that form part of a larger organ.

Summary

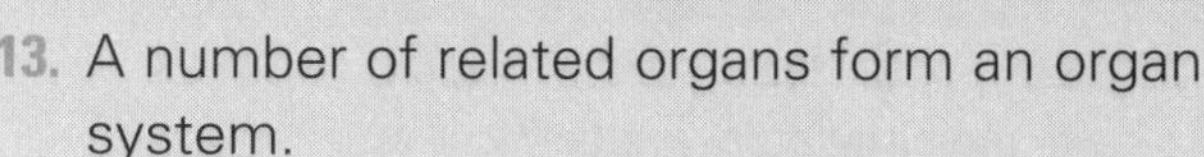

13. A number of related organs form an organ system.
14. The digestive system is the system by which ingested food is acted on physically and chemically, providing the body with nutrients it can absorb and excreting waste products.
15. The digestive systems of different organisms vary, depending on their diets.
16. The circulatory system is composed of the heart, arteries, capillaries and veins, and is responsible for moving blood throughout the body.
17. The excretory system removes excess, unnecessary or dangerous materials from an organism, maintaining a balance within the organism that prevents damage to the body.
18. The skeletal system provides structural support and protection using bones, and is supported and supplemented by ligaments, tendons, muscles and cartilage.
19. The respiratory system's function is to allow gas exchange through all parts of the body. This includes the nose, throat, larynx, trachea, bronchi and lungs.
20. Animals that live in water, such as fish, can obtain oxygen from water through gills.
21. Asexual reproduction involves only one parent and produces offspring that are genetically identical to the parent.
22. Sexual reproduction involves the union or fusion of a male and a female gamete.
23. Cloning is the process of producing populations of genetically identical individuals. This occurs in nature when organisms such as bacteria, insects or plants reproduce asexually, or in tissue culturing.
24. Tissue culturing is a process of growing a plant in the laboratory from cells rather than seeds. Growing seeds is a traditional plant-breeding method.
25. Reproductive technologies involve the application of scientific knowledge to assist in making babies, human or animal, or to prevent conception.
26. Organ transplantation is the process of transferring an organ from one body to another, or from a donor site on the patient's own body. This is done to replace the patient's damaged or absent organ.
27. Bioethics is the study of ethical issues raised by research on living beings and the applications of that research.

Syllabus checklist

Are you able to answer every syllabus question in this chapter? Tick each question as you go through the list if you are able to answer it. If you cannot answer it, turn to the appropriate page in the guide as is listed in the column to find the answer.

For a complete understanding of this topic		Page no.	✓
1	Can I explain how a light microscope can be used to view cells?	4–5	
2	Can I recall what all living things are made up of?	6	
3	Can I make statements about the size, shapes and structures of cells?	6–8	
4	Can I recall the roles of cell membranes?	6–7	

For a complete understanding of this topic		Page no.	✓
5	Can I recall the cellular role of mitochondria?	7	
6	Can I recall the location of chloroplasts and the purpose of chlorophyll in green plant cells?	8	
7	Can I recall the role of chromosomes in a cell's nucleus?	9	
8	Can I explain mitosis?	9–10	
9	Can I explain why cell division is required in a multicellular organism?	10	
10	Can I explain what causes disease in a multicellular organism?	11, 14	
11	Can I recall what cell specialisation in multicellular organisms means?	17	
12	Can I describe what tissues in a multicellular organism are and how they form organs?	17–19	
13	Can I recall what forms an organ system?	17–18	
14	Can I recall the purpose of the digestive system?	19–20	
15	Can I explain how and why the digestive systems of different organisms vary, depending on their diets?	19–22	
16	Can I recall the parts and purpose of the circulatory system?	22–23	
17	Can I recall the purpose of the excretory system?	23–24	
18	Can I recall the parts and purpose of the skeletal system?	24–25	
19	Can I recall the purpose of the respiratory system?	25–26	
20	Can I explain why animals that live in water (e.g. fish) can obtain oxygen from water through gills?	26–27	
21	Can I explain asexual reproduction?	27–28	
22	Can I recall the name of the type of reproduction that involves the union or fusion of a male and a female gamete?	28–31	
23	Can I explain cloning and give examples of where this occurs in nature?	31–32	
24	Can I recall how plant tissue culturing is achieved in the laboratory?	32	
25	Can I recall the purpose of reproductive technologies?	33–34	
26	Can I recall the steps in and purpose of organ transplantation?	34–36	
27	Can I recall the meaning of bioethics?	34	

Chapter test

Go to p. v for *Tips for tests and examinations*

Part A: Multiple-choice questions

(1 mark for each)

1. Select the correct statement concerning a common infectious disease.
   A Tinea is a fungal disease characterised by a red rash that covers most of the body.
   B Malaria and HIV are both infectious diseases caused by protozoans.
   C Measles is a viral disease that responds readily to antibiotic treatment.
   D Tuberculosis is a bacterial disease that causes the formation of lesions in the lungs.

2. Food can be prevented from spoiling by a variety of methods. One of these methods is pickling. In this example the microbes
   A cannot grow as there is insufficient water.

B are killed by the acidic juices.
C are prevented from growing by the dry conditions.
D are killed during the freezing process.

3. An example of a lifestyle disease is
A tuberculosis.
B alcoholism.
C malaria.
D cystic fibrosis.

4. A student uses a monocular microscope with the following lens magnification: eyepiece lens 15×; objective lens: 60×. Calculate the total magnification of the lens system used.
A 75×
B 225×
C 3600×
D 900×

5. Figure 1.60 shows a cross-section of two villi lining the small intestine.

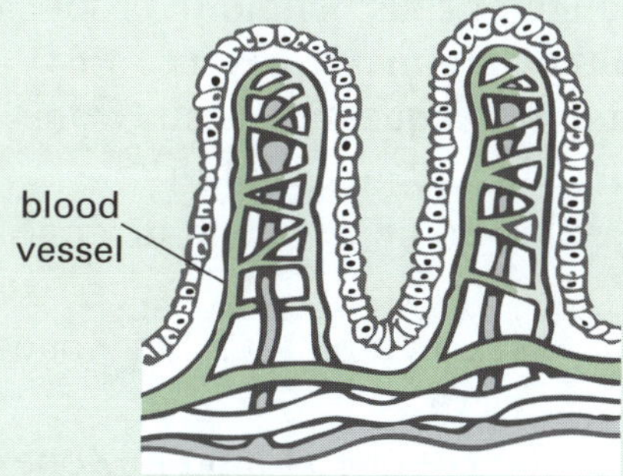

Figure 1.60 Villi cross-section

What is the function of villi?
A to digest food
B to absorb digested nutrients quickly due to a high surface area
C to assist in solid waste elimination
D to release acids to breakdown fats

6. Select the correct statement about reproduction.
A Millions of sperm reach the egg in the fallopian tube.
B One egg is produced from each ovary at the beginning of a cycle.
C Sperm are formed in the scrotum.
D The zygote is the unfertilised egg.

7. Select the *incorrect* statement about reproduction.
A Although more than one sperm reaches the egg, only one actually fertilises it.
B Fertilisation of an egg occurs in the uterus.
C The placenta allows wastes to be removed from the developing foetus.
D The vagina is sometimes called the birth canal.

8. The function of the chloroplast in green plant cells is to
A control the activities of the cell.
B carry out photosynthesis.
C support and strengthen the plant cells.
D store water.

9. The part of a microscope that is used to adjust light intensity is the
A iris.
B objective lens.
C mirror.
D eyepiece lens.

10. With reference to cells, what does the word fission refer to?
A joining of cells to form a tissue
B cell death
C division of a cell into two new cells
D cell enlargement during growth

Part B: Short-answer questions

11. A student used a monocular microscope to examine a 1 mm × 1 mm grid (on clear plastic) under low magnification (see Figure 1.61). The diagram at the left shows the field of view.

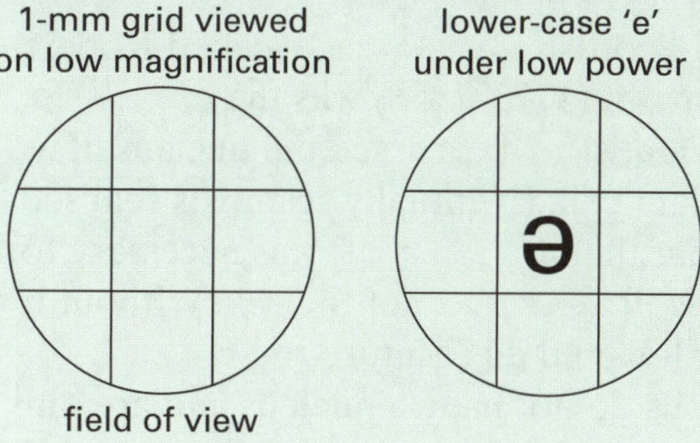

Figure 1.61 Grid under a microscope

a) Estimate the diameter of the low magnification field of view. *(1 mark)*

b) The student then examined the lower-case letter 'e' under the microscope. What do you notice about the image? *(1 mark)*

12. Using the same settings as in the previous question, the student placed a specimen on a glass slide on the grid and viewed it under low power. The field of view showed some round cells (see Figure 1.62). Estimate the diameter of one of these cells. *(1 mark)*

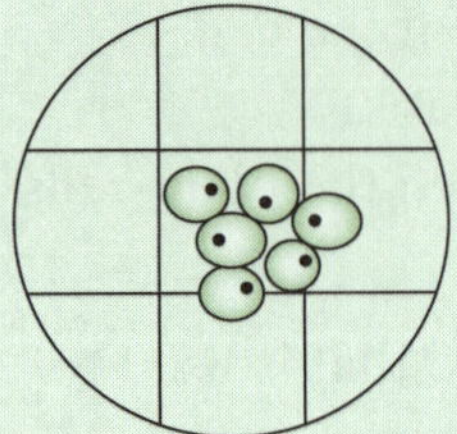

Figure 1.62 Cells

13. Name a structure in the cell for each of the following functions. *(1 mark for each part)*

a) generating energy by a process called respiration

b) storing water and other chemicals

c) controlling the entry of nutrients and removing wastes from the cell

d) containing codes that control cell processes

e) site of photosynthesis in plant cells

14. Various words have been replaced by code letters (a to g) in the following cloze passage. Use the list to identify which words match the code letters: functioning; function; first; replaced; nucleus; scientist; blood. *(1 mark for each part)*

A Scottish a).................... called Robert Brown (1773–1858) was the b).................... scientist to clearly see the nucleus of a plant cell. Eventually scientists realised that the c).................... was necessary for the d).................... of the cell. Without it the cell would die. Mature red e).................... cells do not have a nucleus and are only able to f).................... for a short time. When they die they are g).................... by new red blood cells.

15. The diagrams in Figure 1.63 show an amoeba undergoing binary fission. Which is the correct sequence to these diagrams? *(1 mark)*

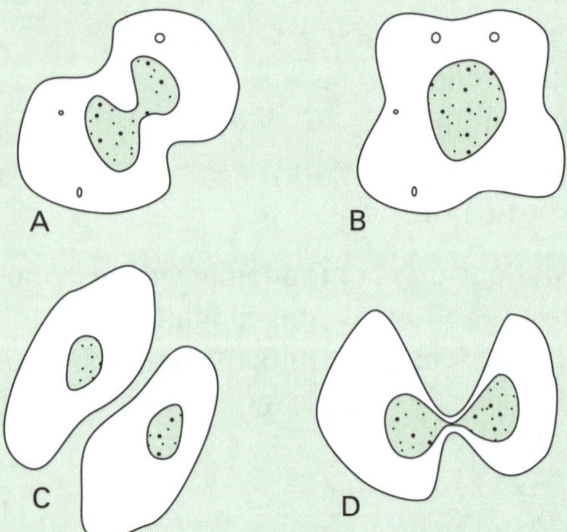

Figure 1.63 Binary fission stages

16. The root tips of plants contain actively dividing cells. The tip of the root is protected by a root cap. Following division, the new daughter cells behind the root cap grow until they are mature. Figure 1.64 shows a jumbled sequence of cells from various zones in a plant's root tip.

a) Which cell(s) could be found in Zone 1? *(1 mark)*

b) Which cell(s) could be found in Zone 2? *(1 mark)*

c) Which cell(s) could be found in Zone 3? *(1 mark)*

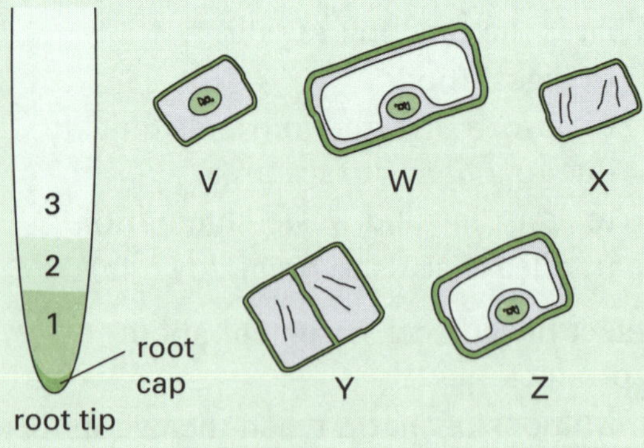

Figure 1.64 Root tip

17. An experiment was performed to assess the effectiveness of three antibiotics on the growth of bacteria in nutrient agar culture plates. The antibiotics (A, B and

C) were impregnated into three paper discs that were placed on the surface of the nutrient agar. The plate was inoculated with three different bacteria (X, Y and Z). After several days in an incubator the plate was examined and the pattern of bacterial colonies noted as shown in Figure 1.65.

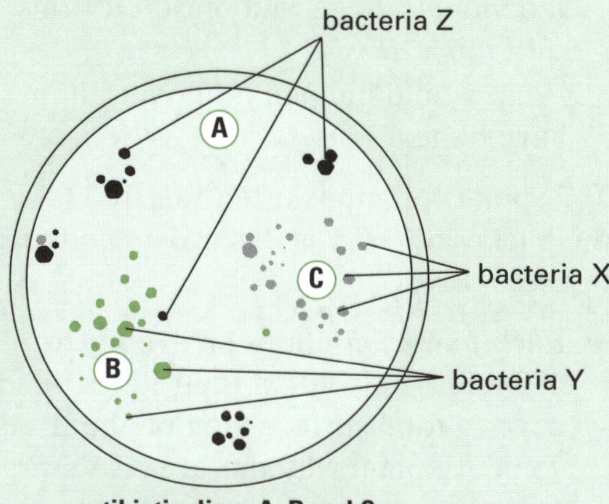

Figure 1.65 Testing antibiotic discs

Discuss the effectiveness of the three different antibiotics on the growth of the three different types of bacteria. *(3 marks)*

18. Many bacteria present in your body reproduce very quickly. Bacterial cells at 37 °C can reproduce by dividing to form two new cells every 20 minutes.

a) Starting with one bacterium at 37 °C, calculate the number of cells formed by the following times (assuming there are no cell deaths). *(3 marks)*

i) 1 hour

ii) 2 hours

iii) 3 hours

b) Explain why the Earth is not overrun with bacteria after several days. *(1 mark)*

19. Figure 1.66 shows how a tendon attaches a skeletal muscle to a bone. Sportspeople can suffer serious problems if they snap their tendons. What is the function of skeletal muscle cells? *(1 mark)*

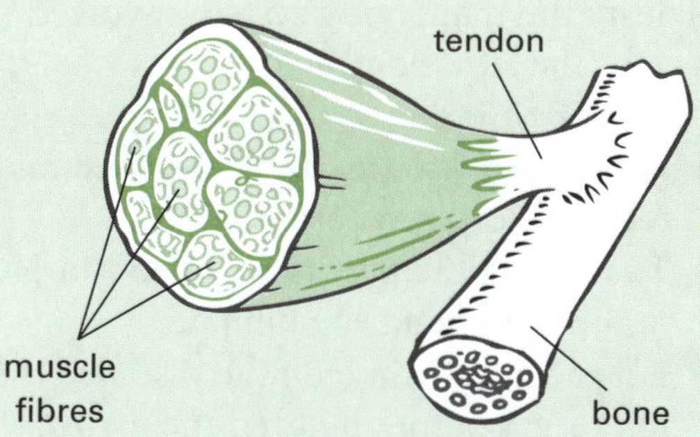

Figure 1.66 Muscle fibres

20. Figure 1.67 shows skin (epidermal) cells in the root of a plant. Some cells are called root hair cells. These cells make close contact with the soil and soil water. Suggest a likely function of root hair cells. *(1 mark)*

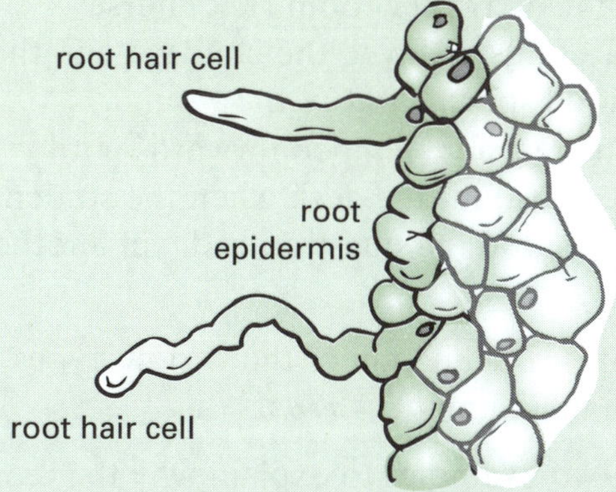

Figure 1.67 Root hair cell

21. What is an advantage of cells in a multicellular organism compared to a unicellular organism? *(1 mark)*

22. To which organ system do the organs in Figure 1.68 belong? *(1 mark for each part)*

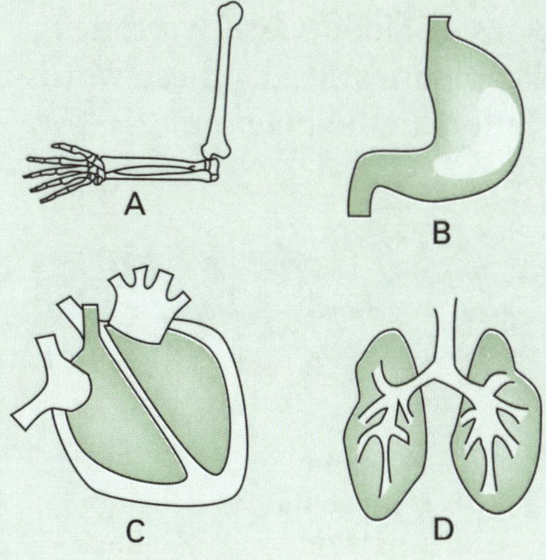

Figure 1.68 Organs

23. Identify the main body system involved in each of the situations below. *(1 mark for each part)*

  a) After the 3-km run, Jodie stopped and took some very deep breaths.
  b) The pin on the brooch stabbed Margaret in the finger and she jumped.
  c) Peter carefully judged the distance to the target before he fired the arrow.
  d) Baby Pattie ate the whole box of chocolates in 30 minutes and then vomited.
  e) After the marathon run, Stefan's legs cramped badly.
  f) The footballer was tackled, fell heavily and broke his collar bone.
  g) The emergency room (ER) nurse carefully removed the bandage and the blood spurted out of the wound.
  h) The teacher slumped forwards until her head hit the desk when the student replied to the question with yet another silly answer.

24. What is the function of the various types of teeth in a human? *(4 marks)*

25. Use two examples to explain why the teeth of animals are different from those of humans. *(2 marks)*

26. How is a cow's stomach different to a human's? Why is this? *(2 marks)*

27. Why is the human heart referred to as a double pump? *(2 marks)*

28. Figure 1.69 shows cross-sections of the various types of blood vessels: arteries, veins and capillaries. Match these words to the code letters in the diagram. *(3 marks)*

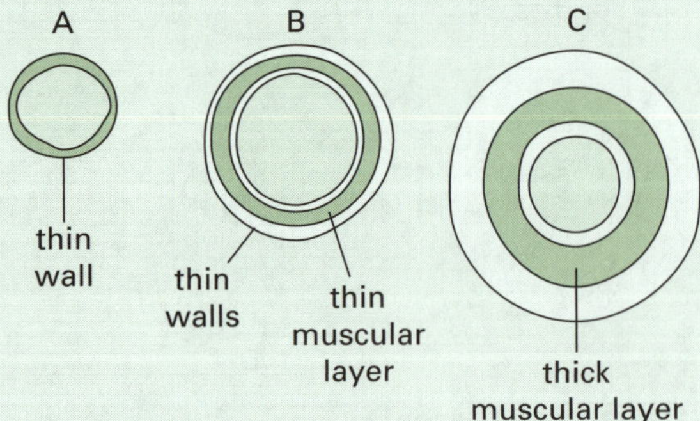

Figure 1.69 Blood vessels

29. Describe the function of our kidneys. What happens when the kidneys are not working properly? *(2 marks)*

30. a) At birth, a newborn baby has over 300 bones, but on average an adult human has only 206 bones. (Numbers can vary slightly from individual to individual.) Suggest a reason for this difference. *(1 mark)*
  b) What would happen if humans didn't have bones? *(1 mark)*

31. The spinal column is often referred to as the backbone. Why is this not a good term to use? *(1 mark)*

32. a) Each human contains an average of 480 million alveoli in their lungs, which cover a total surface area of about 75 m^2. Explain why this area is so large. *(1 mark)*
  b) What happens in the alveoli? *(1 mark)*

33. a) Mudskippers are fish that have the ability to breathe through their skin and the lining of their mouth and throat. How does this help mudskippers live on land? *(1 mark)*
  b) Another important adaptation that aids breathing while out of water is the mudskipper's enlarged gill chambers, where they can retain a bubble of air. How might this assist in living out of water? *(1 mark)*

34. Couch grass can rapidly reproduce by rhizomes (underground stems) that grow just under the soil surface. Shoots and leaves emerge at the soil surface from buds in the rhizome as shown in Figure 1.70.

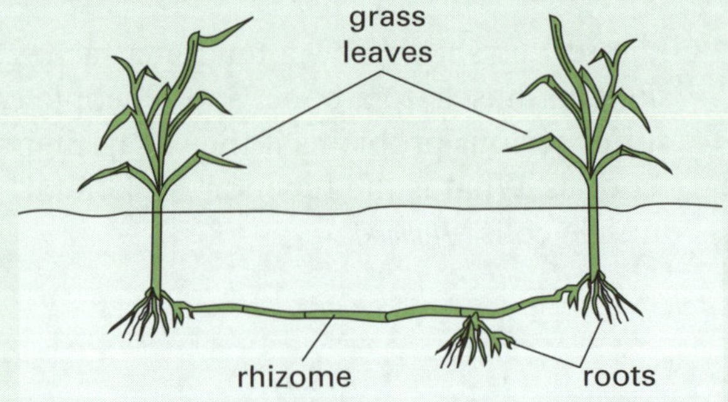

Figure 1.70 Rhizome

Identify the type of reproduction and state whether or not the new plant shoots that emerge are genetically identical to the older parts of the rhizome. *(1 mark)*

35. Figure 1.71 shows the reproductive system of many female birds.

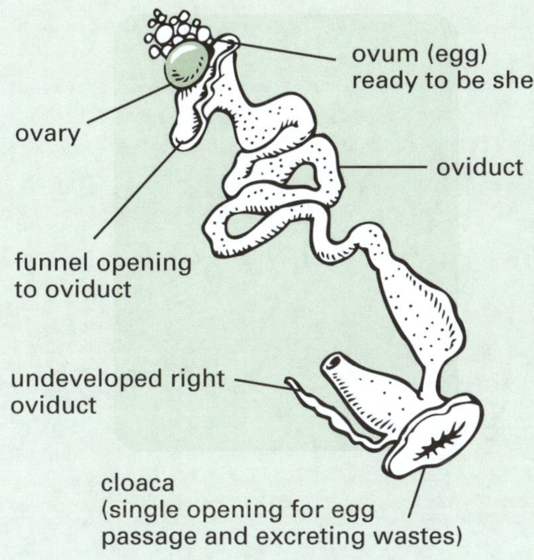

Figure 1.71 Bird reproductive system

a) Suggest a reason why the left ovary is developed but the right one remains undeveloped? *(1 mark)*

b) Birds do not possess a uterus. Why is this? *(1 mark)*

36. Buffalo grass for lawns has been developed that is up to 90% tolerant of shade.

a) Of what use is shade-tolerant buffalo grass? *(1 mark)*

b) Will this grass require more or less watering than buffalo grass grown in full sun? Explain. *(1 mark)*

37. A biologist was interested in finding out about cell membranes. He was researching whether certain substances would pass through the membranes more easily and faster than others. He devised two hypotheses:

- A: If smaller molecules can pass through cell membranes faster than larger ones, then it will take less time for a given number to accumulate inside the cell.
- B: If smaller molecules can pass through cell membranes faster than larger ones, then in a given time a greater number of smaller molecules should pass through.

He made up three solutions containing equal numbers of sucrose molecules (large molecules), glycerol (medium-sized molecules) and urea (small molecules). He placed alga cells in each solution and regularly analysed the contents of some of these cells to determine the concentration of each of these chemicals.

Table 1.11 shows his results. The time given is for the chemical inside the cell to reach half the concentration of that in the solution outside cells.

Table 1.11 Results from experiment

Chemical in solution outside cell	Time (hours)
sucrose	750
glycerol	40
urea	6

a) Why was it important that the three solutions be at the same concentrations (equal number of molecules) initially? *(1 mark)*

b) Given the data he produced, which hypothesis (A or B) did he probably use? *(1 mark)*

c) How can you interpret these results? *(1 mark)*

d) Should this interpretation be the biologist's final conclusion? Why or why not? *(2 marks)*

e) Can these results be applied to all cells, and for all chemicals? *(1 mark)*

f) Another scientist repeated this experiment but also used a fourth chemical, ethyl urethane. This chemical has molecules that are slightly smaller in size than glycerol, but bigger than urea. This scientist found it only took 1 hour for this chemical to reach half the concentration of that in the solution outside the cells. Does this result support the biologist's conclusion? Explain. *(2 marks)*

g) What do the results of these new data suggest about the hypothesis of the biologist? *(1 mark)*

38. Figure 1.72 shows four cells found in the human body. They are not drawn to scale. Use the following list to identify these cells: skin cells; nerve cell; bone cells; muscle cells; white blood cell; red blood cell. *(4 marks)*

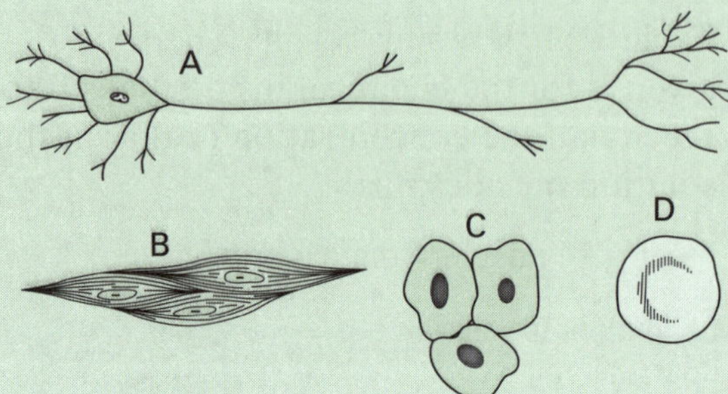

Figure 1.72 Cells

39. At the present levels of technology it is not possible to perform a complete brain transplant. What are some of the ethical issues involved if this could be done? *(3 marks)*

40. Draw up a table with the following headings: Asexual reproduction; Sexual reproduction. Place each of the following statements under the correct heading in the table: *(6 marks)*

- no sex cells produced
- two parents required
- offspring are identical
- sex cells produced
- offspring show differences to the parents
- only one parent.

Go to pp. 186–189 to check your answers.

CHAPTER 2

Elements and compounds

Overview

In this chapter you will learn about:

- solids, liquids and gases
- the physical properties of matter
- changes of state
- elements and the periodic table
- metals, semi-metals and non-metals
- compounds
- molecules and ions
- chemical and physical changes
- simple chemical reactions
- new materials.

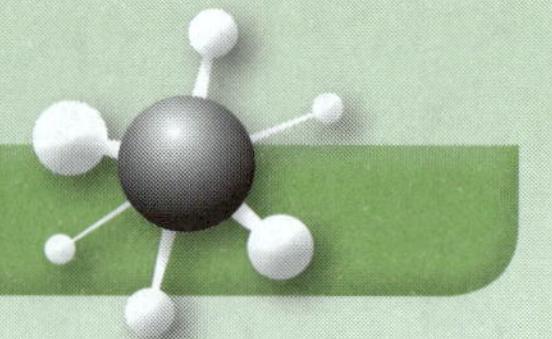

Glossary

Chemical change—a change that leads to the formation of new chemical substances

Compound—a pure substance composed of two or more elements in fixed atomic ratios

Density—the mass of matter in a given volume

Diffusion—the process in which matter spreads out from a region of high concentration to a region of low concentration

Ductile—the ability to be able to be drawn out into thin wires

Ion—a charged atom

Ionic compound—a compound composed of ions of opposite charge

Malleable—the ability to be rolled or hammered into thin sheets

Metal—an element that is lustrous and conducts electricity and heat

Molecular compound—a compound composed of molecules

Non-metal—an element that is dull and does not conduct electricity or heat

Physical change—a change that does not lead to the formation of a new chemical substance

Plastic—a synthetic polymer molecule that can be moulded into various shapes

Polymer—a long chain of molecules formed from thousands of smaller molecules called monomers

Sublime—the conversion of a solid directly into a vapour on heating

2.1 Solids, liquids and gases

Everything around us, including ourselves, is made up of matter. This matter is made up of very small particles, too small to see. This idea that all matter is made from particles is called the particle model of matter.

The idea that matter is made up of particles can be traced back 2500 years to the Ancient Greeks, in particular, Democritus (c. 460 BC–370 BC). Democritus was a natural philosopher who observed the features of the world and then made hypotheses based on these observations. He observed that the sand on beaches was made of tiny sand grains. In themselves, they were minute, but added together, they created a huge structure. Extending this observation, Democritus hypothesised that matter was similarly made of incredibly tiny, indestructible particles which added together to make materials. He called these particles *atomos*, which means 'not able to be sliced apart or divided'. Atoms are so small that they cannot be seen without the use of very high-powered microscopes. Even then they appear only as fuzzy 'balls'. In the 19th century, chemists such as John Dalton became convinced that matter was composed of particles. Experimental data could only be explained satisfactorily in terms of the interaction of these small particles.

The main ideas in the particle model are as follows.

- All matter is made up of tiny particles.
- These particles of matter are continuously moving.
- There are spaces between particles.
- The particles move faster when heated.

In solids, the particles are held together by strong forces (called bonds), so they are not free to move about. But they do vibrate. These particles are usually close together and form regular arrangements.

A solid has a stable, definite shape, with a definite volume. Solids can only change their shape by force, such as when broken or cut, and because the particles are close together, solids cannot be compressed. Their shapes, however, can be modified by grinding, hammering or rolling to produce new shapes. For example, a rock can be crushed to powder, or a piece of metal can be folded or twisted. A cube of solid aluminium, for instance, can be hammered or rolled into very flat sheets. Some sheets are so thin that we call them foils. A block of ice is water in its solid state. If an ice block is hammered it shatters.

Figure 2.1 shows a particle model of a solid and how grinding a solid to form a powdered solid can be achieved using a mortar and pestle. Sugar is an example of one of many solids.

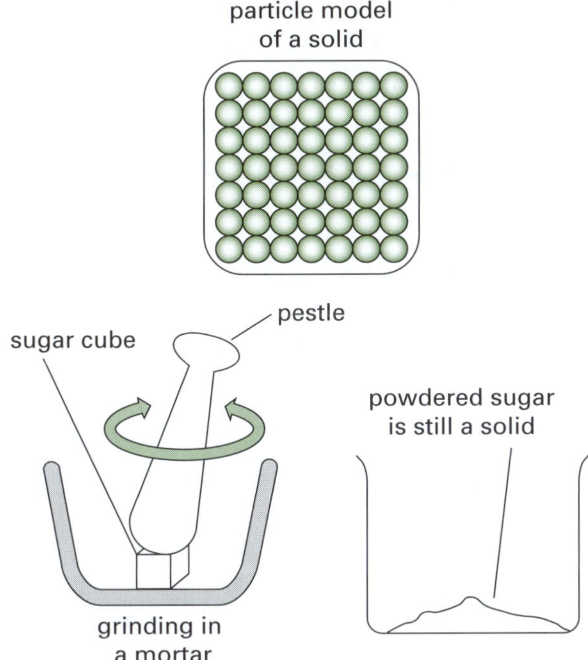

Figure 2.1 Sugar is still a solid after being ground into a powder.

In liquids, the particles are still in close contact but are free to slide or move past each other. The particles do not form regular arrangements. Because the particles are still in close contact, they are virtually incompressible. This is why liquids have fixed volumes, but no fixed shape. They take the shape of the base and sides of the container they are in, as shown in Figure 2.2.

In gases, the particles are much further apart from each other, and so are lighter than solids or liquids. Because the particles are far apart, gases can be easily compressed and completely lack rigidity. They easily fill whatever container they are in, as shown in Figure 2.3. This is why gases have no fixed volume and no fixed shape. Further, they completely mix with other gases. (This is a property that cannot be said of solids or liquids. Some liquids will completely mix with some other liquids, but not with *all* other liquids.)

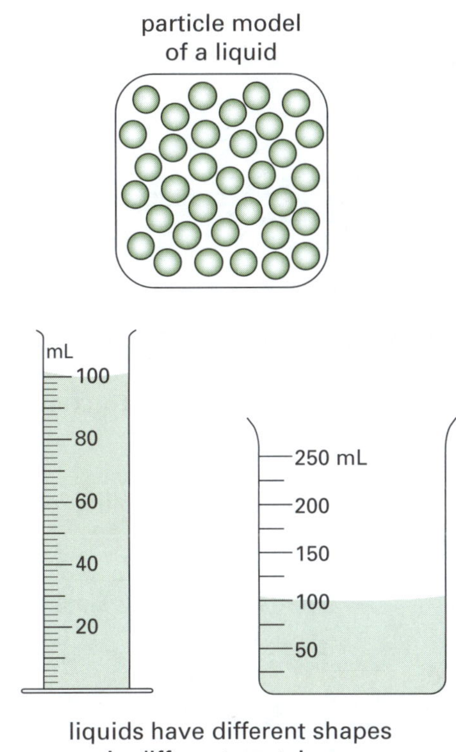

Figure 2.2 Equal volumes of a liquid have a different shape in a different container.

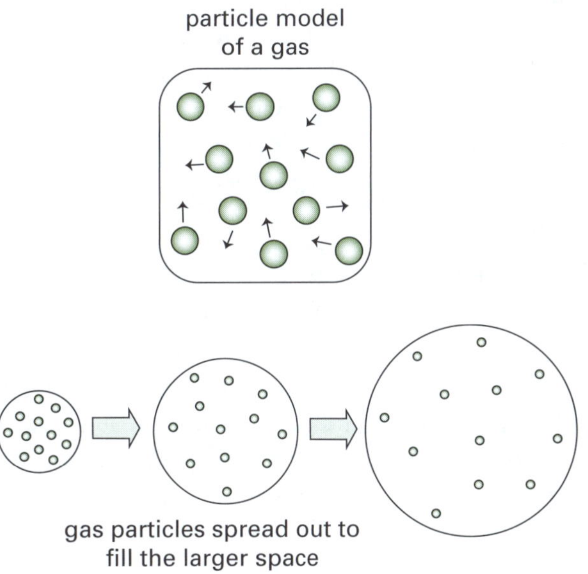

Figure 2.3 Gases fill the containers they are in.

Investigating the properties of matter

Compressibility

Some 50-cm^3 samples of fine sand, water and oxygen were placed in turn in a plastic syringe as shown in Figure 2.4. A fixed force was then applied to the plunger to record the volume to which each sample could be compressed.

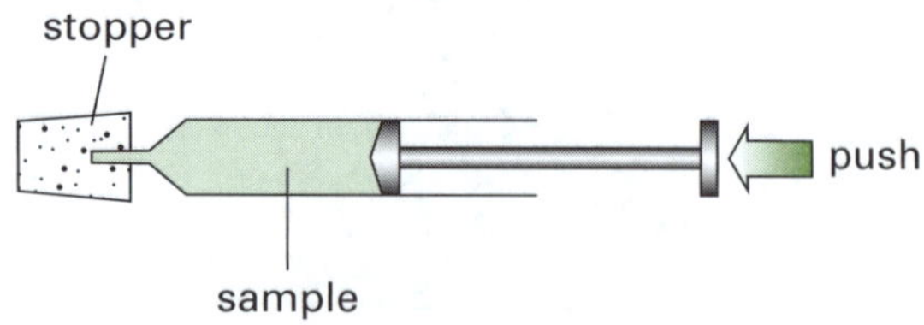

Figure 2.4 Compressibility of matter experiment

The results of the experiment are compiled as shown in Table 2.1.

Table 2.1 Results of compressibility of matter experiment

Sample	Initial volume (cm^3)	Final volume (cm^3)
fine sand	50	50
water	50	50
oxygen	50	10

These and similar experiments show that solids and liquids are incompressible whereas gases like oxygen can be compressed. This result is explained by the particle model of matter. In solids and liquids the particles are very close together with no free space. In a gas there are large spaces between the particles and this allows gases to be compressed.

Diffusion

Diffusion is the process in which matter spreads out from a region of high concentration to a region of low concentration. Solids at room temperature do not diffuse but stay in the original shape. Liquids can diffuse through other liquids that they can mix with. Gases readily diffuse through other gases.

Experiment 1

Diffusion in water

Aim

- To investigate the diffusion of a dye in a liquid

Method

1. A green dye was mixed with alcohol to produce a concentrated green solution.
2. A sample of green alcohol was placed in a syringe. Some of the dyed alcohol was injected into the base of a jar of water.
3. Observe the movement of the dye over time.

Results

Over time the green liquid diffused upwards, eventually throughout all the water, and coloured it a uniform light green as shown in Figure 2.5.

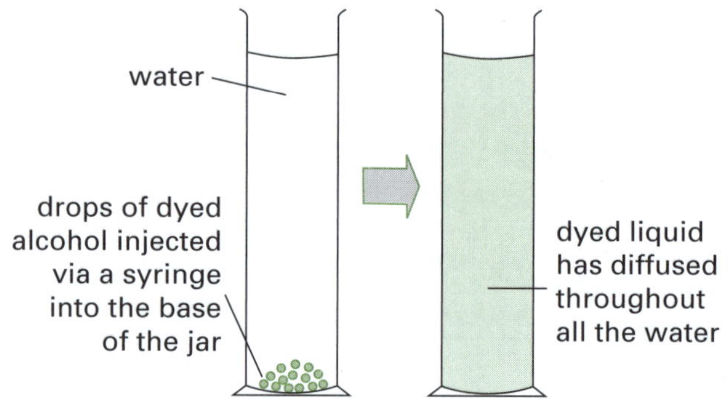

Figure 2.5 Diffusion in liquids

Analysis

Explain your observations. Go to p. 189 to check your answer.

Conclusion

Write a suitable conclusion.

Go to p. 189 to check your answer.

Have you noticed how many cooking odours can be smelled throughout the house? This is due to diffusion of these particles, and can occur reasonably quickly. Smoking has been banned from many indoor areas because poisonous cigarette odours can quickly diffuse to other parts and become hazardous to non-smokers.

Density

Density is defined as the mass of matter present in a fixed volume. For example, 10 mL of water has a mass of 10 g. The density of the water at room temperature can be calculated using the formula:

$$\text{density} = \frac{\text{mass}}{\text{volume}}$$

$$D = \frac{m}{V}$$

For water, $D = 10/10 = 1$ g/mL. Since 1 cm^3 = 1 mL, this can also be written as 1 g/cm^3. Other substances are more or less dense than water.

Experiment 2

Density of glass

Aim

- To determine the density of a glass stopper

Method

1. Determine the mass of the glass stopper using an electronic balance.
2. Determine the volume of the glass stopper by the displacement of water in a measuring cylinder. Lower the stopper by a string into a fixed volume of water in a measuring cylinder. The water level rises. Measure the difference between the initial volume and the final volume as shown in Figure 2.6. This is the volume of the glass stopper.

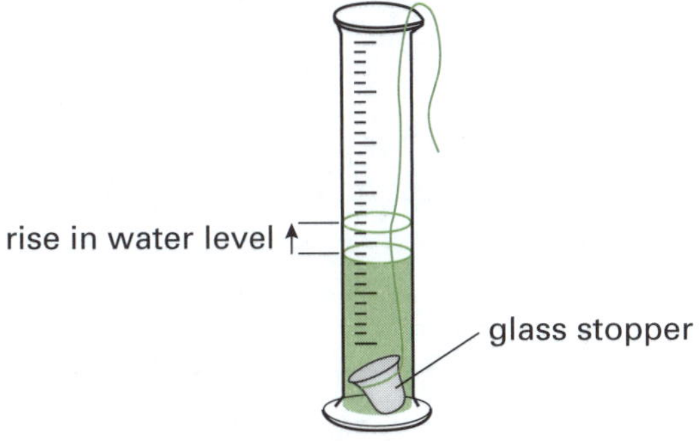

Figure 2.6 Measuring the volume of an irregular solid by displacement of water

Results

Mass of glass stopper (g)	13.2
Volume of water (mL)	40.0
Volume of water + stopper (mL)	45.5

Analysis

Calculate the density of the glass using the formula $D = \frac{m}{v}$. Go to p. 189 to check your answer.

Conclusion

Write a suitable conclusion.

Go to p. 189 to check your answer.

Heating matter

When matter is heated, it expands. Gases show the greatest degree of expansion when heated. When solids are heated, the particles absorb the energy and vibrate more. These vibrations cause the particles to move further apart and so the solid expands (see Figure 2.7).

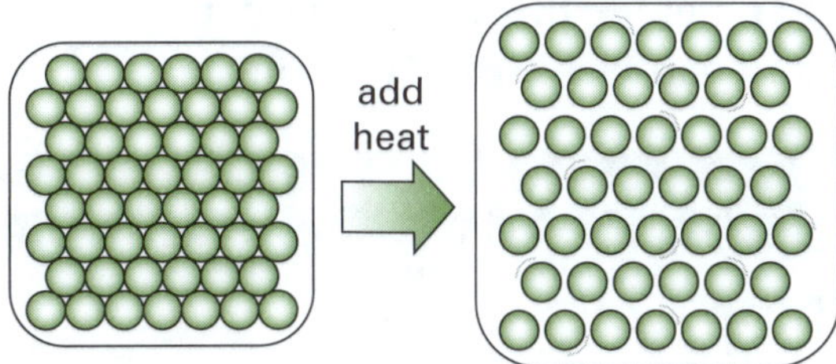

Figure 2.7 Expansion of solids on heating

Liquids expand more than solids when equal masses are heated by the same amount of heat energy. The particles of the liquid vibrate and push each other. The particles become more mobile. This property is made use of in thermometers (see Figure 2.8). The liquid in the bulb expands up the capillary tube and the temperature of the source can be determined.

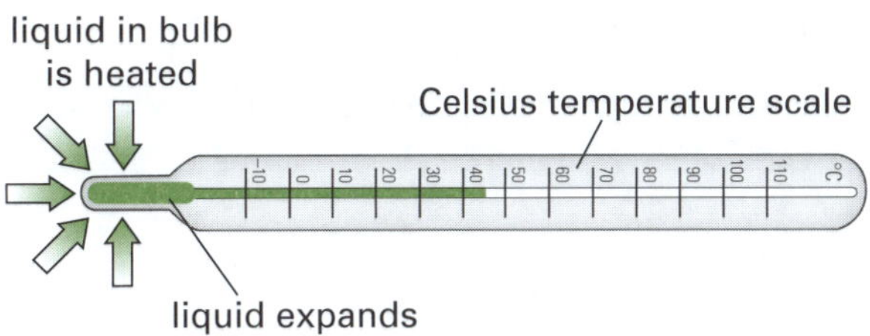

Figure 2.8 Expansion of a liquid in a thermometer

When gases are heated, they expand as long as they are in an open vessel or in a container with flexible walls, such as a balloon. The gas particles move around with increasing speed as they are heated.

Change of state on heating matter

When solids are heated, their temperature rises until a point is reached where they melt and turn into a liquid. Further heating raises the temperature of the liquid until another point is reached where the liquid boils to form a gas.

In order to understand changes in state, look at the motion of the particles involved. When a solid is heated, for example, the energy supplied causes the particles to vibrate faster and faster. This motion becomes so violent that the particles tear away from the crystal and a liquid forms. When a liquid is heated, the added energy causes the particles of the liquid to move around with increasingly greater speed. Some particles near the liquid surface are able to escape from the liquid and enter the air. The hotter the liquid, the more particles escape to form a gas.

Experiment 3

Melting point of wax

Aim

- To measure the melting point of candle wax

Method

1. Place wax flakes into a large test tube containing a thermometer. Clamp the tube in a beaker of water at room temperature. The beaker rests on a hotplate as shown in Figure 2.9.
2. Turn on the heat and allow the hotplate to heat the water and the wax.
3. Monitor the temperature of the wax with time.

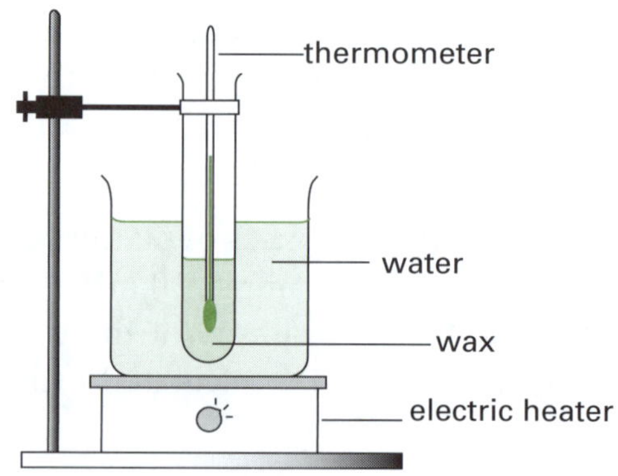

Figure 2.9 Melting wax

Results

The results of this experiment were tabulated and graphed as shown in Figure 2.10.

Time (min)	Temperature (°C)
0	25
2	28
4	31
6	34
8	37
10	40
12	42
14	43
16	44
18	44
20	44
22	44
24	45
26	47
28	50
30	53

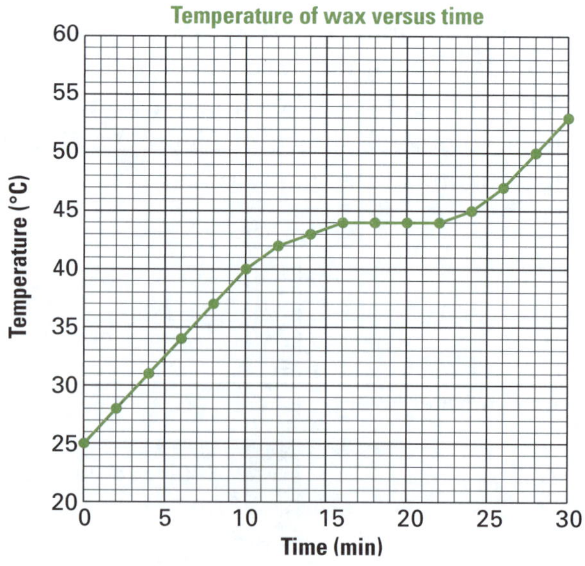

Figure 2.10 Tabulated data and graph of results

Analysis

Analyse the graphical data and explain the shape of the temperature–time graph.

Go to p. 189 to check your answer.

Conclusion

Write a suitable conclusion.

Go to p. 189 to check your answer.

2.2 Elements and the periodic table

Elements are pure substances that are made up of the one kind of atom. They cannot be broken down into a simpler type of matter by either physical or chemical means. They can exist as either single atoms (e.g. helium and neon) or molecules (e.g. oxygen and nitrogen). Carbon atoms, for instance, can combine in different ways with other carbon atoms to form different structures.

Diamond and graphite are two forms of carbon yet both look and feel very different from each other. It is these arrangements of the atoms in the molecules which, in part, give elements their physical properties.

Elements are the basic building blocks of matter. There are about 90 natural elements and about another 28 synthetic elements that have been produced by nuclear processes. Elements can be classified into families on the basis of their atomic substructure and their properties.

Chemical symbols

Each element has been given an internationally recognised symbol. A symbol is like a short version of its name, just like you might use your

Case study 1: Distillation of alcohol

Liquid alcohol can be distilled and the temperature of the distilling alcohol monitored as a function of time. At 30 minutes the heating stopped when half the alcohol had distilled. The results of this experiment are shown in Figure 2.11.

This experiment shows that the temperature of the alcohol increased steadily for the first 20 minutes and then the increase slowed down as the liquid started to boil. As it boiled, the temperature remained constant at 78 °C. This is the boiling point of the alcohol. During the boiling, the heat energy is used to separate the particles of liquid so vaporisation can occur.

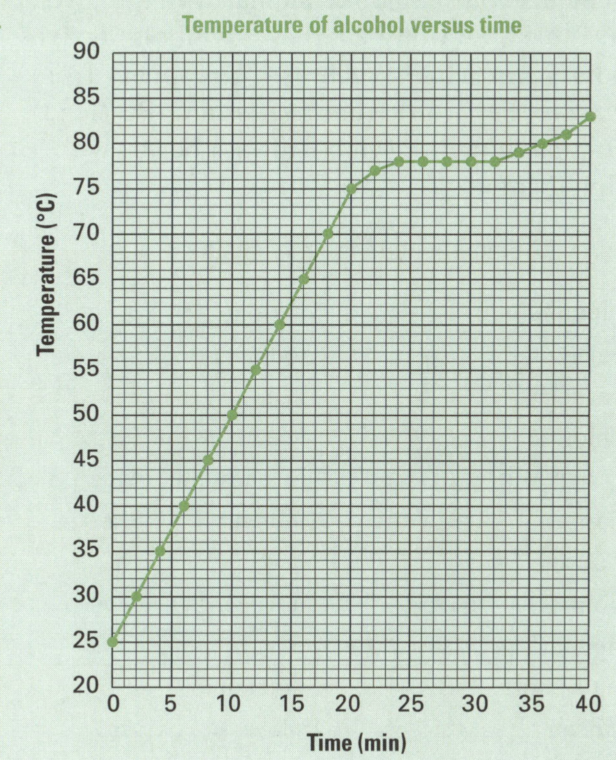

Figure 2.11 Temperature–time graph of boiling alcohol

Case study 2: Sublimation of iodine

Iodine is a non-metal. At room temperature it exists as shiny, violet-black solid flakes. When iodine is heated in a closed vessel it sublimes rather than melts (see Figure 2.12). This can happen just from the heat of sunlight. Sublimation is the process in which a solid turns directly into a vapour on heating. No liquid state is formed. The iodine vapour diffuses and fills the closed container with a violet vapour.

Dry ice (solid carbon dioxide) is often used by outdoor vendors to keep ice cream frozen. Dry ice sublimes at –78.5 °C, that is, it changes from solid to gas at this temperature. Its advantages include a lower temperature than that of water ice and not leaving any liquid residue.

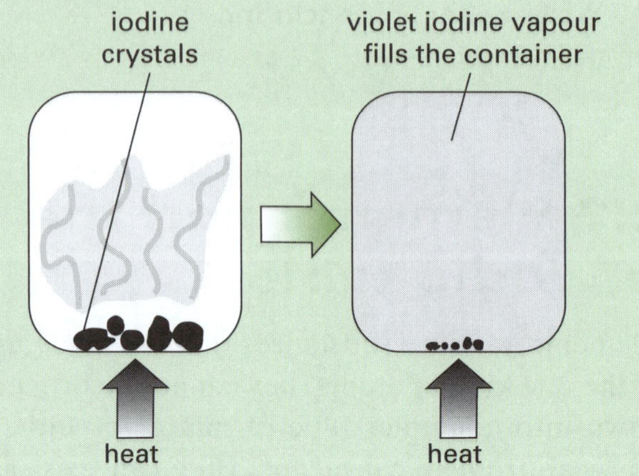

Figure 2.12 Sublimation of iodine crystals

initials. Each symbol consists of a capital letter, or a capital letter followed by a lower-case letter. The symbols for some of the common elements you should be familiar with are given in Table 2.2. Many of these symbols derive from Latin names, such as Cu (copper) from *cuprum*, Fe (iron) from *ferrum* and Pb (lead) from *plumbum*. A symbol always begins with a capital letter.

Table 2.2 Elements and their symbols

aluminium	Al
calcium	Ca
carbon	C
chlorine	Cl
cobalt	Co
copper	Cu
fluorine	F
gold	Au
helium	He
phosphorus	P
silicon	Si
sodium	Na
tin	Sn
zinc	Zn

hydrogen	H
iodine	I
iron	Fe
lead	Pb
magnesium	Mg
mercury	Hg
neon	Ne
nitrogen	N
oxygen	O
potassium	K
silver	Ag
sulfur	S
uranium	U
vanadium	V

Metals, semi-metals and non-metals

The elements can be divided into three categories that have characteristic properties. These are metals, semi-metals and non-metals. Semi-metals are elements that have some characteristics of metals and some characteristics of non-metals. The six semi-metals are boron (B), silicon (Si), germanium (Ge), arsenic (As), antimony (Sb) and tellurium (Te).

There are many more metal elements than non-metal elements. The characteristic properties of metals and non-metals are shown in Table 2.3.

Table 2.3 Properties of metals and non-metals

Properties of metals	Properties of non-metals
shiny (lustrous) surface	dull surface
ductile (able to be drawn into a wire)	brittle
malleable (able to be rolled into a sheet)	non-malleable
good conductor of heat	poor conductor of heat
good conductor of electricity	poor conductor of electricity (except carbon graphite)

Elements have many uses that are related to their properties. Some of these uses are summarised in Table 2.4.

Table 2.4 Uses of elements related to their properties

Element	Use	Properties related to its use
copper (Cu)	electrical wiring	good electrical conductor; ductile; unreactive
magnesium (Mg)	flares	burns in air releasing bright, white light
lead (Pb)	fishing sinkers	high density; malleable
gold (Au)	jewellery	lustrous; unreactive
helium (He)	weather balloons	low density; inert
silicon (Si)	computer chips	semiconductor

Iron is a very useful metal. It is used to make steel and steel is used in building and in manufacturing cars, railway tracks and numerous machines. Iron and steel must be protected from rusting otherwise their properties will change and they will no longer be useful. Painting, lacquering or coating steel with other metals are common methods of protecting it from corrosion.

Metals have such an important role in the history of humans that several archaeological ages have been named after outstanding accomplishments in metallurgy that occurred during those times (see Table 2.5). About 3000 years ago, for example, humans learnt to extract iron from iron ore. From this came steel, an alloy of iron. Steel is used in buildings and vehicles because of its ability to form sheets, rods and wires.

Table 2.5 Archaeological ages related to metals and alloys; bronze is an alloy (mixture) of copper and tin

Archaelogical age	Beginning of age (years ago)
New Stone Age	11 000
Copper Age	7000
Bronze Age	5000
Iron Age	3000

The periodic table

The periodic table is a classification of elements. A full copy of the periodic table is reproduced on the inside front cover of this study guide. Figure 2.13 shows the arrangement of elements in the periodic table.

- The elements are arranged in the periodic table according to their increasing atomic number (Z). The atomic number refers to the number of particles called protons inside the nucleus of each atom.
- The elements are arranged in rows and columns.
- The rows are called periods. There are seven periods. Not all periods have the same number of elements. The first period contains only two elements (hydrogen and helium) whereas the fourth period contains 18 elements. The last two periods are very long because of the presence of two special series of elements: the lanthanoids (Z = 57 to 71) and actinoids (Z = 89 to 103). These are usually extracted and written at the bottom of the table to reduce its width.
- The columns are called groups. Each group represents a family of related elements. The element hydrogen is sometimes placed with Group I, but in some tables it is not allocated to any group.
- The transition metals are a group of metals located between Group II and Group III.

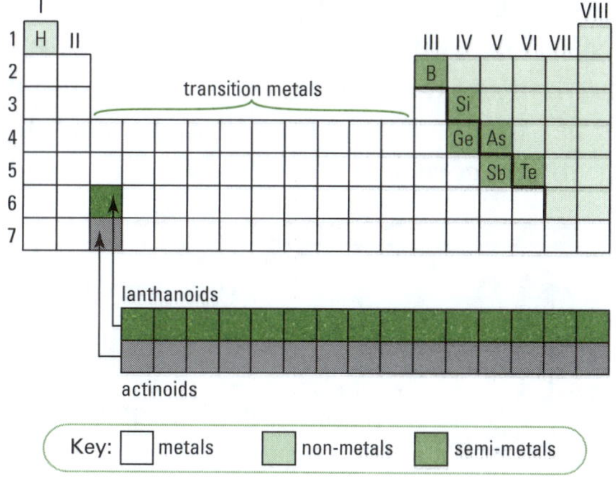

Figure 2.13 Periodic table

Investigating the properties of elements

The properties of various elements can be investigated experimentally.

Electrical conductivity

The electrical conductivity of elements can be determined using a simple conductivity circuit as shown in Figure 2.14. A lamp will shine if the element conducts. An ammeter can be used in place of a lamp.

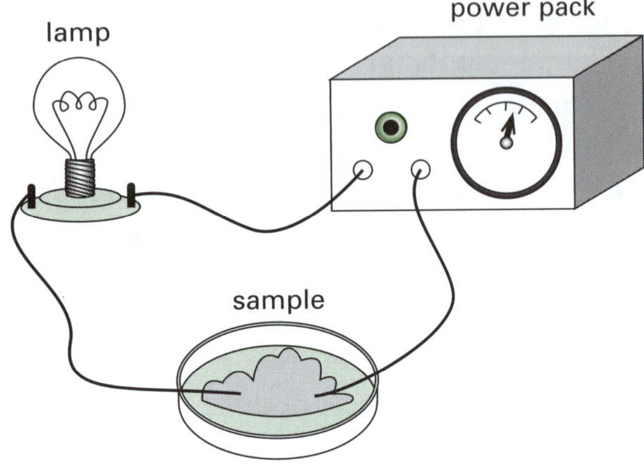

Figure 2.14 Conductivity circuit

Malleability

Samples of each element are provided and students use a hammer to determine whether the substance is brittle (it will shatter) or whether it is malleable.

Heat conductivity

Long rods of various metals are tested for heat conductivity as shown in Figure 2.15. Pieces of chalk are attached to the rod using paraffin grease. The end of the rod is heated with the blue flame of the Bunsen burner. The time that it takes for the grease to melt and the chalk to fall off is recorded.

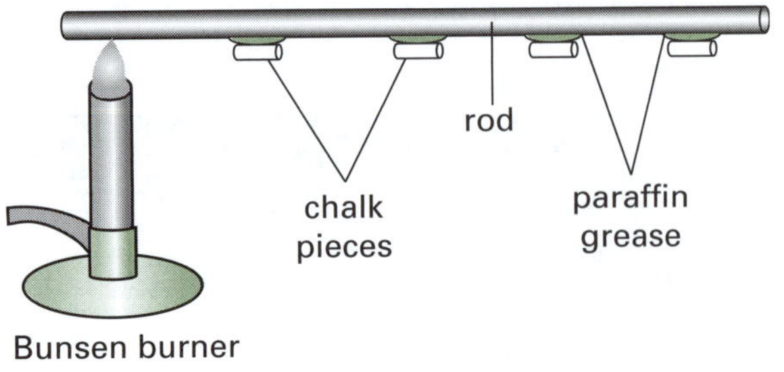

Figure 2.15 Testing heat conductivity

In one series of experiments the following rods were tested: iron, aluminium and copper. The copper rod was shown to have the greatest heat conductivity as the chalk pieces fell off fastest. The aluminium rod was less conductive than the copper and the iron rod had the lowest heat conductivity.

Expansion on heating

Metals expand when heated. The classic experiment called the ball and ring is used to illustrate that solids expand on heating. At room temperature the metal ball will not fit through the metal ring. When the ring is heated, it expands and the room-temperature ball can pass through the ring as shown in Figure 2.16.

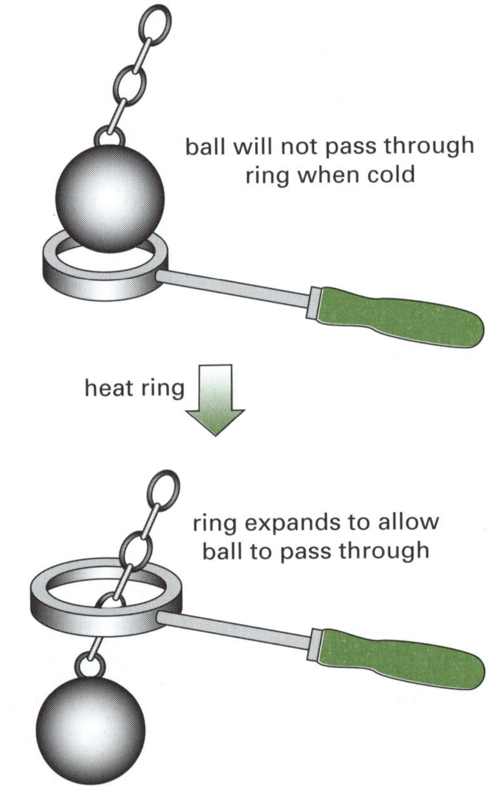

Figure 2.16 Ball and ring experiment

Different metals expand by different amounts when heated. A bimetallic strip consists of two layers of different metals fused tightly together. When heated, one metal expands more than the other and so the bimetallic strip bends. Bimetallic strips are used in such devices as thermostat temperature controls. One of the many uses for bimetallic strips is in electrical circuit breakers where too much current heats the strip, causing it to bend and tripping the switch. This stops the current.

The results of experiments with different bimetallic strips are shown in Figure 2.17.

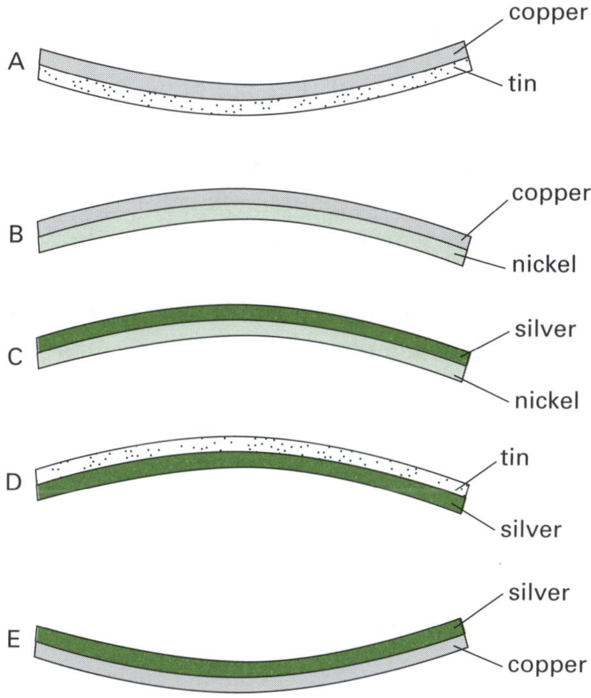

Figure 2.17 Bimetallic strips; each metal expands by a different amount when the strips are heated

- Experiment A shows that tin expands more than copper.
- Experiment B shows that copper expands more than nickel.
- Experiment C shows that silver expands more than nickel.
- Experiment D shows that tin expands more than silver.
- Experiment E shows that copper expands more than silver.

Combining all the results, the decreasing order of metal expansion is: tin, copper, silver and nickel.

2.3 Compounds

A chemist defines a pure substance as having a constant composition. Figure 2.18 shows some particle models to illustrate this definition. Metals such as copper are pure because all the particles (atoms) are identical.

Compounds are also pure because all the composite particles are the same as each other. Substances such as water are pure because each molecule contains one oxygen and two hydrogen atoms. Copper oxide is a solid compound in which there are equal numbers of charged copper and oxygen atoms.

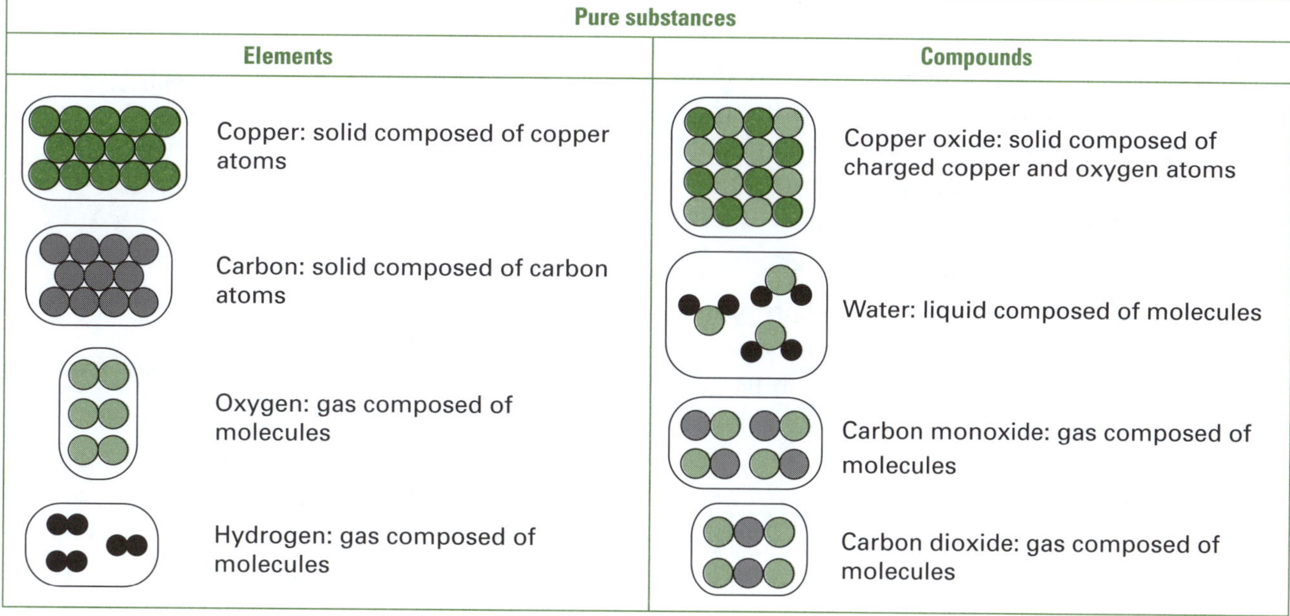

Figure 2.18 Pure substances are either elements or compounds.

Compounds have different properties to the elements from which they are made. Table 2.6 shows some examples of differences in properties between elements and their compounds.

Table 2.6 Physical properties of elements and compounds

Pure substance	Melting point (°C)	Boiling point (°C)	Density (g/cm^3)
aluminium (e)	660	2467	2.70
bromine (e)	–7	59	3.12
aluminium bromide (c)	97	255	3.20
sodium (e)	98	883	0.97
iodine (e)	114	184	4.94
sodium iodide (c)	660	1304	3.67

Key: (e) = element; (c) = compound

In nature compounds are often contaminated with impurities. Water in the ocean, for example, has various salts dissolved in it. These impurities change the properties of the compound. A simple experiment to demonstrate this idea is to measure the boiling point of pure water and water containing salt. A distillation apparatus is used to distil a sample of pure water. The thermometer records a constant temperature of 100 °C as the water boils. The experiment is repeated with a sample of salt water. In this case the salty water does not boil until the temperature is above 100 °C (e.g. 6 g of salt dissolved in 100 mL of water raises the boiling point to 101 °C).

The purity of a compound can therefore be established by measuring a physical property of the substance. If the result does not match the published data then the sample must be impure.

Types of compounds

In nature there are many different compounds. They can be classified into various groups based on the types of particles that are present.

Compounds composed of molecules

Methane is a common molecular compound. It is a major component of natural gas which is used as a fuel. Cows, and other grazing animals, emit a huge amount of methane through belching and flatulence. Methane contributes to global warming, being 23 times more powerful than carbon dioxide.

Chemists have found that methane molecules are composed of one carbon atom and four hydrogen atoms. These atoms are arranged in a tetrahedral geometric shape (triangular pyramid). The carbon atom is joined to the four hydrogen atoms by chemical bonds. In some diagrams of these molecules the bonds are shown by a straight line (—) joining each atom. In reality the atoms are much closer together. Figure 2.19 shows some different ways that chemists represent methane.

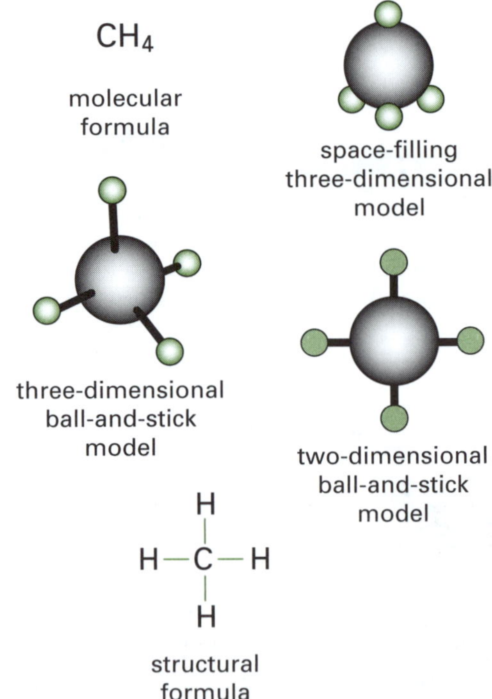

Figure 2.19 Different ways to represent methane molecules

Molecular compounds vary considerably in size (see Figure 2.20). Water molecules have one oxygen atom and two hydrogen atoms. Octane molecules, which are found in petrol, are much larger. They consist of eight carbon atoms and 18 hydrogen atoms.

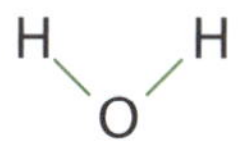

water

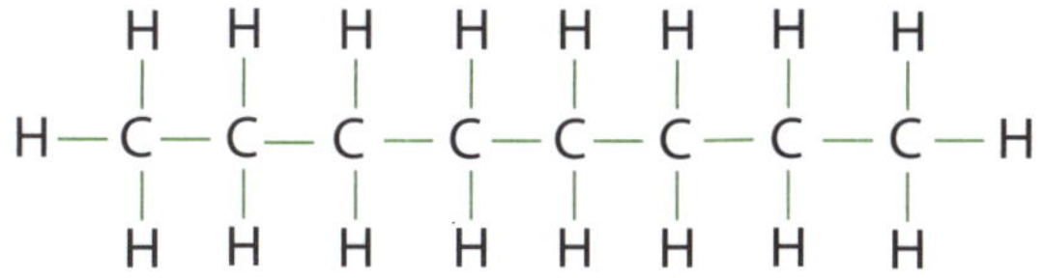

octane

Figure 2.20 Structural formulae of water and octane molecules

Compounds composed of ions

Atoms are electrically neutral. Atoms contain equal numbers of both positively charged particles called protons and negatively charged particles called electrons.

- Atoms can gain or lose electrons.
- An atom with more or less electrons than protons is called an ion.
- A positive ion has more protons than electrons.
- A negative ion has fewer protons than electrons.
- Metals tend to form positive ions.
- Non-metals tend to form negative ions.
- Oppositely charged ions attract one another. This attraction binds the ions together to form a compound.

Example 1: Sodium reacts with chlorine to form sodium chloride

Sodium is a metal. In its neutral atom there are 11 protons and 11 electrons. Sodium can readily lose one electron to form a positive sodium ion. This is represented by the symbol Na^+.

Chlorine is a non-metal. Its neutral atom has 17 protons and 17 electrons. Chlorine can readily gain one electron to form a negative chloride ion. This is represented by the symbol Cl^-.

When sodium and chlorine are heated together they react because sodium gives one electron to chlorine. Positive sodium ions and negative chloride ions form. These ions attract each other and the compound sodium chloride (NaCl) is formed. This compound, therefore, is composed of ions.

sodium + chlorine → sodium chloride

Figure 2.21 shows a model of a small section of a sodium chloride crystal. It is cubic in shape. Next time you use salt on food, look at the shape of the salt crystals. They are cubes because sodium chloride is the major component of salt.

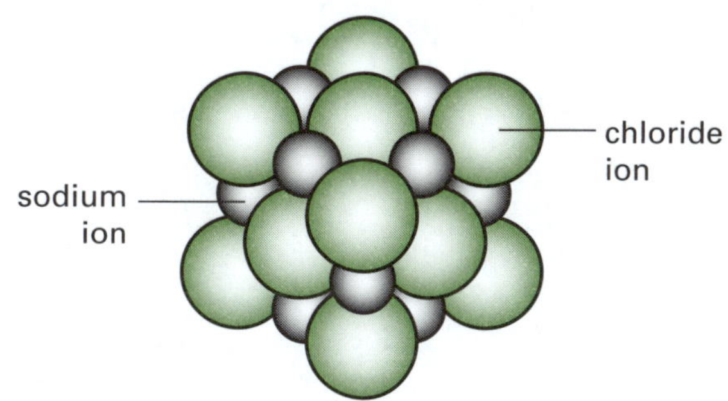

Figure 2.21 Model of a sodium chloride crystal

Example 2: Calcium reacts with oxygen to form calcium oxide

Calcium is a metal. In its neutral atom there are 20 protons and 20 electrons. Calcium can readily lose two electrons to form a positive calcium ion. This is represented by the symbol Ca^{2+}.

Oxygen is a non-metal. In its neutral atom there are eight protons and eight electrons. Oxygen can readily gain two electrons to form a negative oxide ion. This is represented by the symbol O^{2-}.

When calcium and oxygen are heated together they react because calcium gives two electrons to oxygen. Positive calcium ions and negative oxide ions form. These ions attract each other and the compound calcium oxide (CaO) is formed. This compound, therefore, is composed of ions.

calcium + oxygen → calcium oxide

In nature there are many compounds composed of ions. The shiny black mineral called galena is a source of lead metal. It is an ionic compound called lead sulfide (PbS). The mineral bauxite is a source of aluminium metal. It is an ionic compound called aluminium oxide (Al_2O_3).

Uses of common compounds

The chemical properties of a compound are related to its use. Table 2.7 lists some examples.

Table 2.7 Compounds and their uses

Compound	Formula	Property	Use
octane	C_8H_{18}	flammable	vehicle fuel
sulfuric acid	H_2SO_4	corrosive	removing rust from steel sheets
ammonia	NH_3	caustic	dissolves fats in drains
carbon dioxide	CO_2	acidic	dissolves in water under pressure to form soda water
carbon dioxide	CO_2	does not support combustion	fire extinguishers
hydrogen peroxide	H_2O_2	oxidising agent	bleach; kills microbes
magnesium oxide	MgO	thermally stable	high-temperature kilns

Acids and bases

Acids and bases are common examples of compounds. In our homes vinegar contains a weak acid called acetic acid. Lemons and oranges contain a weak acid called citric acid. Soda water contains carbonic acid (H_2CO_3). These weak acids have a sour taste. The school laboratory has much stronger acids including hydrochloric acid (HCl) and sulfuric acid (H_2SO_4).

Bases are chemical compounds that have the opposite properties to acids. Weak bases have a bitter taste. In your home there are some common bases. Glass and floor cleaners contain ammonia (NH_3). Baking soda ($NaHCO_3$) is a weak base used in cooking. Sodium hydroxide (NaOH) is a much stronger base and is used as a drain cleaner.

Great care must be used in handling strong acids and bases as they can damage the eyes and skin. Strong bases are caustic and can dissolve fats and your skin.

Chemical compounds called indicators can be used to distinguish between acids and bases. One common indicator found in the school lab is litmus. This indicator is extracted from a lichen (a living thing in which a fungus and an algae form an association) and its solution in water is purple. When the litmus indicator is added to an acid solution, the indicator turns red. When added to a solution of a base, the indicator turns blue. Other indicators have different colours in acids and bases.

Test yourself 1

Part A: Knowledge

1. Select the true statement about matter. *(1 mark)*
   - **A** Matter particles are stationary.
   - **B** Matter particles vibrate or move faster when heated.
   - **C** The particles of matter are far apart in the solid state.
   - **D** All matter is made of pairs of particles joined together.

2. Which statement is true of liquids? *(1 mark)*
   - **A** Liquids take the shape of the whole container they are in.
   - **B** The particles of a liquid do not move about.
   - **C** The particles of a liquid move and slide around each other.
   - **D** The particles of a liquid are far apart and move rapidly.

3. The density of a solid is 4.0 g/cm^3. Calculate the mass of 40 cm^3 of this substance. *(1 mark)*
   - **A** 40 g
   - **B** 10 g
   - **C** 160 g
   - **D** 44 g

4. Sublimation is the process in which a *(1 mark)*
   - **A** liquid is converted to a solid.
   - **B** solid is converted to a liquid.
   - **C** gas is converted to a liquid.
   - **D** solid is converted to a gas.

5. The set of elements that contains a metal and a non-metal is *(1 mark)*
   A mercury and lead.
   B uranium and silicon.
   C helium and fluorine.
   D nitrogen and zinc.

6. Complete the following restricted-response questions using the appropriate word.
   (1 mark for each part)
   a) Germanium is an element that is classified as a ....
   b) Copper is used for electrical wiring because it is ...., unreactive and a good electrical conductor.
   c) Different metals .... by different amounts when heated.
   d) Water is a compound composed of the elements hydrogen and ....
   e) Methane .... are composed of one carbon atom and four hydrogen atoms.

7. Use the code letters to match the terms or phrases in each column. *(1 mark for each part)*

Column 1	Column 2
A ions	F molecule
B octane	G base
C zinc chloride	H charged atoms
D sodium hydroxide	I gas
E helium	J compound

Part B: Skills

8. Figure 2.22 shows a method for determining the volume of an irregular-shaped solid.

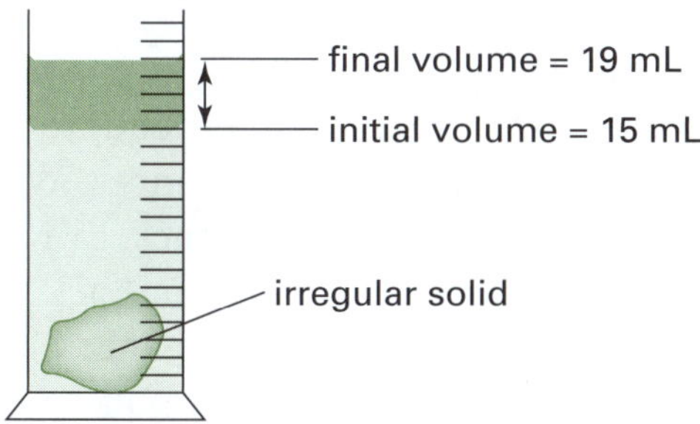

Figure 2.22 Volume of an irregular solid

   a) Calculate the volume of the irregular solid. *(1 mark)*
   b) If the solid has a mass of 32 g, calculate the density of the solid. *(1 mark)*

9. Identify the property of liquids illustrated by the equal volumes of water shown in Figure 2.23. *(1 mark)*

Figure 2.23 Equal volumes of liquids

10. A gas jar of oxygen gas was inverted over a gas jar of a brown gas as shown in Figure 2.24. The apparatus was monitored for several hours. Explain what would be observed. *(2 marks)*

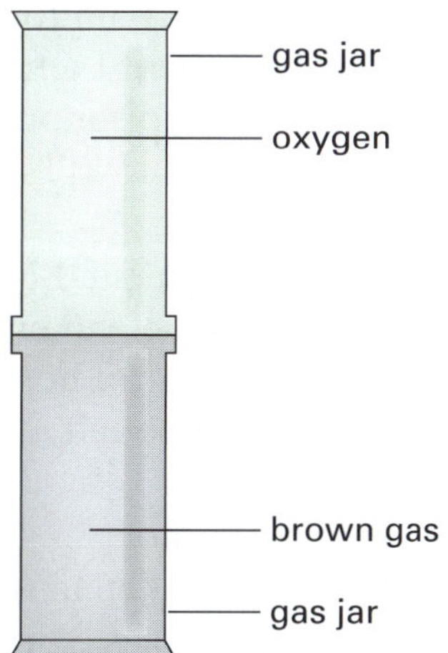

Figure 2.24 Diffusion

11. The particles of matter are too small to be seen with your eyes alone. A student found that the thickness of 500 pages of paper was 5 cm.
   a) How thick is one page? Give your answer in millimetres. *(1 mark)*
   b) If at least 1 000 000 atoms equal the thickness of one page, how big (in millimetres) would you expect atoms to be? *(1 mark)*

12. Two unknown substances, X and Y, were placed in separate syringes until they each occupied a volume of 20 mL. Pressure was applied to the plunger but only X could be compressed easily. Y could be poured easily from one container to another and it took the shape of the lower regions of the container. What states of matter are X and Y? *(2 marks)*

13. Use the periodic table at the back of this book to find the symbols for the following elements and classify them as metals, semi-metals or non-metals. *(5 marks)*
    a) bromine
    b) boron
    c) cobalt
    d) potassium
    e) gallium

14. Classify the following elements as metals or non-metals. *(3 marks)*
    a) I do not occur free in nature. I am found in sea water. As a pure element I am a yellow-green gas. I am very poisonous. I am used in swimming pools to kill bacteria.
    b) I am a soft, silvery-white element. I am less dense than water but react violently with it. I am a good conductor of heat and electricity. I form a compound which is used to season food.
    c) I am black and brittle. I am not malleable. I can be used for sketching in art. As a powder I am used as a dry lubricant. I have a very high melting point and conduct electricity well.

Go to pp. 189–190 to check your answers.

2.4 Chemical and physical changes

Some changes in matter are classified as physical changes and some are classified as chemical changes. In order to understand these differences, consider the following experiment performed by school students.

Example: Properties of freezing water

Problem to solve

- Does freezing water change its properties?

Method

1. Take a sample of pure water.
2. Divide it into two portions.
3. Boil one sample and measure its boiling point.
4. Freeze the other portion, then let it melt and again measure its boiling point.

Results

- Normal water:
  measured boiling point = 100 °C
- Frozen and remelted water:
  measured boiling point = 100 °C

Conclusion

There was no difference between the two measurements. Freezing and remelting the water did not change its properties.

Physical changes

When an ice block is taken from the freezer and placed in a beaker it melts into liquid water over a period of time. If the beaker of water is returned to the freezer, the water turns back into ice. These changes of state are called physical changes; they are changes that do not lead to the formation of new chemical substances.

Other examples of physical changes are:

- evaporation
- sublimation
- condensation
- filtration.

Chemical changes

Chemical changes are different to physical changes; they are changes that lead to the formation of new chemical substances. When you bake a cake and let it cool back to room temperature, the cake does not turn back into the separate ingredients. New substances have formed inside the cake. When you boil an egg, the yolk and white are permanently changed.

Indicators of a chemical change can include some of the following.

- A new substance is formed. Some examples include:
  - a gas is given off that you can see
  - a solid forms in a solution
  - a new odour forms that you can smell.
- An energy change happens. Examples of this include:
  - the temperature rises
  - the temperature drops
  - light or sound energy is given off.
- The original substance disappears.
- A colour change occurs.
- The change is very difficult, if not impossible, to reverse.

2.5 Simple chemical reactions

The synthesis of a compound from its elements is an example of a simple chemical reaction.

Example 1: Burning steel wool in air

A sample of steel wool is teased out to increase its surface area. It is then held by tongs in the blue flame of a Bunsen burner as shown in Figure 2.25. The steel wool quickly sparkles as it reacts. Once the reaction is over, the steel wool has turned a darker grey compared with the unchanged control sample. The grey substance that forms is a type of iron oxide which has formed by the reaction of iron with the oxygen in the air.

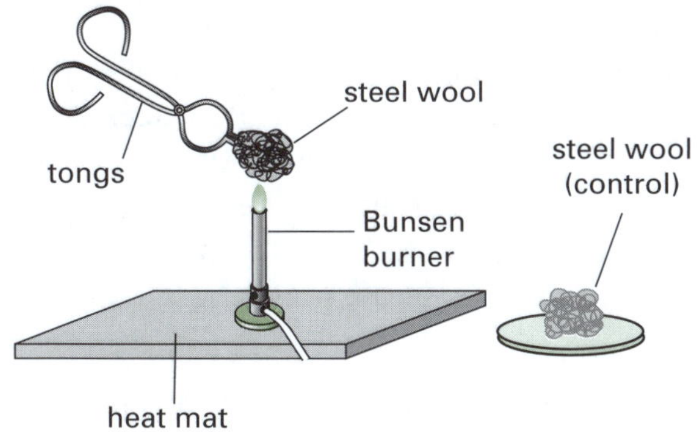

Figure 2.25 Burning steel wool in air

The word equation for this reaction is:

iron + oxygen → iron oxide

Figure 2.26 is a particle model for this reaction. In this reaction the iron oxide that forms has three particles of iron for every four particles of oxygen. In this compound the particles are ions. Other oxides of iron will form if the reaction conditions change.

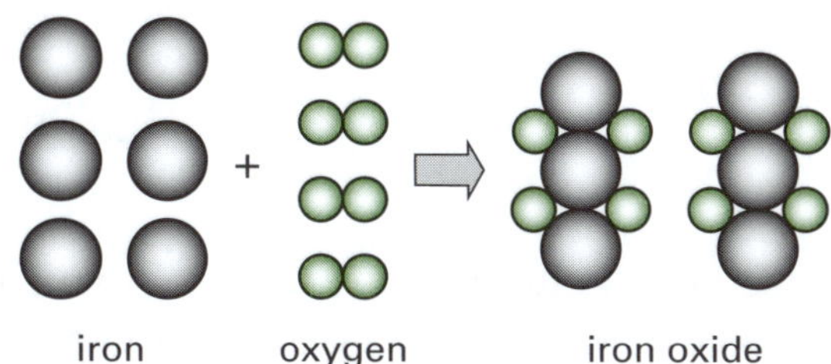

Figure 2.26 Particle model for the formation of iron oxide

Example 2: Burning magnesium in air

A sample of magnesium ribbon is placed in a crucible supported on a pipeclay triangle on a tripod as shown in Figure 2.27. The crucible is heated using the blue flame of a Bunsen burner. The hot magnesium reacts with oxygen in the air and a bright white light is produced as well as considerable heat energy. Once the reaction is over, the grey magnesium ribbon has turned into a crumbly white powder. The white substance that forms is magnesium oxide which has formed by the reaction of magnesium with the oxygen in the air.

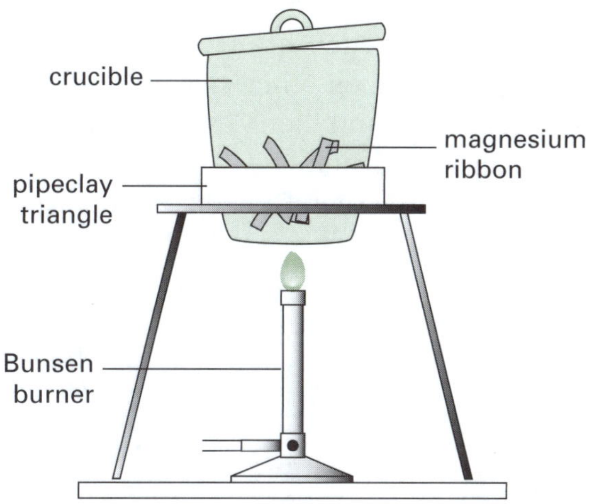

Figure 2.27 Burning magnesium in air

The word equation for this reaction is:

magnesium + oxygen → magnesium oxide

Figure 2.28 is a particle model for this reaction. The particles in the compound are ions.

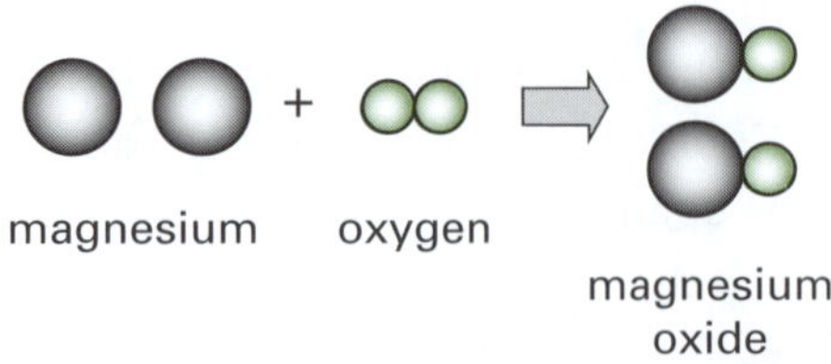

Figure 2.28 Particle model for magnesium reaction with oxygen

The French chemist Antoine Lavoisier discovered that the mass of the products in a chemical reaction is the same as the mass of the reactants. This is now known as the law of mass conservation.

Using simple chemical reactions to determine the atomic weight of elements

If you examine the periodic table reproduced at the back of this book, you will notice that each element has an atomic weight. The atomic weight of each element increases as you move along each row of the periodic table. At the start of the 19th century, John Dalton estimated the relative masses of elements by weighing the reactants and products in a chemical reaction. He soon became convinced that his atomic theory was correct and that atoms of different elements combine in fixed ratios to form compounds.

Dalton knew that hydrogen was the lightest element. If hydrogen atoms are assigned a weight of 1 unit then other elements will have weights greater than 1 unit. Carbon, for example, was found to be 12 times heavier than hydrogen (see Figure 2.29). Therefore, the atomic weight of carbon is 12.

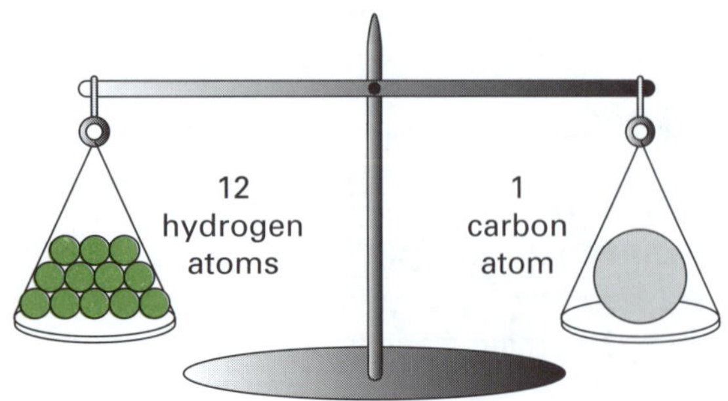

Figure 2.29 Carbon atoms are 12 times heavier than hydrogen atoms.

Comparing the weights of other elements to hydrogen allowed a table of atomic weights of some of the known elements to be established.

2.6 New materials

Chemists have developed a wide range of new materials over the last 100 years. Over the course of history, humans have used natural products such as wood and stone to construct homes and tools. They used natural fibres such as wool and cotton to make clothes and furniture coverings. The development of smelting techniques led to the production of metals and alloys. The discovery of oil deposits led to the petrochemical industry and the production of plastics. Plastics have replaced many wooden and metallic objects in homes and in industry.

Plastics

Plastics are synthetic polymer molecules that can be moulded into various shapes. Depending on the chemical properties of the polymer the final product may be hard or soft and flexible or rigid.

Polymers are long chain molecules formed from thousands of smaller molecules called monomers. These monomers are often obtained from petroleum refining. Polymer chains may be thousands of atoms long.

In the science laboratory, you can model polymers by threading many beads on a long string as shown in Figure 2.30. The beads may be all the same size or different sizes. There may be small lengths of beads hanging off the main chain. This is like a beaded necklace or bracelet with small 'charms' attached along its length. Two chains could also be linked together. These differences can alter the properties and therefore the uses of the polymer.

Common plastics

Polyethylene is an important plastic. It is manufactured from ethylene which is a gas with the chemical formula C_2H_4. Various types of polyethylene can be made which have different properties. These differences in properties are

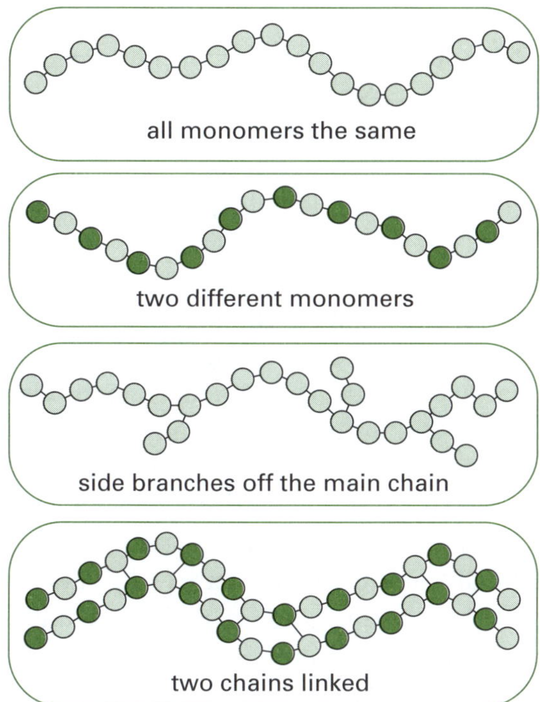

Figure 2.30 Modelling polymer chains like beads on a string

related to the presence or absence of small branches along the polymer chain. High-density polyethylene (HDPE) is a tough, less flexible polymer that is used to make buckets, crates, storm water pipes, wheelie bins and plastic bottles. Low-density polyethylene (LDPE) is a much softer and flexible form of polyethylene due to the presence of side branches along the polymer chain. LDPE is used to make plastic bags, cable insulation and soft squeeze bottles.

Teflon is a heat-resistant polymer. The monomer used to make teflon is derived from ethylene. All the hydrogen atoms are replaced by fluorine atoms and its formula is C_2F_4. Teflon was first used to create a heat-resistant coating on frypans. It is water resistant and it is also used for low-friction coatings in gears and bearings. Table 2.8 outlines the uses of some common plastics.

Nylon was the first synthetic fibre produced. It was initially used to make stockings to replace silk stockings. During World War II it was used to make parachutes. Nylon is commonly used today in carpeting. Often blends of wool and nylon produce durable and soft carpets.

Table 2.8 Uses of some common plastics

Plastic	Chemical name	Common uses
PET	polyethylene terephthalate	soft drink bottles; carpet fibres
HDPE	high-density polyethylene	milk bottles; juice bottles
LDPE	low-density polyethylene	plastic bags; plastic food wrap
PP	polypropylene	industrial fibres; car batteries; carpets
PVC	polyvinyl chloride	non-food bottles; agricultural pipes
PS	polystyrene	DVD cases; foam insulation

Polyester fibres were soon developed by chemists and were used mainly as clothing fibres. Chemically, the polyester used is the same polymer that is used in soft drink bottles which is known as PET. The trade names dacron in the United States and terylene in the United Kingdom were introduced in 1951 to market polyester fabrics.

Blending natural and synthetic fibres allows the best qualities of each fibre produce fabrics with more desirable properties. For example, a blend of cotton and polyester produces clothing fabric that does not crease (like pure cotton) but which is cooler to wear (unlike pure polyester). Wool and dacron blends are often used to make fabrics for suits.

Modern athletes wear form-fitting clothing made of a synthetic fibre called spandex. Fabrics made of 3 to 10% spandex are lightweight, elastic and durable.

Table 2.9 compares the properties of a natural and a synthetic fibre.

Recycling plastics

Rubbish dumps would soon fill up if plastic products were not recycled. Bacteria and fungi do not decompose plastics. Councils provide householders with recycling bins so that plastic materials can be sorted and recycled.

Table 2.9 Comparison of wool and nylon fabrics

Fibre/fabric	Wool	Nylon
useful properties	crease resistant; good heat insulator; does not burn readily; durable; flexible; absorbs dyes well	elastic; dries quickly; strong; good heat insulator; good abrasion resistance; not readily attacked by insects
non-useful properties	shrinks after washing; discolours in sunlight	melts and burns readily; discolours in sunlight; does not allow moisture to escape from the skin

Common plastics have recycling codes (see Figure 2.31). These codes are used to identify the material from which the item is made, and to make recycling or other reprocessing easier. Have a look at various plastic products at home: what type of plastic they are made of?

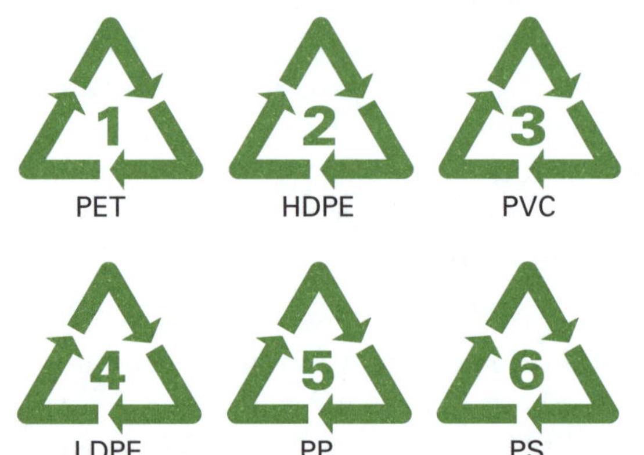

Figure 2.31 Recycling codes: PET = polyethylene terephthalate; HDPE = High-density polyethylene; PVC = polyvinyl chloride; LDPE = low-density polyethylene; PP = polypropylene; PS = polystyrene

Pharmaceuticals

Over thousands of years people have been using extracts from plants and animals to cure human illness. Different cultures made use of the different plants and animals in their region. By 1000 BC Egyptians were using castor oil, garlic, coriander and opium extracts as herbal cures. The extracts contained small amounts of active compounds which helped to reduce pain, reduce swelling or kill microbes. Modern chemists can extract these active compounds and determine their structure. They can then synthesise them in large quantities and sell them as pharmaceuticals. These modern methods are also safer as herbal extracts may also contain other toxic chemicals that could damage the body. Before pharmaceuticals can be legally sold, they undergo rigorous testing on animals and finally on human volunteers to determine appropriate dosage levels and to determine whether there are any short- and long-term problems in its continued use.

Example: Aspirin

Aspirin is the trade name used for acetyl salicylic acid. The salicylic acid was first extracted from the bark of a willow tree (which belongs to the genus *Salix*, hence the chemical name for aspirin). Hippocrates (5th century BC) found that using powder made from the bark and leaves of the willow tree helped heal headaches, pains and fevers. But it is recorded much earlier by the ancient Egyptians that willow and myrtle plants had these curing properties. By 1829 scientists had identified the compound in the willow plants that gave this pain relief.

Aspirin is manufactured from salicylic acid, and it is used to relieve mild aches and pains and to reduce fevers. The structure of aspirin is shown in Figure 2.32. Its molecular formula is $C_9H_8O_4$.

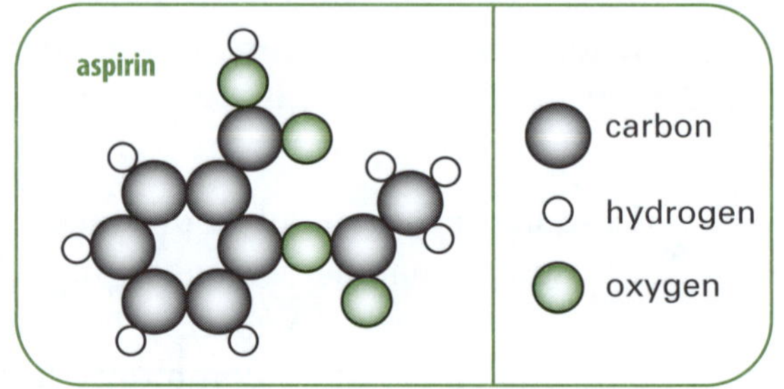

Figure 2.32 Aspirin

Test yourself 2

Part A: Knowledge

1. Which of the following is classified as a physical change? *(1 mark)*
   A melting
   B cooking a cake
   C frying an egg
   D burning a match

2. Identify the product of the reaction between sodium and chlorine. *(1 mark)*
   A chlorine sodide
   B sodium chloride
   C water
   D sodium trichloride

3. Identify the plastic that is a heat-resistant polymer used to coat non-stick frypans. *(1 mark)*
   A PET
   B polyethylene
   C polystyrene
   D teflon

4. Compared with wool, nylon *(1 mark)*
   A melts and burns readily.
   B has poor abrasion resistance.
   C does not turn yellow in sunlight
   D is readily attacked by insects.

5. Plastics have recycling codes. The recycling code for PET is *(1 mark)*
   A 1.
   B 2.
   C 4.
   D 6.

6. Complete the following restricted-response questions using the appropriate word. *(1 mark for each part)*
   a) A blend of cotton and polyester produces clothing fabric that does not .... like pure cotton does.
   b) Polymers are long chain molecules formed from thousands of smaller molecules called ....
   c) The boiling of water is an example of a .... change.
   d) The burning of steel wool in air is an example of a .... change.
   e) When magnesium burns in oxygen the product is called magnesium ....

7. Use the code letters to match the terms or phrases in each column. *(1 mark for each part)*

Column 1	Column 2
A evaporation	F hydrogen
B Dalton	G sodium
C lightest element	H argon
D metal	I physical change
E gas	J atomic theory

Part B: Skills

8. The compound water is made up of two hydrogen atoms and one oxygen atom. Hydrogen atoms are the lightest atom and have a relative mass of 1 unit. Oxygen atoms are heavier and have a mass of 16 units. Calculate the mass of one molecule of water. *(1 mark)*

9. Use the following list of chemical formulae to complete Table 2.10: UO_2; $CuSO_4$; AlN; H_2SO_4; HCl; CuO. *(6 marks)*

Table 2.10 Chemical formulae

Formula	Name	Formula	Name
	hydrogen sulfate		copper sulfate
	hydrogen chloride		uranium oxide
	aluminium nitride		copper oxide

10. Some vinegar was added to a test tube of baking soda as shown in Figure 2.33. Identify two observations that would indicate a chemical change occurred. *(2 marks)*

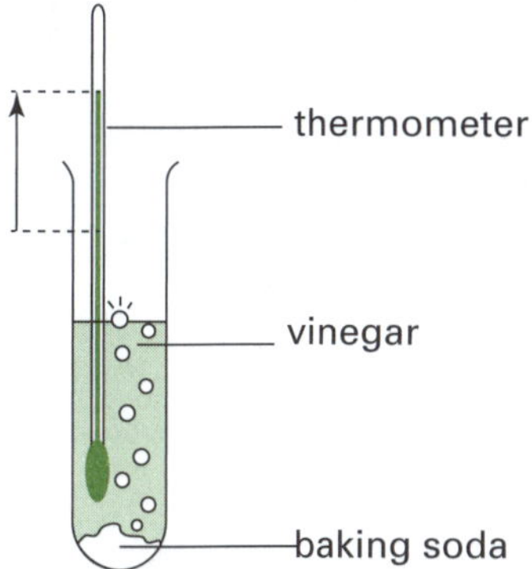

Figure 2.33 Baking soda and vinegar

11. Consider the following experiment and observations.

A test tube was half-filled with white vinegar. One gram of table salt was added to the test tube and then stirred. The salt dissolved. No bubbles were seen. There was no colour change. The solution was tasted and the sour taste of the vinegar and the salty taste were able to be distinguished.

The student concluded that no chemical change occurred. Explain how she came to this conclusion. *(3 marks)*

12. Polymer chains are made up of repeating units called monomers. Figure 2.34 shows a section of a polymer chain. Identify the repeating unit. *(1 mark)*

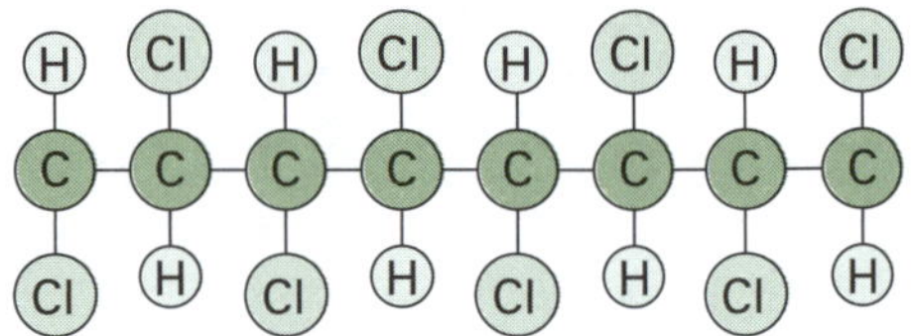

Figure 2.34 Polymer

13. Three structures for sulfur are shown in Figure 2.35.
    - **a)** How many atoms of sulfur are present in each structure? *(1 mark)*
    - **b)** One structure consists of a particle containing a ring of eight sulfur atoms. It is called an octa-atomic molecule of sulfur. Which structure is this? *(1 mark)*
    - **c)** Rings of sulfur atoms can be broken on heating. When this happens, the yellow solid turns into a yellow liquid and then into a brownish treacle-like liquid. Like treacle, the liquid is viscous or sticky due to chains of atoms which tangle in each other. Which structure corresponds to this type of sulfur? *(1 mark)*

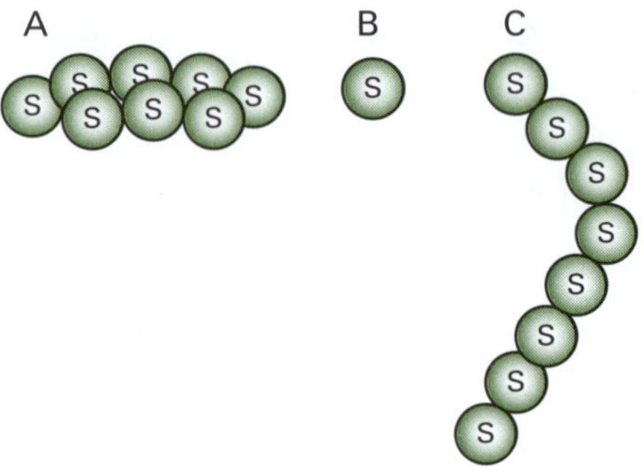

Figure 2.35 Sulfur

14. Figure 2.36 shows a section of the polystyrene polymer chain. Draw a structure for the repeating subsection of the polymer chain. *(1 mark)*

Go to pp. 190–191 to check your answers.

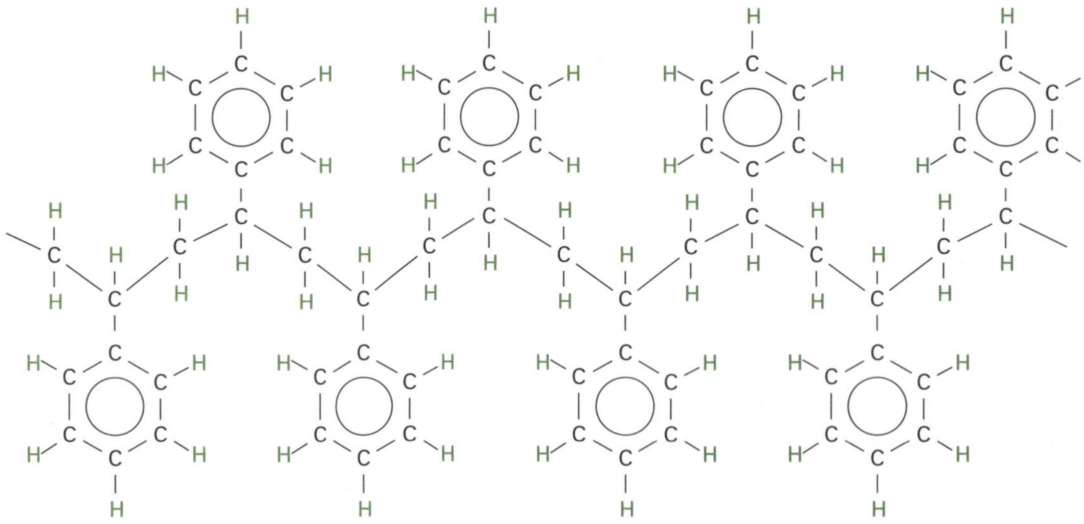

Figure 2.36 Polystyrene

Summary

1. Everything in the universe is made up of matter.
2. Matter is made up of particles.
3. In solids, the particles are held together by strong forces (called bonds), so they are not free to move about.
4. In liquids, the particles are still in close contact but are free to slide or move past each other.
5. In gases, the particles are far apart from each other.
6. When matter is heated, it expands. Gases show the greatest degree of expansion when heated.
7. When solids are heated they reach a temperature where they melt and turn into a liquid. Further heating raises the temperature of the liquid until a point is reached where the liquid boils to form a vapour.
8. Elements are pure substances that are made up of the one kind of atom. They cannot be broken down into a simpler type of matter by either physical or chemical means.
9. Each element has been given an internationally recognised symbol.
10. The elements can be divided into three categories that have characteristic properties: metals, semi-metals and non-metals.
11. The periodic table is a classification of elements. The elements are arranged in rows and columns.
12. Compounds are pure because all the composite particles are the same as each other.
13. Some compounds are composed of molecules and others are composed of oppositely charged ions.
14. The properties of elements and compounds are related to their use.
15. The synthesis of a compound from its elements is an example of a simple chemical reaction.
16. Plastics are synthetic polymer molecules that can be moulded into various shapes.
17. Pharmaceuticals are compounds that are produced as medications for the body. Many pharmaceutical chemicals have been discovered in nature while others have been synthesised by chemical reactions.

Syllabus checklist

Are you able to answer every syllabus question in this chapter? Tick each question as you go through the list if you are able to answer it. If you cannot answer it, turn to the appropriate page in the guide as is listed in the column to find the answer.

For a complete understanding of this topic		Page no.	✓
1	What composes physical substances?	50	
2	Can I state what makes up matter?	50–51	

For a complete understanding of this topic		Page no.	✓
3	Can I recall the term used to describe the strong forces that hold the particles of a solid together?	50	

For a complete understanding of this topic		Page no.	✓
4	Can I explain how the arrangement and motion of particles in a liquid is different to that of a solid?	51	
5	Can I explain how the arrangement and motion of particles in a gas is different to that of a solid or liquid?	51	
6	Can I recall what happens to matter when it is heated?	53–54	
7	Can I recall what happens to solids when they are heated?	53–54	
8	Can I explain why elements are the simplest pure substances?	55	
9	Can I explain why each element has been given an internationally recognised symbol?	55–56	
10	Can I recall the three categories of elements?	56	
11	Can I explain what the periodic table is?	57	
12	Can I explain why compounds are examples of pure substances?	59	
13	Can I state what compounds are composed of?	60–61	
14	Can I give examples of elements and compounds and the properties related to their use?	62	
15	Can I recall what the synthesis of a compound from its elements is an example of?	65	
16	Can I recall what plastics are?	66–68	
17	Can I recall what pharmaceuticals are and where they come from?	68	

Chapter test

Go to p. v for *Tips for tests and examinations*

80 MIN

Part A: Multiple-choice questions

(1 mark for each)

1. The chemical symbols for calcium, iron and tin are

A Ca, Ir, Ti.
B Ca, Fe, Sn.
C Cl, I, T.
D Cu, In, Ti.

2. Which of these is *not* a characteristic of metals?

A usually hard solids at room temperature
B have a lustre when cut or scratched
C usually poor conductors of heat and electricity
D can be bent, or stretched and flattened

3. Elements can be divided into groups called

A compounds and mixtures.
B metals, semi-metals and non-metals.
C pure and impure substances.
D metals and alloys.

4. Which of the following is *not* an example of a physical change?

A condensation
B filtration
C decomposition
D evaporation

5. The word used by Democritus to describe matter was *atomos*. Which of the following is its meaning?

A divisible
B indivisible
C reactive
D unreactive

6. Identify the metal in this list.

A calcium
B nitrogen
C krypton
D arsenic

7. Identify the non-metal in this list.

A barium
B iron

C nickel
D fluorine

8. Identify the semi-metal in this list.
A silicon
B platinum
C helium
D vanadium

9. Identify the set that contains a molecule and a compound.
A sodium chloride and copper oxide
B oxygen gas and water
C nitrogen gas and helium gas
D iron oxide and gold sulfide

10. Identify the element with the highest electrical conductivity.
A silicon
B sulfur
C silver
D germanium

Part B: Short-answer questions

11. Complete the following restricted-response questions using the appropriate word. *(1 mark for each part)*
a) When solids are heated their .... rises until a point is reached where they melt and turn into a liquid.
b) The columns of the periodic table are called ....
c) Chemists have found that .... molecules are composed of one carbon atom and four hydrogen atoms.
d) Carbon dioxide does not support .... and so it is used in fire extinguishers.
e) Lavoisier discovered the law of .... conservation in chemical reactions.

12. Use the code letters to match the terms or phrases in each column. *(1 mark for each part)*

Column 1	Column 2
A plastics	F acidic
B condensation	G polymers
C vinegar	H density
D physical property	I molecular compound
E ammonia	J physical change

13. Rearrange the sentences P to W to produce a paragraph describing the process of melting. *(4 marks)*
P. First the structure is gradually weakened during heating and this expands the solid.
Q. Even though the particles are vibrating, this is not enough to disrupt their structure.
R. Now the solid begins to melt and forms a liquid.
S. So when a solid is heated, the particles gain enough energy to start vibrating faster and faster.
T. Further heating provides more energy until the particles start to break free of the structure.
U. In a solid, the strong attractions between the particles hold them tightly packed together.
V. The particles in a liquid are the same as those in the solid but they have more energy.
W. Now while the particles are still loosely connected, they are able to move around.

14. When you are visiting a friend's place for a barbeque you can often smell the meat sizzling before you enter the front door. Using the particle model, explain why this is so. *(2 marks)*

15. Plants get carbon, hydrogen and oxygen from compounds in their environment. Table 2.11 shows the major trace elements, which plants get from the soil. These trace elements represent 8.12% of the total of all the elements they need.

Table 2.11 Trace elements in soil

Trace element	Percentage of all elements	Percentage of trace elements
nitrogen	3.50	43.1
potassium	2.40	29.5
calcium	1.30	16.0
magnesium	0.40	
phosphorus	0.32	3.9
sulfur	0.20	2.5
total	8.12	100.0

a) Calculate the percentage of magnesium found in the trace elements in soil. *(1 mark)*

b) Draw a divided bar graph of the percentage of trace elements in the soil. *(6 marks)*

16. Copy and complete Table 2.12. *(5 marks)*

Table 2.12 Characteristics of elements

Element	Symbol	Metal, non-metal or semi-metal	Use
tin			
	Hg		
		metal	steel structures
carbon			
	Si		

17. Use the words in the list to copy and complete Table 2.13. Ensure matching statements are opposite each other: malleable; brittle; non-malleable; poor conductor of heat; shiny surface; ductile; dull surface; good conductor of electricity; poor conductor of electricity; good conductor of heat. *(5 marks)*

Table 2.13 Properties of metals and non-metals

Properties of metals	Properties of non-metals

18. Which two of the following changes is an indication of a chemical change rather than a physical change? *(2 marks)*
   A. There is a large drop in temperature when two solids are mixed and stirred together.
   B. A bright flash of white light is observed when two solids are heated together.
   C. The material can be readily converted from a solid to a liquid and then back again.
   D. A solid crystallises out of a solution as the temperature of the solution is lowered.

19. Which of the diagrams in Figure 2.37 could represent copper sulfate ($CuSO_4$)? Each geometric shape represents a different element. *(1 mark)*

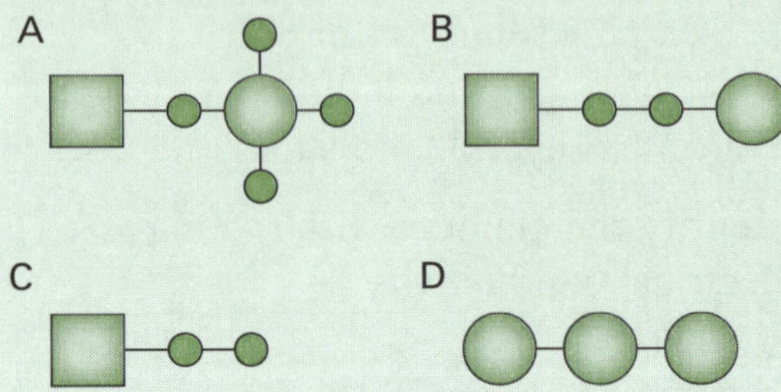

Figure 2.37 Copper sulfate

20. Alloxan is a compound produced by the oxidation of uric acid and causes diabetes when injected into test animals. It has the formula $C_4H_2N_2O_4$. (Oxidation means to react with oxygen.)
   a) How many different elements are present in alloxan? Name them. *(1 mark)*
   b) How many atoms together make one molecule of alloxan? *(1 mark)*
   c) Is uric acid an element or compound? How do you know? *(1 mark)*
   d) We can write a formula for alloxan, but not for petrol. Why is this? *(2 marks)*

21. Jim was performing an experiment on the dining table. He took some laundry bleach and added some acid to it. Within moments the room was filled with a pungent gas. He later found out this gas was chlorine. His family had to clear out of the room for several hours while this gas was ventilated.
   a) What indications are there that a reaction occurred? *(1 mark)*
   b) What were the reactants Jim used? *(1 mark)*
   c) Name a product formed. *(1 mark)*
   d) What precautions should Jim have taken before doing this experiment? *(2 marks)*

22. When the red mineral cinnabar is heated in air, mercury forms at the bottom of the container and a gas, sulfur dioxide, is formed.
   a) How does this indicate that cinnabar is a compound? *(1 mark)*
   b) What elements are present in it? *(2 marks)*

23. Figure 2.38 shows a section of a long-chain plastic. How many different elements form this plastic? *(1 mark)*

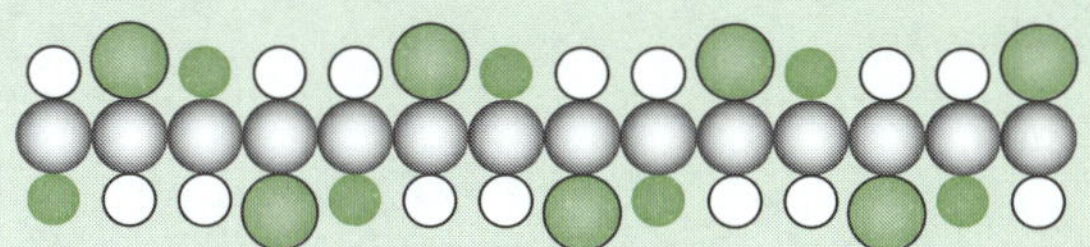

Figure 2.38 Long-chain plastic

24. A compound (lead carbonate) was collected from four different locations and its density determined by measuring the mass and volume of these samples. The results are shown in Table 2.14. Which of the four samples is not a pure sample of lead carbonate? Show your workings. *(4 marks)*

Table 2.14 Four samples of lead carbonate

Sample	Mass (g)	Volume (cm^3)
1	29.70	4.5
2	48.18	7.3
3	42.77	6.5
4	53.46	8.1

25. Figure 2.39 shows the elemental composition of the Earth's crust.
   a) Which is the most abundant metal? *(1 mark)*
   b) Which is the most abundant non-metal? *(1 mark)*
   c) Which is the most abundant semi-metal? *(1 mark)*
   d) Which of the listed elements is found in the first group of the periodic table? *(2 marks)*

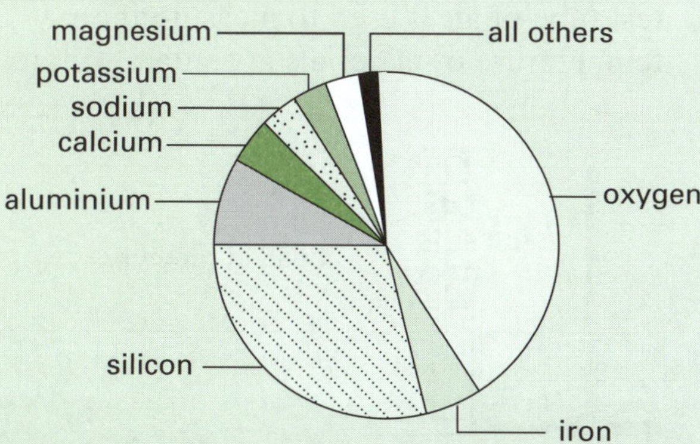

Figure 2.39 Earth's crust composition

26. Twenty millilitres of a brown substance is placed in a container of air with an internal volume of 100 mL. After a while, the brown substance spreads and completely fills the container.
   a) What state of matter is the substance? *(1 mark)*
   b) By what factor has its volume increased by the end of the experiment? *(1 mark)*
   c) How will the final colour of the substance compare to its initial colour? *(1 mark)*
   d) Name the property of matter illustrated by this experiment. *(1 mark)*

27. The density of pure gold is 19.3 g/cm^3, while the density of pure silver is 10.5 g/cm^3. Use the particle theory to explain the difference between the densities of the two solids. *(2 marks)*

28. Calculate the density of alcohol using the following data. *(1 mark)*
   Volume of alcohol = 75.0 mL
   Mass of alcohol = 58.8 g

29. An alcohol thermometer consists of a thin glass bulb and glass capillary tube containing some coloured alcohol.
   a) What happens to the liquid in the tube as the bulb is cooled? *(1 mark)*
   b) What happens to the volume of the alcohol if the whole thermometer is transferred from the laboratory to an incubator held at 60 °C? What happens to the volume of the glass? *(2 marks)*

30. Figure 2.40 shows a gas thermometer. The test tube of air is used to measure the temperature of materials in contact with it.

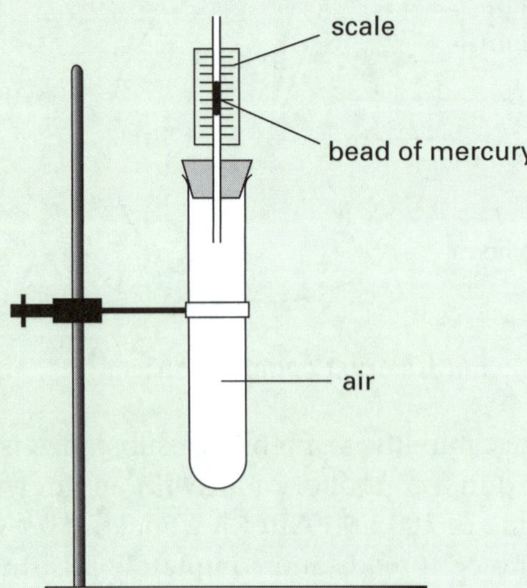

Figure 2.40 Gas thermometer

a) Suggest how this equipment can be used to measure the temperature of a sample of a hot liquid in a beaker. *(1 mark)*

b) Explain in terms of particles how this thermometer works. *(1 mark)*

31. The melting point of mercury is –39 °C and the boiling point is +357 °C.

a) In what state is mercury at 25 °C? *(1 mark)*

b) Some mercury vapour is cooled from 400 °C to 300 °C.

i) What would you expect to see happen during the cooling? *(1 mark)*

ii) What name is given to the process you observe in this experiment? *(1 mark)*

32. For each of the listed elements, name a common use and relate this use to a property of the element. *(6 marks)*

a) aluminium

b) zinc

c) chlorine

33. Refer to the periodic table printed on the inside back cover of this study guide.

a) Name the elements with the following atomic numbers: *(3 marks)*

i) 12; ii) 16; iii) 47.

b) Identify the symbols of the following elements: *(3 marks)*

i) titanium; ii) argon; iii) americium.

34. The malleability of various elements was tested. Identify which elements from the list are malleable and which are brittle: tin; sulfur; nickel; silver; barium. *(2 marks)*

35. Elements and compounds have been modelled in Figure 2.41. Classify each model as an element or a compound. *(6 marks)*

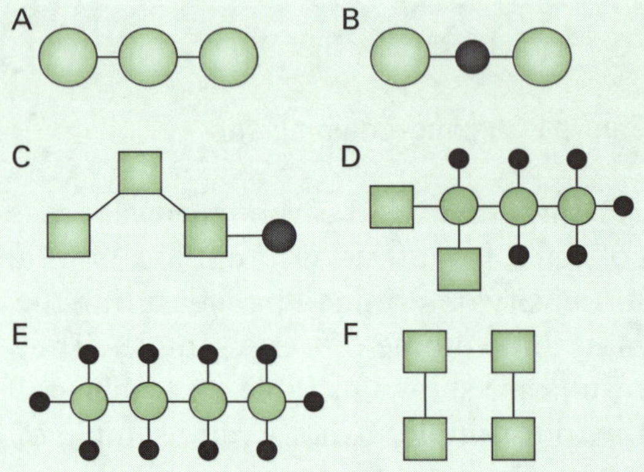

Figure 2.41 Models of elements and compounds

36. Read the following account of an experiment performed by students using an antacid powder called FIZZO. Use this information to explain whether a physical or chemical change has occurred. *(2 marks)*

Problem to solve: Can you get Fizzo powder back after it has been mixed with water?

Method

1. Take a teaspoon of Fizzo and mix it with water.
2. Observe the reaction.
3. Evaporate the resulting solution to produce a dry, white powder.
4. Test the dry powder by adding it to water. Observe what happens.

Results

- At the start, Fizzo fizzed vigorously and a lot of bubbles were released. The final solution was colourless and clear after the fizzing had stopped.
- The white powdered residue formed after evaporating the solution was then

added to another glass of water. It did not fizz at all.

37. Figure 2.42 shows an experiment involving a metal and an acid. The solution that forms after all the metal dissolves is then evaporated on an electric hotplate. White crystals are formed in the evaporating basin.

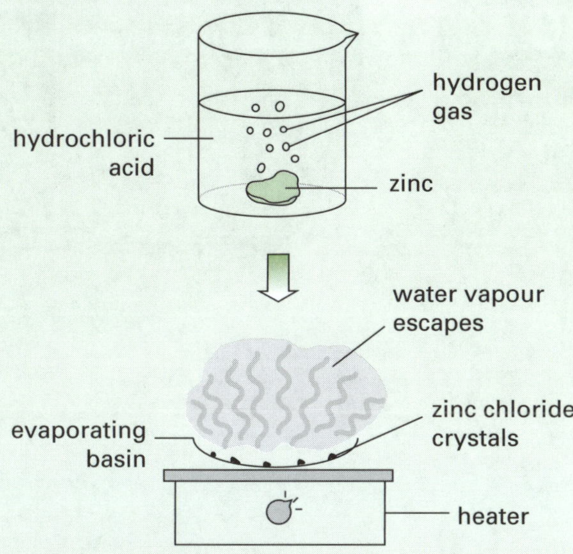

Figure 2.42 Metal and acid

a) Identify the chemical change and the physical change in this experiment. *(2 marks)*

b) Write a word equation for the chemical change. *(1 mark)*

38. Gas jars are used to perform reactions involving gases, such as that shown in Figure 2.43. Some yellow sulfur powder was placed in a deflagrating spoon and the sulfur heated over a Bunsen flame until it caught alight. The sulfur was quickly transferred to a gas jar full of oxygen gas. The sulfur burned brightly with a mauve flame and the jar filled with a misty white vapour.

a) The teacher told the class that the gaseous product of the reaction was toxic. Where is the best place to do this experiment? *(1 mark)*

b) The product of the reaction is sulfur dioxide. Write a word equation for this reaction. *(1 mark)*

c) The sulfur dioxide molecule consists of one sulfur atom and two oxygen atoms. The molecule is V-shaped. Draw a particle diagram of the structure of sulfur dioxide. *(1 mark)*

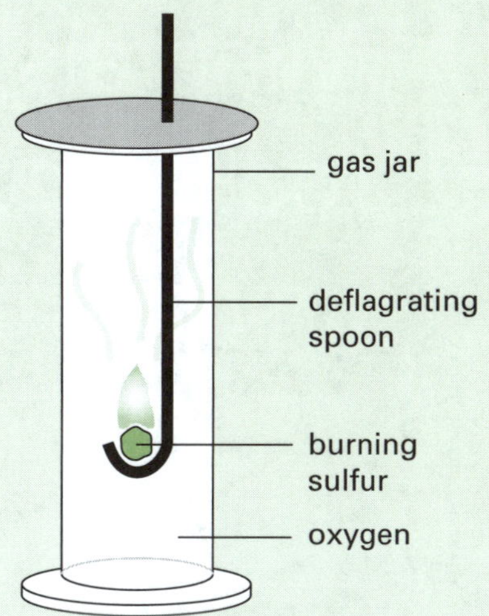

Figure 2.43 Burning sulfur

39. Vinegar is a 4% solution of a weak acid called acetic acid. The acidic nature of vinegar makes it very useful. Name and explain a common use of vinegar that is related to the food industry. *(2 marks)*

40. Pure sodium chloride has a density of 2.20 g/cm^3. A student was provided with two samples of crystalline sodium chloride and asked to experimentally determine which sample was pure and which was impure. The results of the experiment are:

Sample A: mass = 34.5 g; volume = 15.0 cm^3

Sample B: mass = 33.6 g; volume = 15.3 cm^3

Determine whether each sample is pure or impure. *(3 marks)*

Go to pp. 191–193 to check your answers.

CHAPTER 3

Minerals, rocks and the environment

Overview

In this chapter you will learn about:

- the structure of the Earth
- minerals
- weathering and erosion
- sedimentation and sedimentary rocks
- igneous rocks
- metamorphic rocks
- the rock cycle
- ores and metals
- environmental issues and recycling
- improving farming yields and sustainability
- Aboriginal people, ecosystems and fire.

Glossary

Crystalline—a solid material, whose constituent atoms, molecules or ions are arranged in an orderly repeating pattern that extends in all three spatial dimensions

Exoskeleton—an external skeleton that supports and protects an animal's body, including those of insects and crustaceans

Gangue—the commercially worthless material that surrounds, or is closely mixed with, a wanted mineral in an ore deposit

Lithosphere—the solid outer layer of the Earth; includes both the land area and the land beneath the oceans and other water bodies

Longshore drift—the transport of sediments along a coast at an angle to the shoreline, depending on the wave action and prevailing wind direction

Native metals—metals found naturally in the Earth not chemically combined with other substances. Only two metals, gold and platinum, are found principally in their native state. Silver, copper, iron, osmium and several other metals also occur in the native state.

Ore—a type of rock that contains minerals with important elements including metals and is economical to mine

Sustainable agriculture—this can provide high food, feed or energy crop yields without destroying the environment or undermining current productivity

Tectonic—the structures within the lithosphere of the Earth including the forces and movements that operate in a region to create these structures

3.1 Minerals and Earth structure

Structure of the Earth

Our Earth is composed of four main layers, which are called the crust, mantle, outer core and inner core. The crust on which we live is the very thin outer layer of the Earth. On average it is about 20 km thick. The crust under the oceans is not as thick as the continental crust.

The lithosphere includes the Earth's crust and the upper part of the mantle, as shown in Figure 3.1. The rocks which make up the lithosphere do not remain static but may undergo changes. Most of these changes occur very slowly, so over one person's lifetime it appears as if little is happening. Wind and water are continually shaping the landscape, wearing away rocks and carrying away the broken-down debris to other locations.

Meteorites have hit Earth for millions of years. But because of nature, the craters they form have often been destroyed or covered over by layers of sediment or vegetation. Contrast this with the Moon where, because of the absence of life and the altering power of wind and water, the craters remain unchanged.

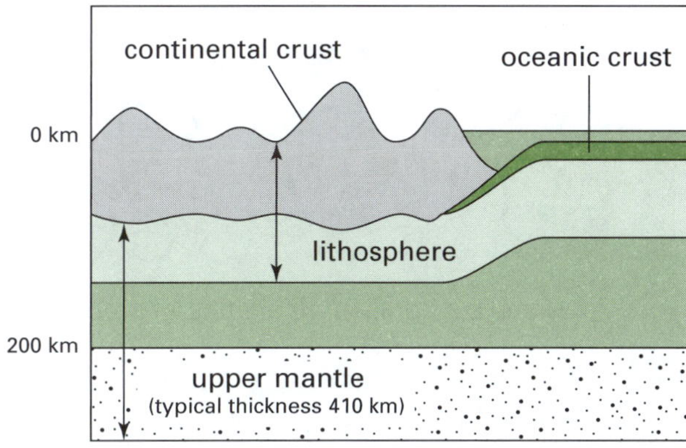

Figure 3.1 The extent of the lithosphere

Minerals

All rocks are made up of minerals. They are the most common solid material found on Earth. There are over 4000 different kinds of minerals, though only about 100 are common. The mineral quartz, for example, is found in many rocks such as granite and sandstone.

A mineral is a naturally occurring inorganic solid material that has a definite chemical composition. While people in different contexts often use the term mineral, mineralogists (scientists who study minerals) use the term for a substance with the following features:

- it is found in nature, and not synthetic
- it is made up of substances that were never alive
- it has the same chemical make-up wherever it is found, and
- the atoms in a mineral are arranged in a regular pattern forming a solid unit called crystal.

Minerals are used to make many products. Cement, steel, aluminium, fertilisers for farming, and chemicals used in manufacturing are all examples of minerals at work.

As Figure 3.2 shows, there is a great variety in the appearance and feel of minerals. Some minerals have a glass-like appearance and sparkle; others are dull and may feel greasy. The hardest mineral can easily scratch glass, while the softest can be easily scratched with a fingernail.

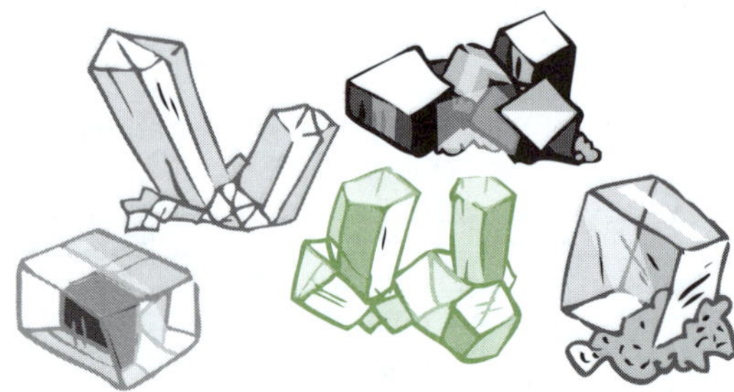

Figure 3.2 Minerals come in all shapes and colours

Mineralogists use a number of characteristics to identify minerals, such as the following.

- Colour. Many minerals are coloured, while some are colourless. Pure quartz is colourless and clear, but impurities often make it look white, pink, green, purple or even black.
- Crystal shape. Some crystals have regular geometric shapes (e.g. cubes) but the crystals may be deformed and so shape is not always a good way of identifying it.
- Lustre. Minerals may be shiny, dull, pearly, resinous, glassy or metallic
- Hardness. Minerals range in hardness. The Mohs scale rates the hardness of minerals on a 10-point scale. Diamond is the hardest natural mineral (10) and talc is the softest (1).
- Cleavage. This is the way a crystalline mineral splits into pieces with flat surfaces. Some cleave into flat sheets and others along the planes of geometric shapes.
- Streak. This is the colour of the powdered mineral. This is often different to the unpowdered mineral.

There are other tests that can be used to identify minerals. Some of these are chemical, such as seeing whether the mineral reacts with certain substances. For example, calcite is a mineral of calcium carbonate ($CaCO_3$). Scientists know that carbonates react with dilute acids. So putting a few drops of an acid (e.g. vinegar) onto a piece of calcite will produce fizzing, as the acid reacts with the carbonate and carbon dioxide gas is produced. Other tests are physical and involve things such as general appearance and density.

3.2 Weathering and sedimentary rocks

Landscapes change with time. This change may be so slow during one person's lifetime that very little alteration is noticed. Sometimes changes can be dramatic that they occur almost instantly. Volcanic eruptions, earthquakes, mudslides and storms sometimes show their fury very dramatically.

But most changes in the landscape are due to slow physical and chemical changes. Destruction of the landscape occurs as soon as it is formed. Weathering is the term used to describe the breakdown in the Earth's crust. This is the mechanical breakdown and chemical alteration of rock materials during exposure to air, moisture and organic matter. The main agents in nature doing this breaking down are wind, water, ice and the sea.

Physical weathering

Physical weathering involves breaking down the rock into smaller and smaller pieces without any change in their chemical composition. The following are some examples.

- Temperature variation. During the day, rocks are heated and they expand. At night they cool and then contract. This repeated heating and cooling occurs on the surface of the rock, as rocks are poor conductors of heat. Eventually the strains are enough to crack the rock and begin its breakdown to rubble.
- Abrasion. Rocks rub against one another, chipping and breaking small pieces off and thereby changing their physical characteristics. Abrasion occurs in glaciers, stream beds and windblown sediments.
- Plant action. Soil and dust collects in rock crevices over time, and if a seed is blown there by the wind, a plant may grow as well. As the roots of the plant grow, they push on either side of the rock's crevice and cause it to become larger. (See Figure 3.3.)

Figure 3.3 As the diameter of the root grows, it exerts a force powerful enough to wedge rocks apart.

- Frost wedging (or, ice cracking). Water may collect in rock crevices. On cold nights, especially in alpine regions, this water freezes. Water expands on freezing, thus putting further pressure on the cracks in rocks. (See Figure 3.4.)

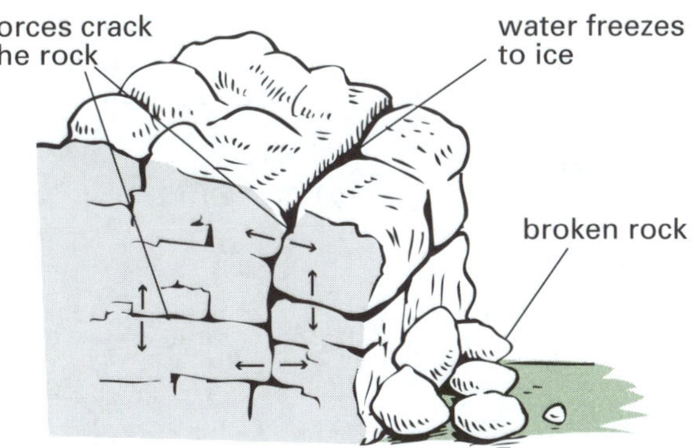

Figure 3.4 In frost wedging water seeps into crevices and, on freezing, pushes rocks apart.

Chemical weathering

Chemical weathering involves chemically altering the composition of the minerals in the rocks. The altered chemicals may not be able to hold the rock together and so it crumbles. Because moisture and heat promote chemical reactions, weathering generally goes deeper in moist and warm climates, rather than in dry and cold ones.

The following are some examples of chemical weathering.

- Oxidation. Oxygen from the air, or dissolved in water, can act on the minerals in rocks producing new, softer substances. For example, iron in some rocks is converted to a rusty brown iron oxide that is soft and crumbly.
- Acids. Rainwater percolating through the ground or over rocks carries with it some dissolved gases, such as carbon dioxide. These gases produce weak acids in the water. For instance, when carbon dioxide dissolves it forms a weak solution of carbonic acid. As the water moves further through the soil, its acid strength increases many times by the addition of carbon dioxide created via the bacterial decay of vegetation. The acid can react on rocks with high carbonate content, such as limestone, and produce limestone caverns such as the Jenolan Caves in south-eastern NSW (see Figure 3.5). Acids can react with potassium feldspar turning it into clay.

Figure 3.5 Limestone caves, a feature of chemical weathering

- Hydration. Some rocks that contain clay minerals swell when they absorb water. This changes its chemical structure and may become a softer material. The repeated hydration and drying out can cause weathering.
- As water moves through rocks, it carries away (leaches) soluble matter. While plants can use some of this leached material, a lot of it finds its way to the sea.

Erosion

The weathered materials are recycled to form new landscapes. Erosion is the movement of the broken-down and dissolved rock material from one place to another on the Earth's surface. The agents of weathering mount a constant assault that even the most resistant rocks cannot withstand. Mountain peaks may erode by only a few centimetres each year and, unless there are movements in the Earth's crust to rebuild them, will eventually be reduced to rubble and transported away.

The main agents of erosion include the following.

- Water. Moving water (as in rivers, rain or sea waves) can generate a lot of force to transport boulders, as well as grains of rock and dissolved minerals, to new locations. This material is deposited when water slows such as in lakes, lagoons and estuaries.
- Glaciers. Rivers of ice carve out U-shaped valleys as they carry debris downhill. Some of the rocks they carry rub along the walls or floor of the valley, further weathering them physically. (See Figure 3.6.)

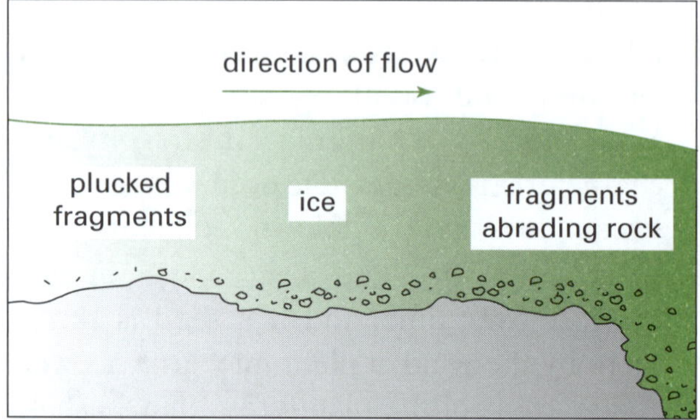

Figure 3.6 Plucking and abrading rock as the glacier moves

- Wind. Broken down (weathered) particles are whipped up by the wind and carried some distance before they are dropped. As these particles are blown along, they can rub against other rocks and sandblast them. A type of wind-deposited sediment is called loess. This is silt, usually accompanied by some clay and fine sediment. This occurs around the world and can form deposits many metres thick.
- Gravity. Landslides, rockslides and mudflows all occur under the influence of gravity. The material can no longer be supported because material below it is broken down or removed. As the rock, soil or mud moves downhill, it can collect more and more, destroying everything in its path (see Figure 3.7). Some mudflows have been known to travel at 20 to 30 cm per second, killing hundreds of people in their paths. Clearing the land of trees (as has occurred in many underdeveloped countries) has increased the chances of this happening, resulting in immense damage to life and property.

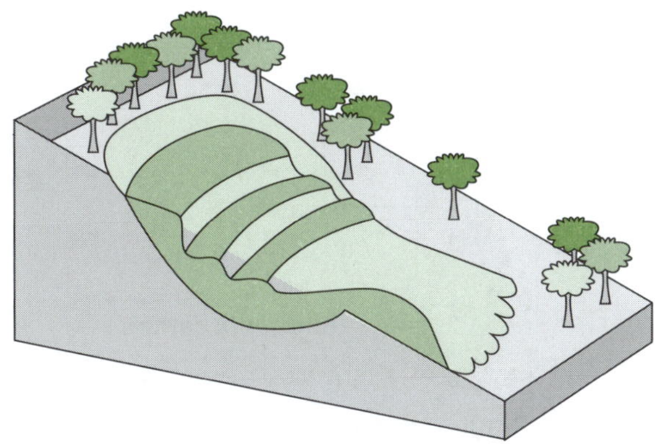

Figure 3.7 Landslides and mudflows occur when heavy rainfall or high levels of groundwater trigger soil or sediments to move causing extensive erosion.

Water is the most powerful agent of erosion. Some 100 billion tonnes of sediment are washed into the oceans every year.

Experiment 1

Modelling erosion along the coast

Aim

- To construct a model of a coastal river in a stream tray and determine how the speed of water flow affects the transport of sediment out to sea

Method

1. Set up a stream tray as shown in Figure 3.8. Create a mountain area, a coastal plain and a coastline using the sand, silt and pebbles.

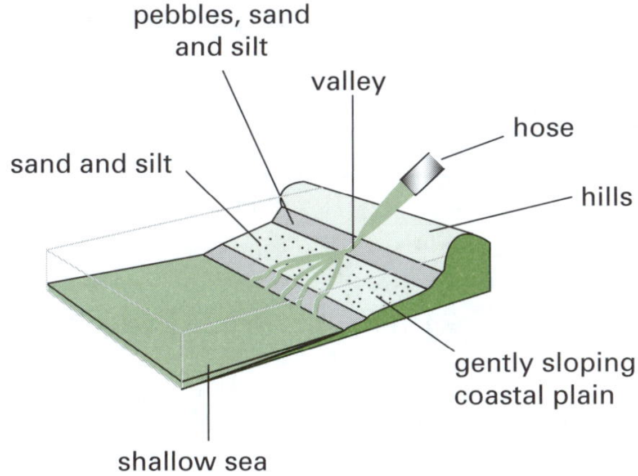

Figure 3.8 Modelling erosion along a coastal plain

2. Add water at one end to create a shallow sea. Create a valley in the mountain area and allow a stream of water from a hose to run down the valley.
3. Investigate the way the running water carves out new channels. When the speed of the water is low, observe what type of sediment is carried furthest. Identify what sediment is carried into the sea.
4. Find out what happens to the sediments and the extent of erosion if the speed of the stream increases.

Results and analysis

When the flow rate of the water is low, the rate of erosion is lower and only the silt and some sand are washed across eroded channels in the plain into the shallow sea. As the water flow rate increases, more sand and some pebbles are washed along the channels towards the sea.

Analyse these results and explain any differences in erosion and transportation of sediments. Go to p. 193 to check your answer.

Conclusion

Use short sentences to state what conclusions can be made from this experiment.

Go to p. 193 to check your answer.

Sedimentary rocks

The sediments carried by water and wind are eventually deposited as horizontal layers. These sediments are pressed and cemented together to form new sedimentary rocks. This is a slow process taking many millions of years.

When first deposited, the sediments are loose as there is plenty of water between the particles. As more sediment is deposited on top, the growing overburden compresses the particles which fuse as water is squeezed out. Water slowly oozing through the various layers deposits mineral matter around the grains. The mineral matter acts as 'glue' cementing the particles together.

The deeper layers are older than the upper layers. This process is shown in Figure 3.9.

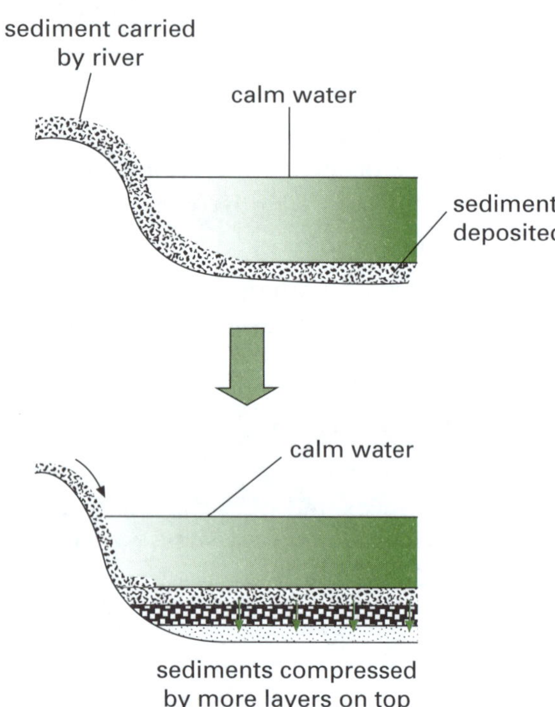

Figure 3.9 Formation of sedimentary rocks

Experiment 2

Particle size and sedimentation

Aim

- To identify how particle size and mass affects the deposition of sediments

Method

1. Place equal quantities of each sediment sample (silt, sand, mud, small pebbles) in an empty soft drink bottle so that the bottom is filled to a depth of about 5 or 6 cm.
2. Fill the bottle with water. Replace the lid.
3. Tip the bottle upside down several times and shake it.
4. Stand the bottle on your benchtop and observe what happens to the samples over a period of 20 minutes (see Figure 3.10). Leave the sediment to settle.

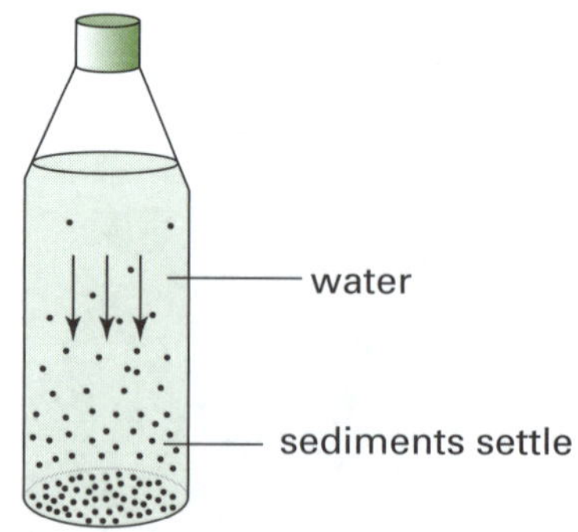

Figure 3.10 Settling of sediments

Results and analysis

The large pebbles were observed to settle first followed by the sand/silt and the fine clay/mud particles settle last. Some fine mud and clay does not settle but remains suspended.

Analyse these results. Go to pp. 193–194 to check your answer.

Conclusion

Write a short conclusion for this experiment.

Go to pp. 193–194 to check your answer.

Sedimentary rocks form when sediments settle out of water. Less commonly, they also form when the sediments deposit out of the air or ice. Very fine particles settle out and compact to produce siltstones or mudstones. Coarser particles can produce sandstone. When some larger pieces of rock are trapped, conglomerates can form. Table 3.1 shows how various sedimentary rocks are classified according to the particle size of the sediments.

Table 3.1 Rock formation

	Coarse	Medium	Fine
loose particles	boulders, pebbles, gravel, rubble, shingles	sands	mud, clay
cemented rock	conglomerate, breccia	sandstone	mudstone, shale

Limestones are deposits which were formed from dissolved material, generally seawater. One way is where seawaters are warmed and evaporated. This increases the concentration of calcium carbonate (the chemical present in shells) in seawater to a point where it is precipitated. Tiny grains of calcium carbonate or calcite ($CaCO_3$) serve as centres where more calcium carbonate deposits. Sometimes microscopic organisms aid the deposition. Often the hard remains (shells) and other fragments of sea animals dominate limestone. Adding drops of dilute acid can show the presence of calcite in a rock. A fizzing reaction and gas evolving indicates calcite.

Coal is formed from the remains of plant material which fell into muddy swamps hundreds of millions of years ago. As more sediments fell on top, it compressed the plant remains before they had time to decay and moisture was squeezed out. Slowly coal was formed.

Fossils

Sedimentary rocks cover about three-quarters of the world's land area and are often used as building materials. Sandstone has been used more than any other kind of natural stone for building purposes. Besides building, ground-up limestone is the main constituent of cement.

Some sedimentary rocks contain fossils. These are the naturally preserved remains or traces of animals or plants. Scientists studying sedimentary rocks and fossils can work out the environment that must have existed at the time when the fossil was being formed.

Generally, fossils found in rock are made of the hard parts of animals, such as bone, shell or chitin. Chitin is the tough outer part of insects. When an animal dies, it starts to decay, scavengers eat it and the weather helps break it down. After a few months or years there may be very little trace on the ground of where the animal died. This is what happens to many organisms; these organisms fail to form fossils. For a fossil to form, conditions must be right at the time the animal dies. Oxygen must be excluded so that decay is slowed. Suppose an animal dies and soon becomes buried in fine sediment such as mud, sand or silt. The hard parts of the body are not destroyed by scavengers or the weather and stand a much better chance of being preserved. Animals and plants that live in water, especially seas, have a far better chance of being preserved than those that live on the land. For animals and plants on land to become fossils, they generally need to be swept into lakes, rivers, swamps or peat bogs and become quickly buried in sediment before decay begins. Over thousands and millions of years, the sediment becomes compacted and cemented into rock and the remains of the organism become a fossil.

There are a number of different types of fossils:

1. unaltered soft parts (e.g. frozen mammoths in Siberia)
2. unaltered hard parts (e.g. insect fossils in amber)
3. altered hard parts: chemical and physical changes occur (e.g. petrified wood, mineralisation of skeletons, carbonised leaves)
4. trace fossils (e.g. shell moulds and casts, dinosaur footprints, animal burrows).

Figure 3.11 shows examples of fossil types.

Figure 3.11 Some fossils

Table 3.2 shows the environments in which various types of sedimentary rocks form.

Table 3.2 Sediments and sedimentary rocks

Sedimentary rock	Sediment	Forming environment
mudstone	clay/mud	sediments carried by quiet, slow-moving water and deposited in lakes, lagoons and on the ocean floor
shale	silt/mud	sediments carried by quiet, slow-moving water and deposited in lakes, lagoons and on the ocean floor
sandstone	sand	sediments carried by faster-moving streams and rivers or ocean waves and deposited on the ocean floor
conglomerate	gravel/ pebbles	heavy sediments carried by fast-moving streams and rivers and deposited in these environments
limestone	crushed shells/ coral	coral and also marine invertebrates with shells grow in shallow, warm seas; after death their exoskeletons are crushed and cemented together
coal	plants	swamp plants die and are buried in mud on the floor of the swamps

3.3 Igneous rocks

Igneous rocks are formed from molten rock inside the Earth (magma). Igneous rocks can be classified into groups based on their crystal size and rate of cooling. When the magma cools slowly deep inside the Earth, the rock that forms has large crystals. Such rocks are called plutonic (intrusive) igneous rocks. Magma that flows out onto the Earth's surface to form lava cools rapidly and forms rocks with very small crystals. Such rocks are called volcanic (extrusive) igneous rocks. Intermediate-sized crystals can form in rocks that solidify more rapidly than plutonic rocks. Figure 3.12 shows the different types of igneous structures.

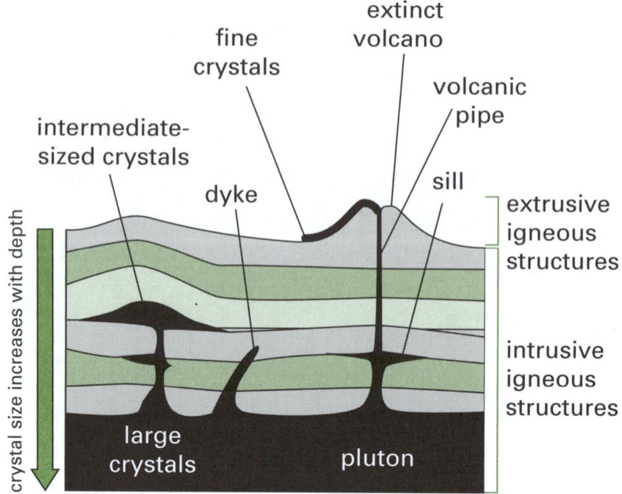

Figure 3.12 Igneous structures

Table 3.3 classifies common igneous rocks according to crystal size and composition.

Table 3.3 Classification of some common igneous rocks

Type of igneous rock	Crystal size and rate of cooling	Examples
plutonic (intrusive)	large crystals due to very slow cooling of magma	• granite: rich in light-coloured minerals such as feldspar and quartz • gabbro: rich in dark minerals such as hornblende and biotite mica
volcanic (extrusive)	fine crystals (or none) due to rapid cooling of lava	• rhyolite: rich in light-coloured minerals • basalt: rich in dark minerals • pumice: light-coloured lava froth filled with gas cavities • obsidian (volcanic glass): no crystals

Over 700 types of igneous rocks have been described. Most of these rocks formed beneath the surface of Earth's crust. Magma can intrude into existing strata to form various structures such as plutons, dykes and sills. The presence of such intrusions can complicate the geological history of an area.

3.4 Metamorphic rocks

Changes in rock strata can also occur due to prolonged periods of heating and pressure caused by igneous activity or tectonic activity in the crust. This leads to the formation of metamorphic rocks.

Any rock may be subjected to heat and pressure causing profound physical and/or chemical changes. Three main ways that metamorphic rocks form are:

1. simply being deep beneath the Earth's surface, and subjected to the high temperatures and great pressure of the rock layers above it
2. tectonic processes such as continental collisions, causing horizontal pressure, friction and distortion
3. rock is heated up by the intrusion of hot molten rock called magma from the Earth's interior.

Studying metamorphic rocks that are finally exposed at the Earth's surface, following erosion and uplift, provides information about the temperatures and pressures that occur at great depths within the Earth's crust. Metamorphic rocks are classified by texture and by their chemical and mineral content.

During the process of metamorphism the particle sizes of the rock may change. For instance, the small crystals of calcite (a mineral) in the sedimentary rock limestone change into larger crystals in the metamorphic rock marble. Similarly, in metamorphosed sandstone, the original quartz sand grains recrystallise resulting in very compact quartzite. Both high temperatures and pressures contribute to recrystallisation. High temperatures allow the atoms and ions in solid crystals to migrate and reorganise themselves in the crystals, while high pressures make the crystals dissolve within the rock at their point of contact.

Figure 3.13 shows one way metamorphic rocks can form.

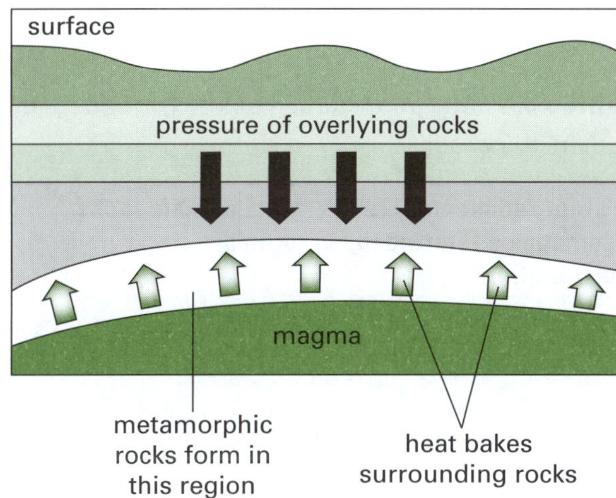

Figure 3.13 One way metamorphic rocks form

Metamorphic rocks can be classified into the following two groups based on the way the process occurs.

1. Contact metamorphism takes place when magma is injected into the surrounding solid rock (country rock). The changes that occur are greatest wherever the magma comes into contact with the rock because the temperatures are highest at this boundary. This is shown in Figure 3.14.

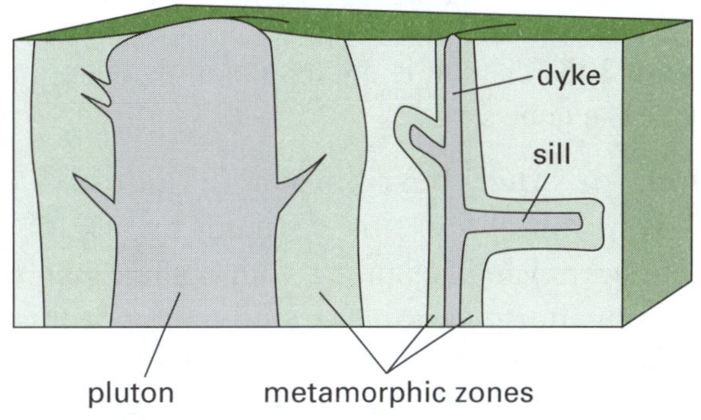

Figure 3.14 In contact metamorphism the alteration of the surrounding rocks is caused by heat and fluids originating from the intruding igneous body. This alteration zone can vary from a few centimetres to several kilometres in width, depending on the size and temperature of the intrusion.

2. Regional metamorphism occurs in great masses of rock over a wide area. Simply being deep within the Earth's surface and subjected to high temperatures and pressures can metamorphose rocks.

The metamorphic rock that forms depends on a number of factors, including the parent material (see Table 3.4).

Table 3.4 Metamorphic rocks formed from sedimentary rocks

Parent sedimentary rock (sometimes referred to as protolith)	Metamorphic rocks formed
quartz sandstone	quartzite
limestone	marble
shale	hornfels, slate, schist, gneiss

Identifying metamorphic rocks is assisted by the fact that they often have banding (foliation). Foliation is common in rocks affected by regional metamorphism (see Figure 3.15). Contact metamorphic rocks are recrystallised and rarely show foliation.

Figure 3.15 Foliation is often seen in metamorphic rocks like gneiss.

One of the best-known examples of metamorphic rocks is formed by regional metamorphism. Beginning with a shale parent, a sedimentary rock, increasing metamorphism produces a sequence of metamorphic rocks that goes through from slate to phyllite, schist and gneiss. It can be hard to imagine that all these very different-looking rocks can come from the same parent rock.

shale → slate → phyllite → schist → gneiss

increasing temperature and pressure →

Test yourself 1

Part A: Knowledge

1. The layer of the Earth which is second-closest to the surface is *(1 mark)*
   A inner core.
   B outer core.
   C crust.
   D mantle.

2. The lithosphere includes the *(1 mark)*
   A crust.
   B crust and upper mantle.
   C crust and mantle.
   D mantle and outer core.

3. The hardest mineral on the Mohs scale is *(1 mark)*
   A talc.
   B quartz.
   C diamond.
   D calcite.

4. An example of the chemical weathering of rocks is *(1 mark)*
   A oxidation.
   B frost wedging.
   C temperature variation.
   D abrasion.

5. The sediment that forms shale rock is *(1 mark)*
   A sand.
   B silt and mud.
   C gravel.
   D clay and pebbles.

6. Complete the following restricted-response questions using the appropriate word. *(1 mark for each part)*
   a) Coal is ultimately formed from the remains of .... that have been buried and compressed over long periods of time.
   b) Basalt is an example of an .... igneous rock.
   c) Limestone undergoes metamorphic changes and turns into ....

d) Granite is a light-coloured rock rich in .... and quartz minerals.

e) When lava cools rapidly, the volcanic rock that forms has .... crystals.

7. Use the code letters to match the terms or phrases in each column. *(1 mark for each part)*

Column 1	Column 2
A limestone	F coral
B sedimentary rock	G obsidian
C igneous rock	H schist
D metamorphic rock	I U-shaped valleys
E glaciers	J mudstone

Part B: Skills

8. Examine Figure 3.16 concerning weathering.

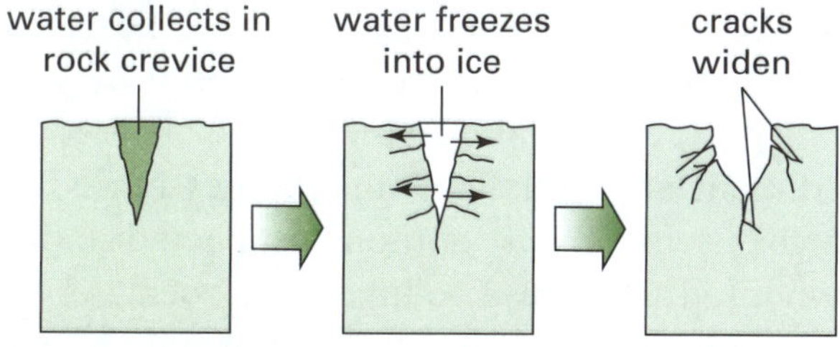

Figure 3.16 Weathering

a) Identify the type of weathering. *(1 mark)*

b) Name the property of water that produces this type of weathering. *(1 mark)*

9. Sediments can be classified as gravel, sand or mud according to the size of their particles. Gravel particles are always larger than 2 mm in diameter. Examine the information about gravels shown in Table 3.5.

Table 3.5 Information about gravels

Size (mm)	Classification
2.0–4.0	granule
4.0–64	pebble
64–256	cobble
> 256	boulder

a) Classify sediments with the following particle sizes.

i) 200 mm *(1 mark)*

ii) 10 mm *(1 mark)*

b) Which type of gravel is most readily transported by a fast-flowing river? *(1 mark)*

c) Sand particles lie in the range 0.0625 mm (very fine sand) to 2.0 mm (very coarse sand). Mud particles lie in the range < 0.0039 mm (clay) to 0.0625 mm (coarse silt).

i) What type of particle is one with a diameter of 0.01 mm? *(1 mark)*

ii) Which type of sediment is carried furthest out to sea by a coastal river? Why? *(1 mark)*

10. Three minerals are calcite, fluorite and apatite. Fluorite scratches calcite but doesn't scratch apatite. Which are the hardest and softest of these minerals? *(2 marks)*

11. Beach sand contains plenty of the mineral quartz. Why should you not put your plastic sunglasses down on the sand at the beach? *(1 mark)*

12. Look at the rectangular block diagram of a cross-section of the countryside shown in Figure 3.17.

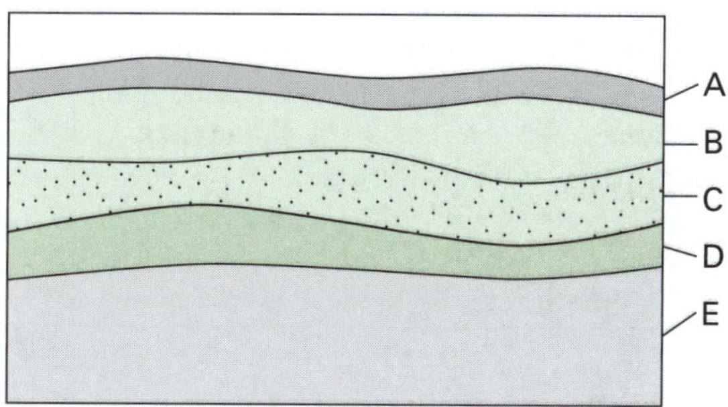

Figure 3.17 Cross-section of rock layers

a) What feature of the diagram suggests that the rock beds shown could be sedimentary? *(1 mark)*

b) Which layer shows the oldest rocks? *(1 mark)*

c) Which layer shows the youngest rocks? *(1 mark)*

13. The following steps describe the formation of sedimentary rocks. The statements are out of order. Rewrite them in your workbook in the correct order. *(4 marks)*

A. This weathered material is constantly being swept from bedrock, carried away and eventually deposited in lakes, river valleys, seas and many other places.
B. By compacting and cementing the sediments together, the material is transformed into solid rock.
C. While being transported, the particles are broken down even further.
D. Weathering begins the process of wearing down the landscape.
E. After the material is deposited, it is called sediment.
F. One way the sediments are cemented together is by mineral matter being deposited in the spaces between particles.
G. The particles are carried to a new location where they are deposited.
H. Erosion agents such as wind, ice, waves and running water remove the products of weathering.

14. Table 3.6 shows the percentage of each type of rock in the total volume of the crust. Use this information to draw a pie chart. *(4 marks)*

Table 3.6 Percentage of each type of rock

Rocks	% abundance
igneous	65
metamorphic	27
sedimentary	8

3.5 Rock cycle

Rocks are constantly broken down and reformed. This transformation of the three kinds of rocks (sedimentary, igneous and metamorphic) is one of the fundamental pathways through the rock cycle, which is shown in Figure 3.18. It is not a quick process, with some steps taking millions of years. It is transformations of this kind that have resulted in all the great diversity of rocks we find on the Earth.

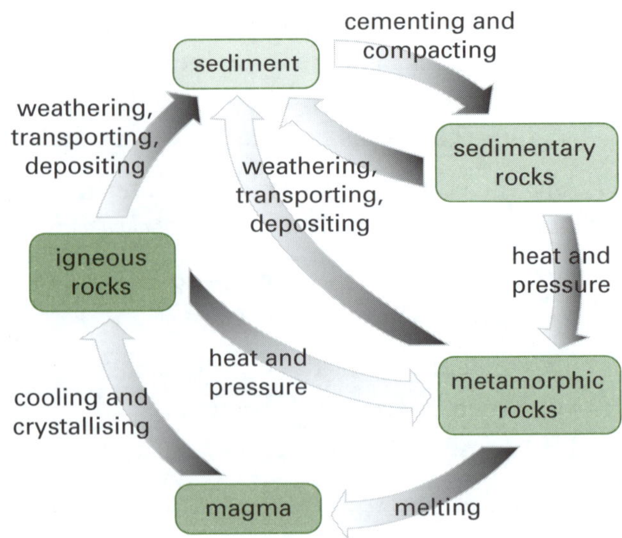

Figure 3.18 The rock cycle

In the rock cycle, magma cools and crystals form. This results in an igneous rock. On the Earth's surface weathering will break down these, and other, rocks. The weathered material may be transported by water, wind or ice. This loose sediment accumulates and becomes buried by further sediment. It is then compacted and/or cemented to form a sedimentary rock. As the sedimentary layers become increasingly buried, the resultant heat and pressure begins to reshape the crystals, forming a metamorphic rock.

However, in reality, most rocks never pass through the complete rock cycle in a continuous manner, such as that outlined in Figure 3.18. Within this overall cycle there are numerous alternate pathways that a rock may pass during its history. For example, a buried igneous rock may undergo metamorphosis without ever being exposed to the surface for erosion to occur.

The oldest rocks that have been found are more than 3.9 billion years old. The Earth itself is at least 4.5 billion years old, but rocks from the beginning of Earth's history have changed so much from their original form that they have become new kinds of rock. In this time some rocks have changed many times, others only a few. The youngest rocks are those obtained when lava from erupting volcanoes solidifies and can be only a few days or weeks old.

3.6 Ores

Most of our non-biological resources are obtained from the ground. Many rocks contain minerals with important elements, especially metals. Sometimes there is not enough of the metal present (too low a grade) to make it economical to remove the metal from it. When there is enough concentration of the metal, and when it can be removed economically, the mineral is called an ore. An ore is a mineral that can be mined for a profit.

The grade or concentration of an ore, as well as where and how it occurs, will directly affect the costs associated with mining the ore. If it is too expensive to mine or extract the metal, it may not be profitable and so not worth mining. Sometimes mines can be closed up for years until the price of the metal increases or more efficient methods of mining and processing are developed.

Metal ores are generally oxides, sulfides, silicates or native metals (not usually forming compounds) such as native platinum and gold. The ores must be processed to extract the metals of interest from the waste rock and from the ore minerals. Table 3.7 outlines the details of some important ores.

Table 3.7 Some important ores

Ore	Chemical formula	Metal extracted
bauxite	Al_2O_3	aluminium
cassiterite	SnO_2	tin
chalcocite	Cu_2S	copper
chalcopyrite	$CuFeS_2$	copper
cinnabar	HgS	mercury
galena	PbS	lead
hematite	Fe_2O_3	iron
ilmenite	$FeTiO_3$	iron, titanium
magnetite	Fe_3O_4	iron
pitchblende	UO_2	uranium
sphalerite	ZnS	zinc
wolframite	$(Fe, Mn)WO_4$	tungsten

How mineral ores form

Ore minerals tend to be concentrated in small, localised rock masses that form by geologic processes. These processes include the following.

- Magmatic deposits. As the magma cools, mineral crystals settle and concentrate at the bottom of the magma chamber.
- Hydrothermal deposits. Hot aqueous solutions, which carry dissolved minerals, deposit them where they cool.
- Mechanical accumulation. On the Earth's surface flowing water transports mineral grains and deposits them elsewhere in quieter bodies of water.
- Beach placers. Waves reaching a beach at an angle produce a longshore drift. This occurs parallel to the beach and can produce a beach placer. Beach placers are a major source of ilmenite and rutile from which titanium metal is extracted.
- Placer deposits. A typical placer deposit involves liberating gold or silver from source regions by mechanical erosion, transporting downstream, and depositing the material in slowed waters, at sharp breaks in slope, or at areas of increased roughness (see Figure 3.19).

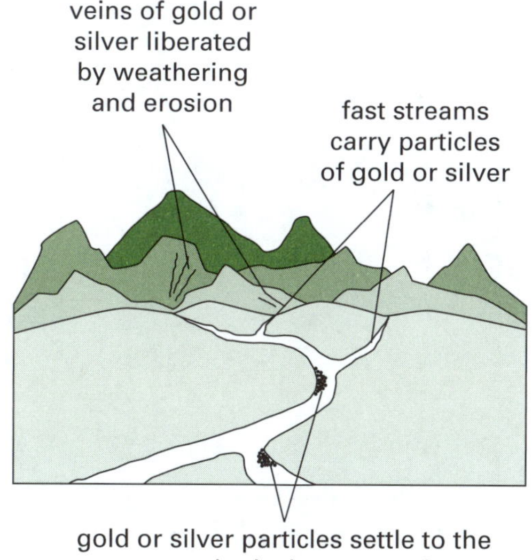

Figure 3.19 Placer deposits can occur along rivers or streams.

No deposit consists entirely of a single ore mineral. There are always mixtures of valueless minerals, collectively called gangue. The more concentrated an ore mineral, the more valuable the mineral deposit. The ores are extracted through mining. These are then refined to extract the valuable element(s).

3.7 Environmental issues and recycling

Many of the non-living resources we use are not renewable. With increasing world population and with an increased demand for materials there is a need to keep the materials we have lasting longer. One way to do this is through recycling. Another is to maintain the developed

Case study 1: Iron ore

Iron is the basic ingredient for steel and, mixed with other metals, makes a whole variety of useful alloys. We rely on iron (as steel) to make almost everything we need for living in the 21st century!

Iron isn't often found on its own but is normally combined with other chemicals in rocks. Australia has huge reserves of iron ore (mostly in the Pilbara region of Western Australia), supplying about an eighth of the world's iron ore needs. Figure 3.20 shows the location of Australia's operating iron ore mines.

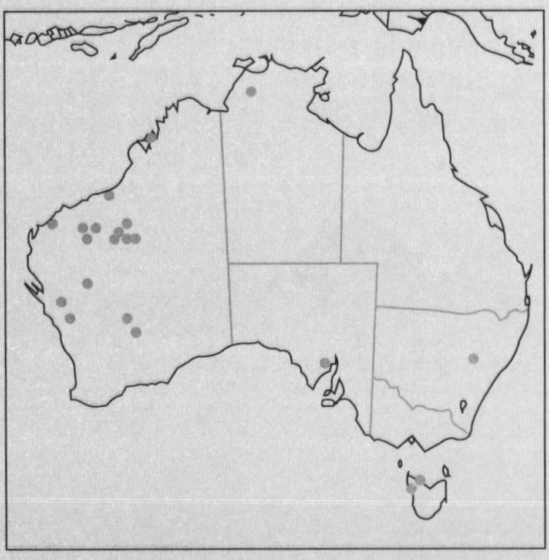

Figure 3.20 Operating iron ore mines in Australia

The rock bearing iron ore is first blasted and dug up from open pit mines, crushed, smelted (heated with other substances including carbon) in huge furnaces, then combined (alloyed) with metals like nickel, chromium, manganese or titanium. Alloying gives steel special properties like electrical resistance, and resistant to wear, rust, impact, shock or expansion when heated. The cooled steel is shaped and can be coated with tin, zinc or paint to help protect it from rusting.

The following are some iron facts.

- Around 6000 years ago Egyptians and Sumerians used iron from meteorites to make beads, ornaments, tools and weapons.
- Some ancient peoples used finely ground hematite, an ore of iron, as rouge. Some Australian Aboriginals still use this blood-red colouring for ceremonial face painting.
- The Iron Age (~1000 BC) began when the Hittites from the Middle East produced iron that was harder than bronze.
- With iron cannons and cannon balls becoming more popular during the Middle Ages, iron overtook copper and bronze as the most widely used metal.
- In the late 19th century wooden ships gave way to steel, steel machinery came to the factories and the railroad was invented.
- Tin cans are really steel cans with a thin tin coating to prevent rusting.
- About 60% of steel available for recycling goes back into making new steel.

Case study 2: Copper ore

While copper is sometimes found in its pure (native) form, it is generally locked together with other minerals. Australia produces around 20% of the world's copper. Australia's largest copper production is at Mt Isa in Queensland while Olympic Dam in South Australia is mining out one of the largest copper-bearing ores in the world. At Olympic Dam, copper-bearing rock (ore) is:

- blasted underground
- scooped up by front-end loaders
- taken in large trucks to underground crushers
- hoisted to the surface in skips up shafts
- crushed further at the surface,
- mixed with water and other special chemicals to remove the waste (gangue) rock
- froth-floated and the copper ore skimmed off
- then heated and chemically treated in other ways to purify the copper and remove it from any other metals.

Figure 3.21 shows the location of Australia's operating copper mines.

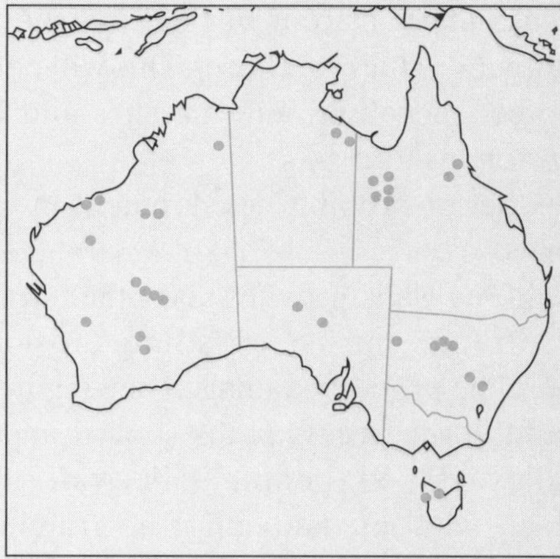

Figure 3.21 Operating copper mines in Australia

The following are some copper facts.

- Copper was the first metal used in place of stone as it was far easier to shape. It began being used about 9000 years ago.
- Bronze, a hard alloy, was discovered around 4000 BC by mixing in some tin with copper. This marked the beginning of the Copper–Bronze Age.
- Copper's resistance to rusting made it popular for roofing on important buildings in the past.
- The first major discovery of copper in Australia was in South Australia in 1842.
- Copper wires were used in the first electric telegraph and telephone.
- Copper and brass are easily recycled. Some 70% of the copper now used has been recycled at least once.
- The average family home contains more than your weight of copper. (Can you think of where?)
- There is about 1.8 tonnes of copper in a Boeing 747-200 jet plane.

As with iron ore, there are many areas of science involved in mining and extracting copper ore. Geologists and geophysicists need to survey and find the ore, mining engineers are responsible for extracting it, and chemists and chemical engineers are then responsible for refining it. All the while the industry, though employing a variety of scientists, is responsible for not polluting the environment through its actions. And once the ore has been removed and the mine closed, the land needs to be returned to how it looked before, so it doesn't remain an eyesore. This is the job for biologists, ecologists and horticulturalists.

and natural environments from deteriorating, especially with the greater demands more and more people on this planet put on it.

Recycling

Recycling helps our society because it:

- reduces the waste in landfills
- takes less energy to recycle something than to make the same item from raw materials.

For example, recycling scrap aluminium requires only 5% of the energy used to make new aluminium. The process involves simply re-melting the metal, which is far less expensive and energy intensive than creating new aluminium through the electrolysis of aluminium oxide (Al_2O_3), which must first be mined from bauxite ore and then refined.

Australians are among the world's best recyclers. However, we produce more waste than any country except the United States. By not buying unnecessary items or preferring items with less throwaway packaging, using the recycling bin and by composting organic matter we can reduce what goes to the tip.

Strategies to maintain the local environment

The Tank Stream provided Captain Arthur Phillip and the new settlement at Sydney Cove in 1788 an appropriate campsite and a reliable source of water. For over 40 000 years previously, Aboriginal people used this stream for food and water, and had learned to live in harmony with it. Excavations near the original course of the stream revealed a high concentration of flake stone Aboriginal artefacts made from water-worn pebbles.

As the settlement grew, Phillip realised that he had to protect the fresh water source. Anyone caught polluting the stream was severely punished. It remained the main source of water for 30 years. But when Phillip left Australia this protection was forgotten and it quickly became polluted. Eventually it turned into an open sewer and became a major source of smell and illness. Starting in 1860, the Tank Stream was progressively covered over and is now a storm water channel running under Sydney.

This story illustrates what can happen to the environment if people abuse it.

With more people living in close proximity to each other in an area, there are stresses set up on environments such as bushland, beaches, lakes or rivers. Here are some strategies to protect a shoreline.

- Maintain/restore wetlands; preserve habitat for vulnerable species; preserve coastal land/development.
- Remove shoreline structures such as bulkheads and other engineered structures to allow for shoreline migration and maintain sediment transport.
- Plant sea grasses to stabilise sediment and reduce erosion.
- Create marsh by planting the appropriate species (typically grasses, sedges or rushes) in the existing substrate. This will help preserve the habitat for vulnerable species.
- Create dunes along the backshore of the beach; this includes planting dune grasses (such as marram and spinifex grasses) and sand fencing to induce settling of wind-blown sands.
- Install rock sills and other artificial breakwaters in front of tidal marshes along energetic estuarine shores. This will naturally protect shorelines and marshes and inhibit erosion inshore.
- Restrict or prohibit development in erosion zones.
- Build sea walls, groynes and other structures to protect against coastline erosion and shoreline property damage. (See Figure 3.22.)
- Build beach access paths (board and chain walkways) so people can walk to the beach without damaging the fragile dune environment. (See Figure 3.23.)

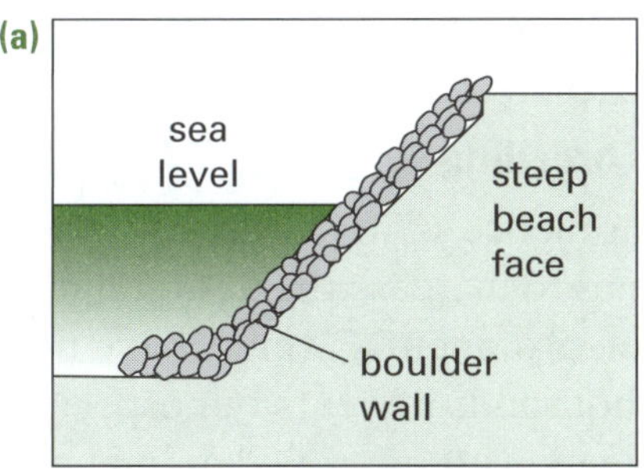

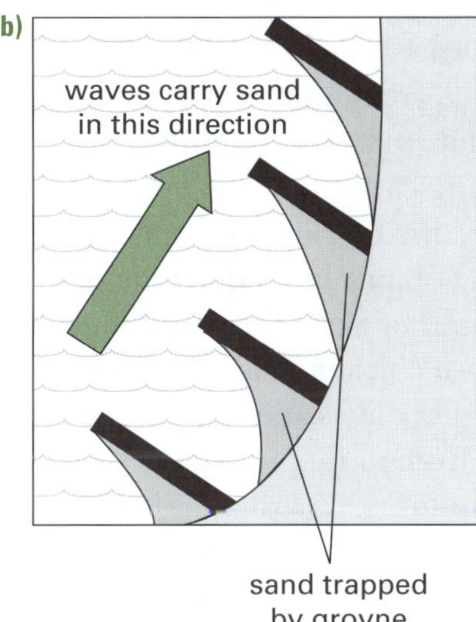

Figure 3.22 Some beach stabilisation structures. **(a)** Boulders or interlocking cement blocks can be built along a shoreline or steep facing slope to protect it against wave or current erosion. **(b)** Groynes are concrete and/or rock barriers or walls jutting out to the sea from the shoreline.

Figure 3.23 Some examples of shoreline sand dune management measures

Sand dune systems are dynamic and potentially fragile coastal habitats. They need to be maintained by natural accumulation of sand and erosion, allowing dunes and bare sand to be mobile and recolonise naturally with minimum impact from humans. Ideally, natural processes should be allowed to proceed without hindrance. However, this can often be difficult especially where buildings and roads have been constructed in or near the dunes and are affected by any windblown sand.

Improving farming yields and sustainability

Agriculture has changed significantly, especially since the end of World War II. New technologies, mechanisation and increased chemical use has allowed food and fibre production to soar. In developed countries these changes allowed for fewer farmers with reduced labour demands to produce the majority of the food and fibre.

Farmers live on and work the land not for short-term profit, but to maintain and have it productive for generations to come. Sustainable agriculture involves using ecology (relationships between organisms and their environments) and other scientific practices in the farming methods.

Rotating crops

Growing and harvesting crops removes nutrients from the soil. Not replenishing these causes the land to suffer from depletion, either becoming unusable or producing fewer crops in the following years. Sustainable agriculture involves replenishing the soil while minimising the use of non-renewable resources, such as fertilisers and pesticides.

One method of sustainable agriculture is crop rotation. This is the practice of growing a series of dissimilar types of crops in the same area in following seasons. For instance, legumes such as alfalfa, clover, peas, beans, lentils, soy and peanuts have nodules on their roots that contain bacteria and can use nitrogen from the air (see Figure 3.24). It therefore makes good agricultural sense to alternate them with cereals and other plants that require nitrates. A common crop rotation practice is to alternate soybeans and corn, or cotton and rice.

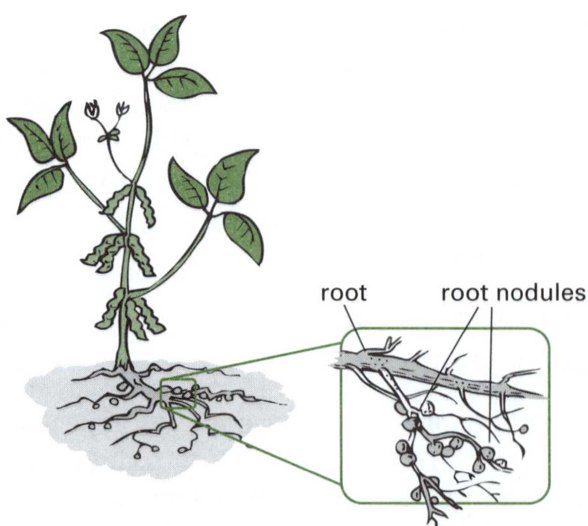

Figure 3.24 Legumes contain bacteria (rhizobia) in their root nodules, which can fix atmospheric nitrogen. This reduces farmers' fertiliser costs, and allows them to be part of a crop rotation program replenishing soil that has been depleted of nitrogen.

The benefits of crop rotation include:

- avoiding the build-up of pathogens and pests that often occurs when only one type of plant is continuously grown
- improving soil structure and fertility by alternating deep-rooted and shallow-rooted plants; soil enrichment produces healthier plants that resist disease.
- balancing the fertility demands of various crops to avoid depleting soil nutrients; replenishing nitrogen by using green manure or legumes in sequence with cereals and other crops can do this.

Green manure is a type of cover crop grown mainly to add nutrients and organic matter to the soil. Typically, a green manure crop is grown for a specific period, and then ploughed under and incorporated into the soil.

A farm that is viable may have negative effects on the environment elsewhere. This is not sustainable agriculture. For example, over-applying synthetic fertilisers or animal manures can improve farm productivity but it can also pollute nearby rivers and coastal waters. On the other hand, destroying rainforests (as in the case of slash-and-burn farming), can result in low crop yields because the nutrients in the soil become depleted.

Selective breeding

Another practice is to selectively breed plants and animals to increase the desired qualities in a suitable environment. Farmers have been doing this for thousands of years but especially in the last 150 years with improved animal nutrition.

This is done for the following reasons.

- Produce greater quantities of the marketable product more efficiently. For example, breeding cattle that grow quickly with good quality meat; cows that produce a greater volume of milk containing the right nutrients; increased egg production in a flock of White Leghorn chickens; selecting for increased oil content in olives and canola; and greater crop yields from more productive strains of wheat, corn and rice.
- Produce strains and races of economically important plants and animals that are resistant to a wide variety of diseases. For example, introducing genes into plants allowing them to combat infections from viruses or bacteria, or to produce toxins that will kill plant-eating insects.
- Better suit the animal or plant to the environmental conditions. For example, to overcome the problems of traditional dairy breeds performing at reduced levels under hot, humid and tick-infested conditions of northern Australia, the Australian Milking Zebu cattle was selected for heat tolerance, tick resistance and improved milk production.

Experiment 3

Soils and water

Aim

- To determine the ability of three soils to hold water

Method

1. Line three flowerpots with paper towel to prevent the soil escaping through the drain

holes. Half-fill each pot with one of the soils provided (sand, clay and loam soils).

2. Place each pot in its own plastic container. A beehive shelf will help to raise it off the base.
3. Measure out 500 mL of water in the measuring cylinder and pour it into the pot with the clayey soil as shown in Figure 3.25. Don't allow water to run out over the top of the pot. Repeat with the other two soils.
4. Let the pots stand in their container trays for about 10 minutes.
5. Pour the water collected in the tray back into the measuring cylinder and read off how much water ran out of the pot.

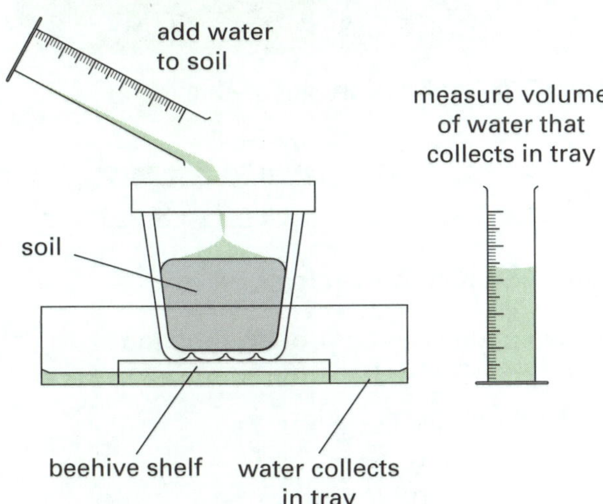

Figure 3.25 Soils and water

Results

The volume of water collected in each experiment is:

- sandy soil: 405 mL
- loam soil: 195 mL
- clay soil: 110 mL.

Analysis

Analyse these results in terms of the water holding capacity of each soil. Go to p. 194 to check your answer.

Conclusion

Write a suitable conclusion for this experiment.

Go to p. 194 to check your answer.

Aboriginal people, ecosystems and fire

Aboriginal people have been in Australia for perhaps 60000 years. In that time fire was used for warmth, hunting, communication, ceremonies, cooking, warfare, encouraging plant re-growth and providing a fire-safe environment (such as clearing tracks through the bush). Different patterns were used in different places, and probably at different times.

By trial and error they learned to use fire to:

- flush animals from grass and make them easier to hunt
- signal between one Aboriginal group and another
- encourage kangaroos and other prey animals, which favoured a more open countryside, to congregate on the areas of fresh vegetation sprouting on burnt areas
- encourage the regrowth of eucalypt trees and edible plant food such as bracken, from which the roots, young leaves and shoots could be eaten
- clean up the countryside since a pattern of low-intensity fire burns maintained the food supply as trees and shrubs were not totally killed.

Fire recycles nutrients and promotes new growth, especially in the nutrient-poor Australian soils, and this would subsequently attract herbivores.

Many plants have adapted to regular, low-intensity burning by being fire tolerant or fire resistant. Some orchids and lilies flourish after fire, as they often possess underground storage organs or tubers, which were eaten by the Aborigines. Studies show that burning burrawangs (*Macrozamia communis*) could increase the productivity of cones two to three times. The female cone seeds of the burrawang (see Figure 3.26) have very high starch contents and were eaten by the Aboriginal peoples after being washed to remove toxic compounds.

Figure 3.26 The female cone seeds of the burrawang

Fire-tolerant *Banksia* species regenerate from both lignotubers and epicormic shoots located along the trunk as shown in Figure 3.27. The trunk is usually protected by bark 1 to 3 cm thick. Also, seeds are released earlier and quicker from cones exposed to high fire temperatures (see Figure 3.28). If good rainfall follows, a very large proportion of seeds will germinate.

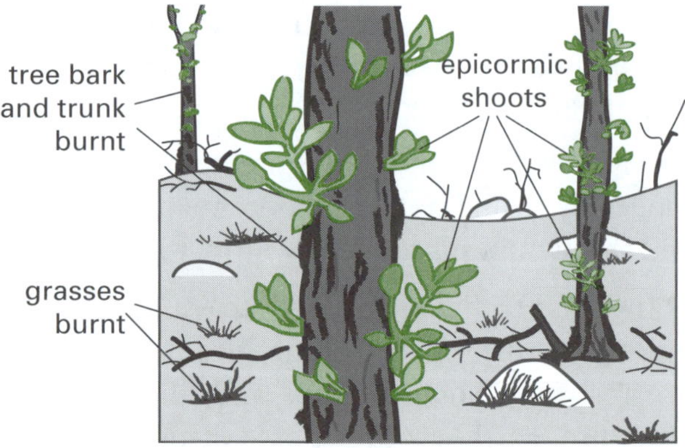

Figure 3.27 Epicormic shoots develop from the stem of some fire-tolerant plants. This is one sort of adaptation for surviving environments that experience fire.

While Aboriginal people used fire as a tool for increasing the productivity of their environment, European settlers saw it as a threat. However, without regular low-intensity burns, leaf litter accumulates and the resulting intense fires can destroy everything in its path.

Figure 3.28 Banksias are adapted to fires.

Test yourself 2

Part A: Knowledge

1. Sediments can be produced by the weathering and erosion of *(1 mark)*
   A sedimentary rocks.
   B igneous rocks.
   C metamorphic rocks.
   D all types of rocks.

2. A low-grade ore body *(1 mark)*
   A contains large amounts of metallic minerals.
   B contains very low amounts of minerals that may make it uneconomic to mine.
   C is commonly found in coastal areas where extensive erosion has occurred.
   D always contains sulfide minerals.

3. The metal that can be extracted from bauxite ore bodies is *(1 mark)*
   A copper.
   B iron.
   C tin.
   D aluminium.

4. Most iron ore mines are located in which Australian state? *(1 mark)*
   A South Australia
   B Queensland
   C Western Australia
   D Tasmania

5. Crop rotation is a common agricultural practice. This involves *(1 mark)*
   A growing a series of dissimilar types of crops in the same area in following seasons.
   B growing the same crop in the same paddock each season.
   C extensive use of fertiliser in the soil.
   D improving soil structure and fertility by growing different deep-rooted plants each season.

6. Complete the following restricted-response questions using the appropriate word. *(1 mark for each part)*
   a) Many plants have adapted to regular, low-intensity .... by being fire tolerant or fire resistant.
   b) Green manure is a type of cover crop grown mainly to add .... and organic matter to the soil.
   c) Sea walls, groynes and other structures are built to protect against coastline ....
   d) Recycling scrap aluminium requires only 5% of the .... used to make new aluminium.
   e) Hydrothermal mineral deposits are formed when hot .... solutions which carry dissolved minerals are allowed to cool.

7. Use the code letters to match the terms or phrases in each column. *(1 mark for each part)*

Column 1	Column 2
A gold	F lead
B galena	G magnetite mineral
C magma	H metamorphic rock
D heat and pressure	I molten rock
E iron	J placer deposit

Part B: Skills

8. Sea walls are sometimes built to protect property. Sea walls are strong barriers built right up against the properties or shoreline roads. Waves lose most of their energy as they roll up the coast, but a sea wall deflects the wave (and its energy). As the wave bounces back, it erodes the beach. As the beach erodes even further, more force is unleashed against the sea wall. No matter how strong the sea wall is, it is only a matter of time before it, too, is eroded away by the sea. Figure 3.29 shows a newly erected sea wall on a beach. Use the information provided in this paragraph to draw a new diagram showing the destructive effects on the beach over time. *(2 marks)*

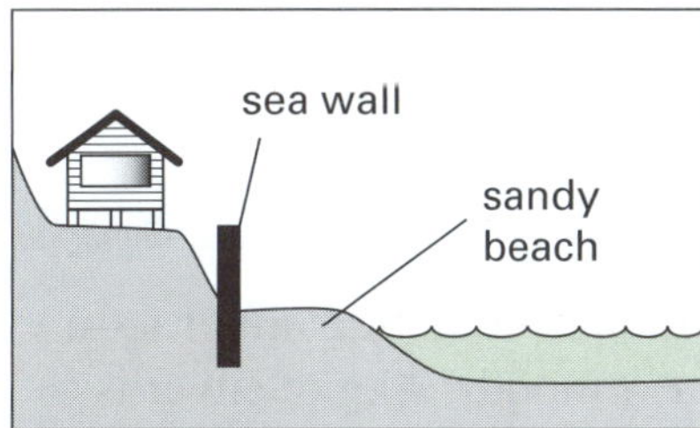

Figure 3.29 Sea wall

9. A statement can be changed into a question and answer. For example, the statement:
   Sand dune grasses tend to grow on top of the dune.
   can be turned into the following question and answer:
   Q: Where do sand dune grasses tend to grow?
   A: On top of the sand dune.
   Convert the following statement into a question and answer. *(2 marks)*
   Coastal dunes are formed when sand dumped onto the shore dries out and is then blown further to the back of the beach.

10. When marram grass rather than spinifex grass is planted on an eroded primary sand dune, the shape of the dune changes from broad, level and low to hilly, uneven and steep.

Use the following information to explain these observations.

- Spinifex has a root system of runners from which new shoots develop.
- Marram has a root system of both horizontal and vertical root-like stems. This grass grows quickly. *(2 marks)*

11. Table 3.8 is an analysis of a typical sample of iron from a blast furnace.

Table 3.8 Analysis of an iron sample from a blast furnace

Element	Percentage by weight
iron	94.0
carbon	
silicon	0.7
manganese	0.4
others	0.5

a) What is the percentage by mass of carbon in this iron? *(1 mark)*

b) Draw a pie graph of this data. *(6 marks)*

12. Argentite is a silver sulfide mineral. In a particular ore body, 2 kg of silver is obtained from each tonne of ore.

a) What percentage by weight of silver is present in this ore? *(1 mark)*

b) If 50 000 tonnes of ore is mined each day, what mass of silver would be produced? *(1 mark)*

13. The steps involved in producing pure copper from copper ore are listed below but they are not in the correct order. Use the code letters to place these steps in the correct order. *(2 marks)*

A. The copper compounds are smelted to produce metallic copper.

B. The copper minerals are roasted to produce copper sulfide and oxide compounds.

C. Froth flotation separates the copper mineral from the waste rock.

D. Metallic copper is refined electrolytically to produce pure copper.

E. Mined ore is crushed into a fine powder.

14. The Hunter Valley north of Sydney is one location where iron and aluminium are produced from their ores. The Hunter Valley is rich in coal. Explain how coal is useful in the production of iron and aluminium. *(2 marks)*

Go to pp. 194–195 to check your answers.

Summary

1. The lithosphere includes the Earth's crust and upper part of the mantle.
2. A mineral is a naturally occurring solid formed through geological processes that has a characteristic chemical composition with a highly ordered atomic structure and specific physical properties.
3. Minerals can be identified from a number of characteristics such as colour, crystal shape, lustre, cleavage, hardness and streak.
4. Weathering is the breakdown and changes in rocks and sediments at or near the Earth's surface produced by biological, chemical and physical agents or some combination.
5. Erosion is the process by which the Earth's surface is worn away by the action of water and wind.
6. A sedimentary rock is a rock type that has been created by the deposit and compression of sediments. This type of rock is created over millions of years.
7. When molten rock cools, igneous rocks are formed. Magma that solidifies within the Earth's crust cools slowly and results in coarse-grained (large crystal) rock.

Summary cont.

8. A metamorphic rock is any type of rock that changes in texture or composition after its original formation, as a result of extreme heat, pressure or chemically active fluids.
9. The rock cycle describes the dynamic transitions through geologic time among the three main rock types: sedimentary, igneous and metamorphic.
10. An ore is a rock from which metal or minerals can be extracted at a financial profit.
11. Recycling involves processing used materials into new products, preventing the waste of potentially useful materials, reducing the consumption of fresh raw materials, reducing energy usage and reducing air and water pollution.
12. The exploitation of the environment, especially in recent years, has led to incessant degradation of the natural world. Humans need to take responsibility for caring for the ecology and surroundings, thus preserving the environment that has already been depleted.
13. Sustainable farming involves farming that avoids soil erosion and pollution and does not destroy the land for future generations.
14. Australian Aboriginal peoples have used fire in a responsible and sustainable manner in daily activities and in a way that is beneficial to the environment.

Syllabus checklist

Are you able to answer every syllabus question in this chapter? Tick each question as you go through the list if you are able to answer it. If you cannot answer it, turn to the appropriate page in the guide as is listed in the column to find the answer.

For a complete understanding of this topic		Page no.	✓
1	Can I recall what the lithosphere consists of?	79	
2	Can I recall what a mineral is?	79–80	
3	Can I recall the characteristics used to identify minerals?	80	
4	Can I explain what weathering is and what causes it?	80	
5	Can I explain what erosion is?	80–83	
6	Can I recall how a sedimentary rock is created and how long this process takes?	83–85	
7	Can I describe how igneous rocks form and the relationship with crystal size?	86	
8	Can I recall what a metamorphic rock is?	87–88	

For a complete understanding of this topic		Page no.	✓
9	Can I recall what the rock cycle describes?	90	
10	Can I explain why metal or minerals can be extracted from ore at a financial profit?	91	
11	Can I recall what recycling involves and its benefits?	94	
12	Can I explain what has led to the incessant degradation of the natural world?	92, 94	
13	Can I explain what sustainable farming involves?	95–96	
14	Can I explain how Australian Aboriginal peoples have used fire in their daily activities and explain how this is beneficial to the environment?	97–98	

Chapter test

Go to p. v for *Tips for tests and examinations*

80 MIN

Part A: Multiple-choice questions

(1 mark for each)

1. Minerals may look shiny or dull. These descriptions can also be referred to as their
   A colour.
   B streak.
   C cleavage.
   D lustre.

2. A sedimentary rock formed from the remains of shelled sea animals is
   A sandstone.
   B limestone.
   C siltstone.
   D mudstone.

3. Which of the following is not a mineral?
   A mica
   B quartz
   C gabbro
   D feldspar

4. At which point on the river in Figure 3.30 would a placer deposit of a heavy mineral occur?

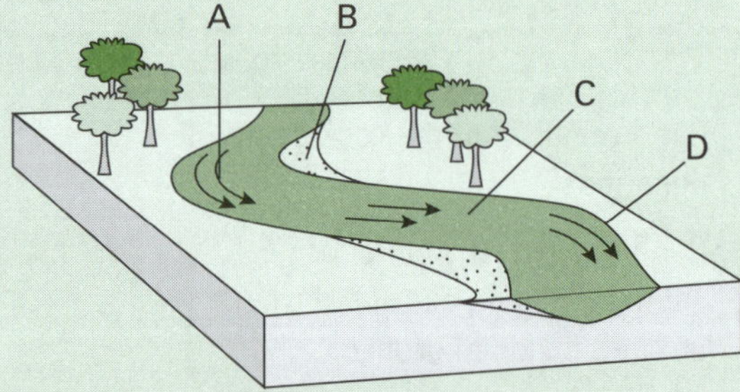

Figure 3.30 Placer deposit

5. Aboriginal peoples used fire to
   A kill kangaroos for food.
   B burn the landscape so they could place their camps.
   C clear the land so they could plant their crops.
   D allow plants to regenerate and grow new shoots.

6. About 6000 years ago the Egyptians used iron from meteorites to make
   A tools.
   B boats.
   C buildings.
   D money.

7. The term weathering refers to
   A the breakdown of the Earth's crust.
   B moving small particles from one location to another.
   C the action of rain only.
   D the action of wind and rain only.

8. The streak of a mineral refers to
   A the colour of its crystals.
   B the colour of the powdered crystals.
   C the way it splits when struck by a hammer.
   D its lustre.

9. Conglomerate is an example of a
   A mineral.
   B metamorphic rock.
   C igneous rock.
   D sedimentary rock.

10. Select the statement that is true of granite.
    A It is a metamorphic rock.
    B It is an extrusive igneous rock.
    C It is an intrusive igneous rock.
    D It is composed of very fine crystals.

Part B: Short-answer questions

11. Complete the following restricted-response questions using the appropriate word.
    (1 mark for each part)
    a) A .... is a naturally occurring inorganic solid material that has a definite chemical composition.
    b) .... refers to the breakdown of rocks and minerals in the Earth's crust.
    c) .... is the movement of the broken-down and dissolved rock material from one place to another.
    d) .... rocks are formed by particles pressed and cemented together.

e) .... are the impressions of dead plants or animals that can be found in sedimentary rocks.

f) Igneous rocks cooling deep with the Earth have .... crystals.

g) .... rocks have been subjected to heat and pressure causing profound physical and/or chemical changes.

h) An important ore of aluminium is ....

12. Use the code letters to match the terms or phrases in each column. *(1 mark for each part)*

Column 1	Column 2
A lithosphere	H splitting
B igneous	I mined mineral
C water	J main eroding agent
D conglomerate	K crust and upper mantle
E cleavage	L volcano
F ore	M sedimentary
G recycling	N reusing material

13. a) Does ice fulfil the definition of a mineral? Explain. *(3 marks)*

b) Explain why each of the following is not a mineral.

i) an artificial diamond *(1 mark)*

ii) coal *(1 mark)*

iii) sand *(1 mark)*

iv) calcium and phosphorus found in milk *(1 mark)*

14. Figure 3.31 shows a cross-section from a mountain to a lake, and describes the early stages in the formation of sedimentary rocks.

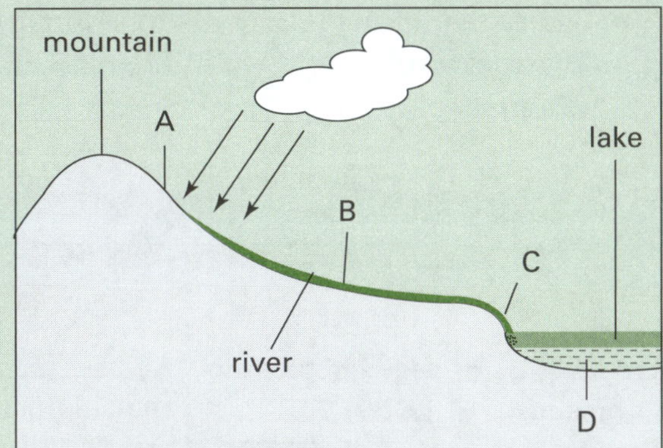

Figure 3.31 Early stages of formation of sedimentary rocks

Select the appropriate captions from the following list to replace the letters in the diagram: sediments turn to rock; deposition; transportation; weathering and erosion.
(1 mark for each part)

Use the following information to answer the Questions 15 to 17.

Igneous rocks form when molten rock cools and solidifies (crystallises). Molten rock trapped underground is called magma. If magma is allowed to cool slowly over thousands or millions of years (think of a magma chamber as a well-sealed oven), large crystals form because the crystals have had a long time to grow before the crystallisation process is complete. This produces coarse-grained igneous rocks like granite and gabbro. Generally the minerals making up the rock are easily seen as they crystallise separately. When molten rock erupts at the surface, lava is produced. Lava cools quickly and rocks formed on the surface are composed of tiny, microscopic crystals (fine grained) as the crystals have not had time to grow. This process happens over a few seconds to perhaps several days. Rocks formed on the Earth's surface include rhyolite and basalt. Basalt has the same chemical composition as gabbro. The distinguishing feature is the size of the crystals. Igneous rocks are generally divided into three compositional groups.

i) Felsic: these mostly contain quartz and orthoclase felspar and are generally white to pink in colour.

ii) Intermediate: mostly contain the minerals plagioclase with some quartz and sodium-rich pyroxene felspar. They are usually light to dark grey in colour.

iii) Mafic: these mostly contain the minerals calcium-rich pyroxene felspar and olivine. These rocks are usually black to dark green in colour.

15. a) What are igneous rocks? *(1 mark)*

b) How do igneous rocks differ from sedimentary rocks? *(3 marks)*

c) Why do igneous rocks that have formed deep within the Earth have large crystals? *(1 mark)*

16. a) Basalt and gabbro have the same chemical composition. How can you tell them apart? *(2 marks)*
b) Intermediate and mafic rocks both contain pyroxene felspar. Is this the same felspar in each case? *(3 marks)*
c) Sedimentary rocks sometimes contain fossils. Would igneous rocks contain fossils? *(2 marks)*
d) Obsidian is an igneous rock that looks like broken dark glass. Would this rock have formed within the Earth or on the surface? Why? *(1 mark)*

17. Table 3.9 shows a classification of some igneous rocks.

Table 3.9 Classification of igneous rocks

	Slow cooling (coarse grained)	Fast cooling (fine grained)	Very fast cooling (glassy)
mafic (dark colour)	gabbro	basalt	scoria
intermediate (intermediate colour)	diorite	andesite	
felsic (light colour)	granite	rhyolite	Pumice, obsidian

Describe a difference between:
a) gabbro and granite. *(2 marks)*
b) pumice and granite. *(2 marks)*
c) diorite and obsidian. *(2 marks)*

18. Use the following information to answer the questions.
Metamorphic rocks form from heat and/or pressure changing the original, or parent, rock into a completely new rock. There are three ways metamorphic rocks can form.
- *Contact metamorphism occurs when magma comes into contact with an already existing body of rock. When this happens, the temperature of the rock rises and fluids from the magma can infiltrate it. The area affected is usually small (1 to 10 km). Examples of these rocks are marble, quartzite and hornfels.*
- *Regional metamorphism occurs over a much larger area and is caused by such processes as mountain building. Rocks produced show cleavage due to tremendous heat and pressure applied to them. Examples include gneiss and schist.*
- *Dynamic metamorphism is also due to mountain building with tremendous pressures and heat causing rocks to bend, fold, crush, flatten and shear. Metamorphic rocks are generally harder than sedimentary rocks and usually as hard as igneous rocks.*

a) Give a simple definition of metamorphic rocks. *(1 mark)*
b) Are the parent rocks from which metamorphic rock is derived sedimentary, igneous or metamorphic? *(2 marks)*
c) Give one difference between regional metamorphism and contact metamorphism. *(1 mark)*
d) Figure 3.32 shows some magma pushing its way into three layers of sedimentary rocks.

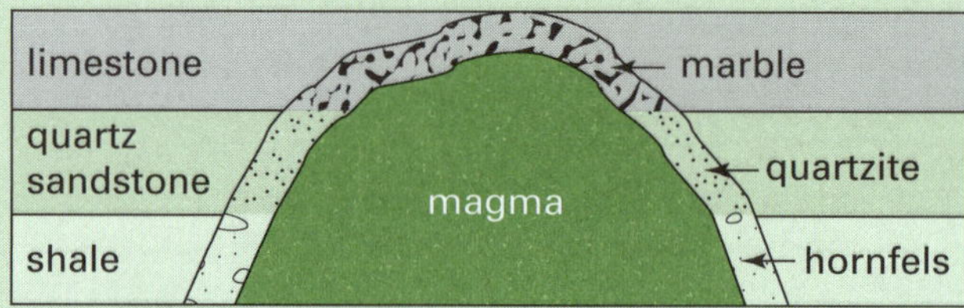

Figure 3.32 Metamorphism

i) What metamorphic rocks are produced by each of these rocks? *(3 marks)*
ii) Which type of metamorphism produced these? *(1 mark)*

19. a) Write five brief statements to summarise the rock cycle as shown in Figure 3.33. *(5 marks)*

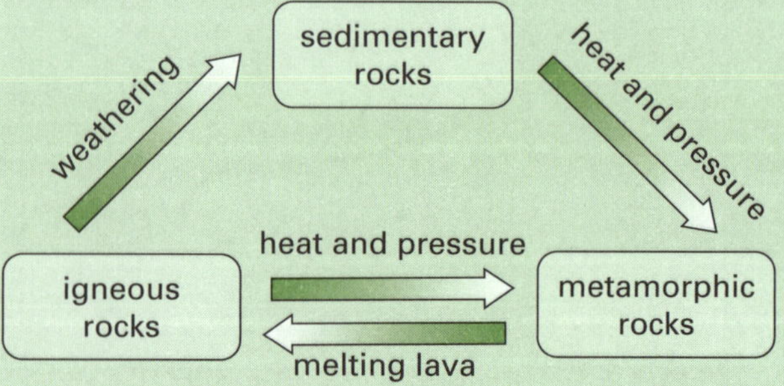

Figure 3.33 Rock cycle

b) Where on Figure 3.34 would you find contact metamorphism? *(1 mark)*

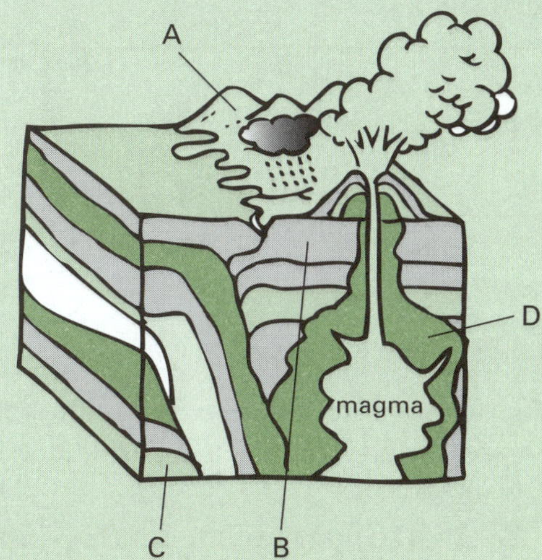

Figure 3.34 Contact metamorphism

20. Table 3.10 gives Mohs scale of hardness for minerals.

Table 3.10 Mohs scale of hardness for minerals

Mineral	Hardness	Common tests
talc	1	scratched by fingernail
gypsum	2	
calcite	3	scratched by copper coin
fluorite	4	scratched by window glass or knife blade
apatite	5	
feldspar	6	scratches a knife blade or window glass
quartz	7	
topaz	8	
corundum	9	
diamond	10	scratches all common materials

a) Which of the following will scratch quartz: topaz; calcite; apatite; diamond? *(2 marks)*

b) Feldspar will scratch all of the following except *(1 mark)*

A talc.
B fluorite.
C gypsum.
D corundum.

c) Galena scratches gypsum, but is scratched by calcite. Its hardness is closest to *(1 mark)*

A 2.
B 2.5.
C 3.
D 3.5.

d) Which of the following will *not* scratch galena? *(1 mark)*

A knife blade
B window glass
C copper coin
D fingernail

21. The use of firestick farming by Aboriginal peoples in Australia led to significant changes in the distribution of plant species over time.

a) Explain what this statement means. *(2 marks)*

b) How were these small, regular burns good for the environment? *(2 marks)*

22. Figure 3.35 shows the grain yield of wheat in a country compared to the amount of fertiliser applied.

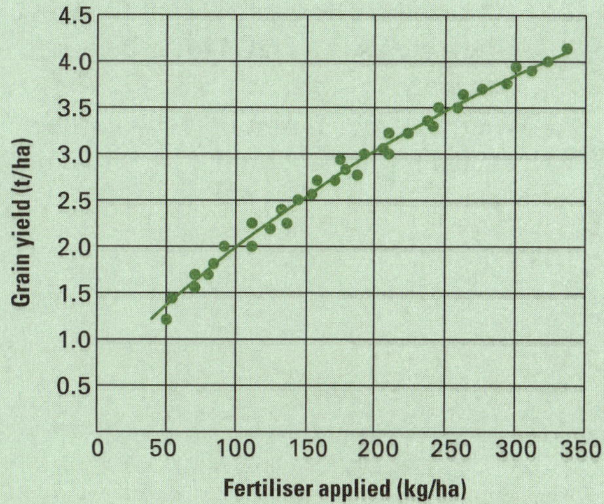

Figure 3.35 Grain yield

a) Describe the relationship between grain yield and the amount of fertiliser applied. *(1 mark)*

b) Will applying more and more fertiliser keep improving the wheat yield? *(1 mark)*

c) What are two disadvantages to applying more fertiliser? *(2 marks)*

23. A lignotuber is a starchy swelling of the root crown possessed by some plants as a protection against destruction of the plant by fire (see Figure 3.36).

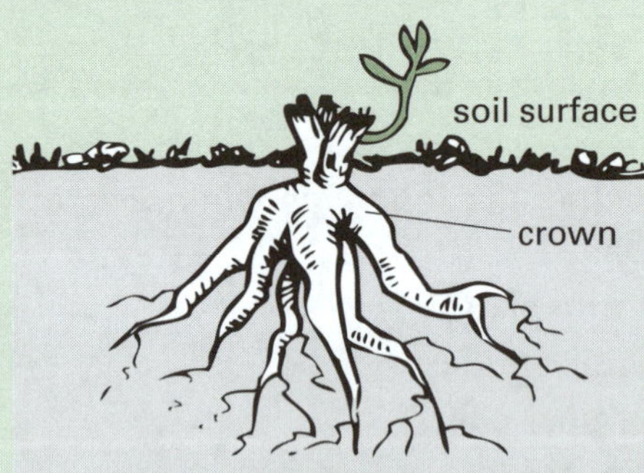

Figure 3.36 Lignotuber

a) How can this feature prevent the plant from being totally destroyed? Select *two* of the following responses.
   i) The roots and lignotuber are not killed if the fire only burns the leaves down to ground level.
   ii) The starch in the lignotuber will not burn.
   iii) The lignotuber is fire resistant.
   iv) The crown of the lignotuber is a site where new buds can sprout to form new stems and leaves after the fire.
   (2 marks)

b) What must be present in the lignotuber to sustain the plant until it can grow new leaves? *(1 mark)*

24. Figure 3.37 shows household recycled materials collected from the average Australian.

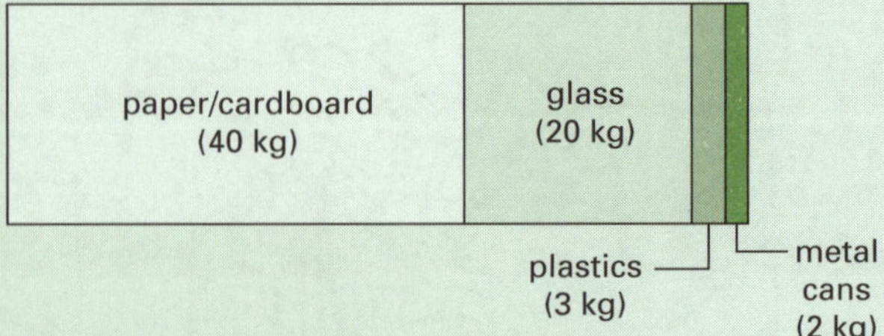

Figure 3.37 Household recycled material

a) How much recycled material altogether is collected from each Australian every year? *(1 mark)*

b) What percentage of the total is glass? *(1 mark)*

c) State two advantages of recycling waste materials. *(2 marks)*

25. Figure 3.38 shows the beginning of an ore mining operation.

a) What steps in ore mining are shown in the figure? *(2 marks)*

b) Aluminium is the most abundant metallic element in the Earth's crust (8.3%). Yet it is only mined in certain places. Why is this? *(1 mark)*

26. Figure 3.39 shows a sequence involving a sedimentary rock. Use the following terms to replace the code letters: sand grain; sandstone strata; quartz crystal; sandstone. *(2 marks)*

Figure 3.38 Ore mining

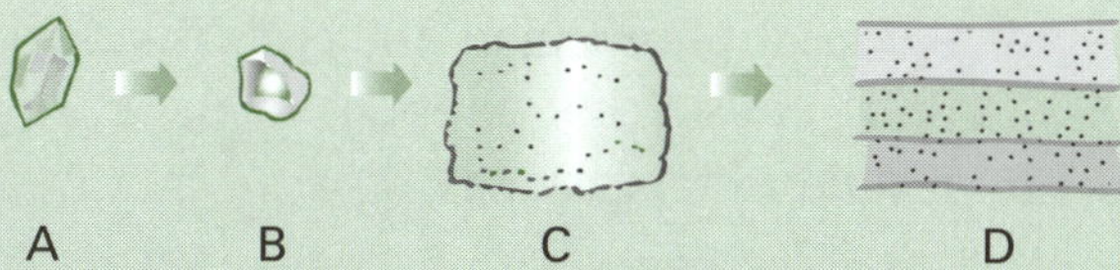

Figure 3.39 Sedimentary rock sequence

27. Coastal rivers carry rock material towards the sea. The river carries the material in several ways.

- X: materials dissolved in the water
- Y: materials suspended in the water
- Z: materials rolled and pushed along the river bed

Classify the following materials into one of the above three groups. *(3 marks)*

a) pebbles
b) mineral salts
c) clay
d) boulders
e) fine sand
f) very coarse sand

28. Use the key in Figure 3.40 to name the rock described below.

The rock is made of a mixture of large and some small particles cemented together. The particles have different colours. The rock does not fizz when dilute acid is added. *(1 mark)*

29. Use the following descriptions of five minerals to construct a written key to identify them. *(5 marks)*

- *Calcite. This is a very soft mineral with a hardness of 3. Its crystal and streak are both white. Its lustre is dull.*
- *Chalcopyrite. A metallic, yellow mineral that has a black streak. Its hardness is 4.*
- *Fluorite. This is a soft mineral with a hardness of 4. Its crystals are vitreous and white. Its streak is white.*
- *Magnetite. A black, metallic mineral that is black when powdered. Its hardness is 6.*
- *Pyrite. A yellow, metallic mineral that is green when powdered. Its hardness is 6.*

30. Each year the average Australian household creates 1.3 tonnes of landfill waste. Up to 40% of household waste can be composted, including food scraps, garden cuttings, paper and card. How many kilograms of landfill waste can be composted? *(1 mark)*

31. What are some factors that mining companies need to consider when determining whether a mineral deposit is worth mining as an ore? *(3 marks)*

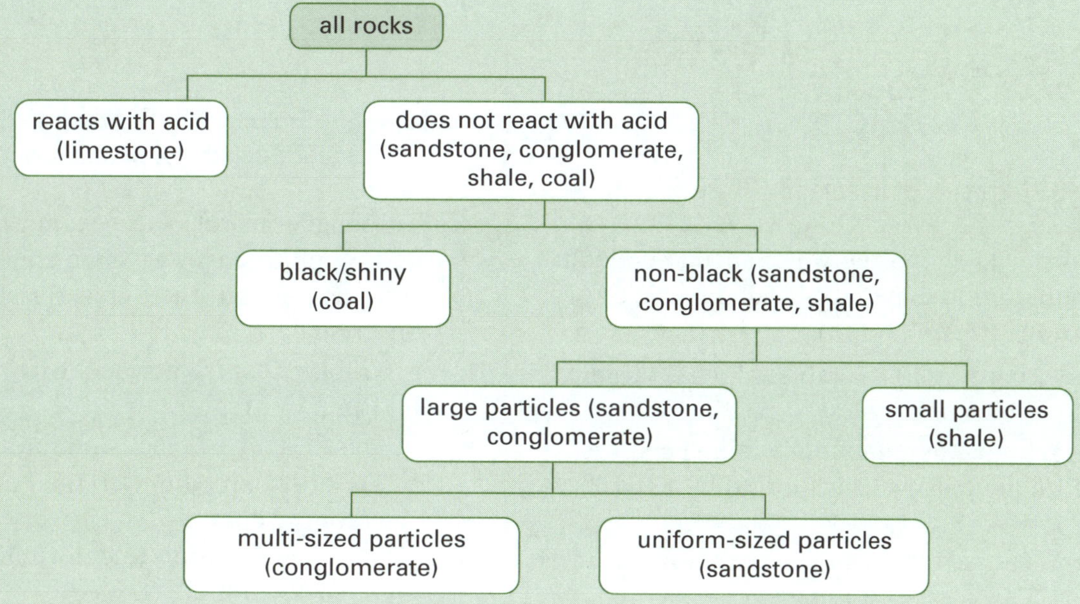

Figure 3.40 Diagrammatic key

32. Figure 3.41 shows the erosion of a headland along the shore. Write a series of captions to go with each diagram. *(4 marks)*

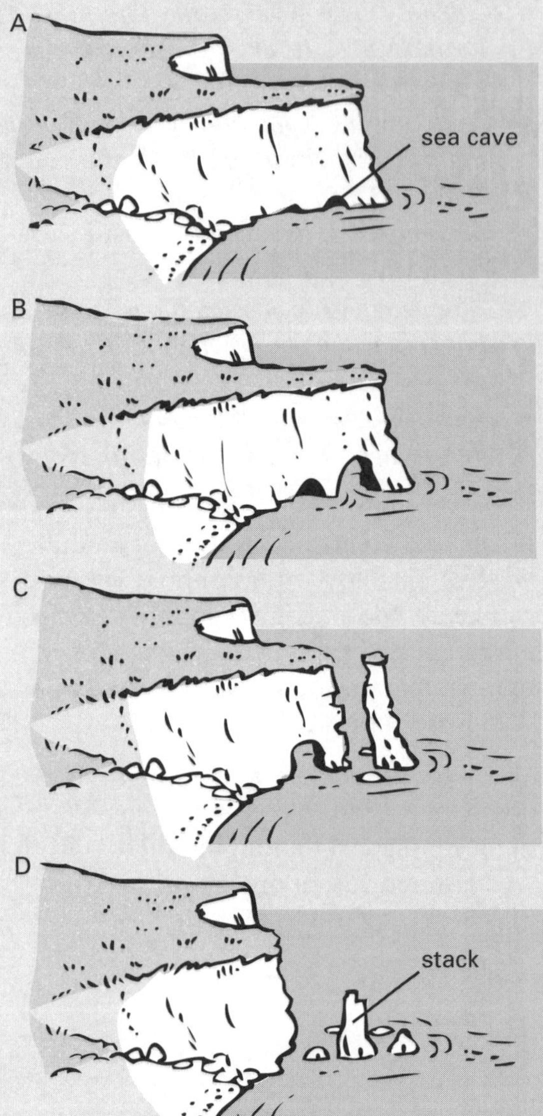

Figure 3.41 Erosion of headland

33. Table 3.11 shows the water speed needed to maintain movement of eroded particles of various sizes.
    a) According to the table, which is larger: pebbles or cobbles? *(1 mark)*
    b) What is the minimum water speed for boulders to be transported by a river? *(1 mark)*
    c) What particles can a stream flowing at 1.5 m/s carry? *(1 mark)*
    d) A stream flowing at 100 cm/s is slowed to 45 cm/s. What happens to the particles it might be carrying? *(1 mark)*

Table 3.11 Water speed needed to transport particles

Size of particle (mm)	Name	River speed needed to transport particle (m/s)
> 256	boulder	≥ 3.00
64–256	cobble	> 1.70
4–64	pebble	> 0.75
2–4	granule	> 0.60
0.06–2	sand	> 0.40
0.004–0.06	silt	> 0.25
< 0.004	clay	< 0.25

34. Figure 3.42 shows two layers of sedimentary rocks (P and Q).
    a) Which layer is older? *(1 mark)*
    b) Which layer is harder? *(1 mark)*
    c) Outline the events that might have formed this scenery. *(4 marks)*

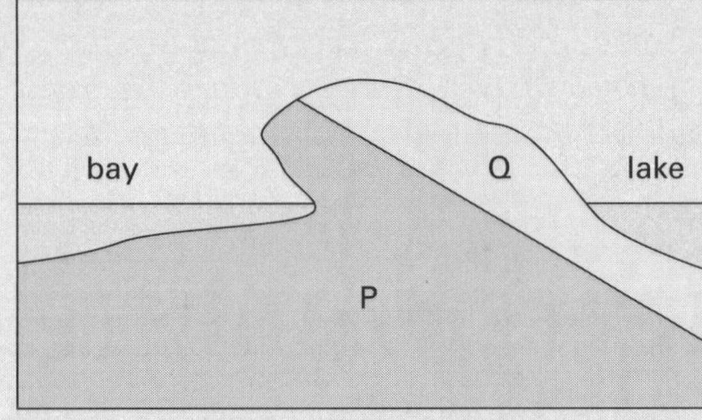

Figure 3.42 Sedimentary landscape

35. Three igneous rocks are examined and found to have the same chemical composition, but their crystal sizes are different.
    - Gabbro: coarse grained, with crystal size 1 mm or greater.
    - Basalt: grey to black and fine-grained.
    - Dolerite: a medium-grained dark-coloured rock.

    Use this information to determine where these rocks formed. *(3 marks)*

36. A student is provided with a rock to identify. She knows it is one of the rocks in the following list: coal; limestone; obsidian; pumice; sandstone; slate.

She recorded the following features of the rock in her notebook. Using the diagrammatic key in Figure 3.43 below, identify the unknown rock from the following description.

The rock is grey-black. It is hard and forms flat sheets. The rock is fine grained. It does not burn when held in a Bunsen flame. It does not fizz when dilute acid is added.

(1 mark)

37. Use the information in Table 3.12 to identify the unknown mineral from its properties. Give your reasons.

Unknown mineral's properties:

- *colour: black*
- *streak: white*
- *hardness: 3.8*
- *lustre: resinous.*

(2 marks)

Table 3.12 Properties of some minerals

Mineral	Colour	Streak	Hardness	Lustre
apatite	w, gr, y, p	w	4.5–5	r
chlorite	g	w, gr	2.2–5	pe
gypsum	w, c, g, y, r	w	1.5–2	pe to v
haematite	g	r, br	5–6	m
limonite	br, y, bl	br, y, w	4.5–5	d to v
quartz	c, p, r, y	w	7	v
sphalerite	y, br, r, bl, w	w	3.5–4	r

Abbreviations:
Colour and streak: b = blue; bl = black; br = brown; c = colourless; g = grey; gr = green; p = purple; r = red; w = white; y = yellow
Lustre: d = dull; m = metallic; pe = pearly; r = resinous; v = vitreous (like glass)

38. The idea of a rock cycle began with the 18th-century 'father of geology', James Hutton. He used the famous quote: 'no vestige of a beginning, and no prospect of an end'. What do you think he meant by this? *(1 mark)*

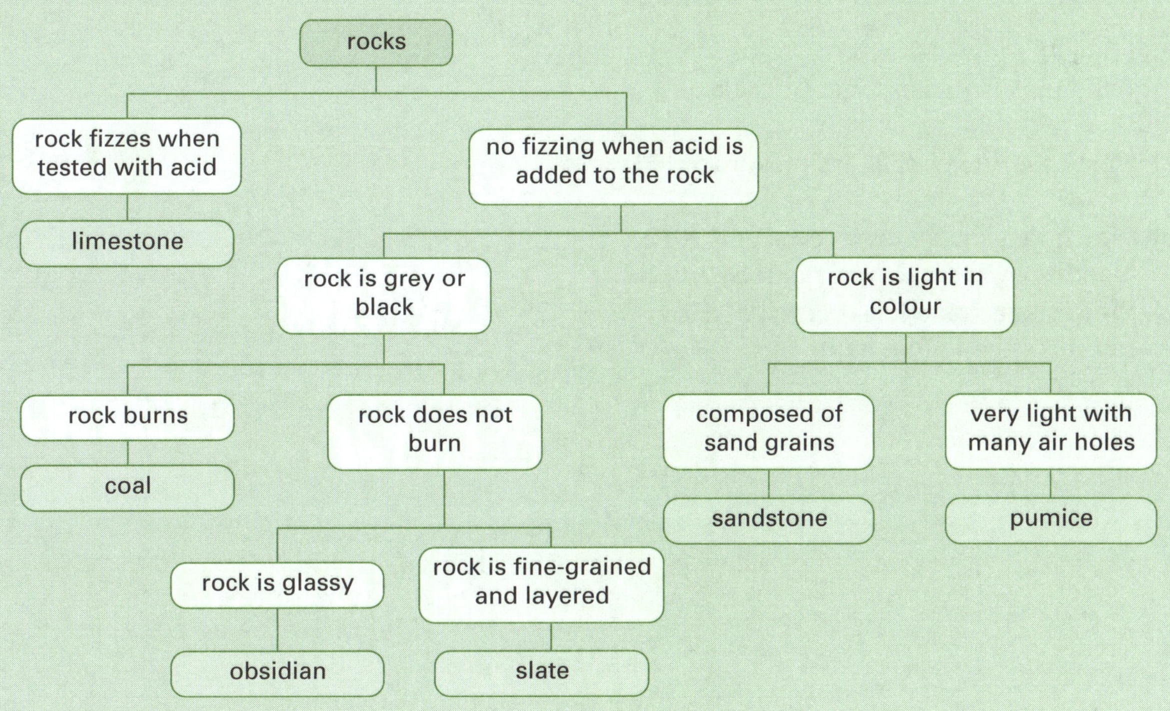

Figure 3.43 Rock key

39. Figure 3.44 shows the mass of iron ore exported from Australia.

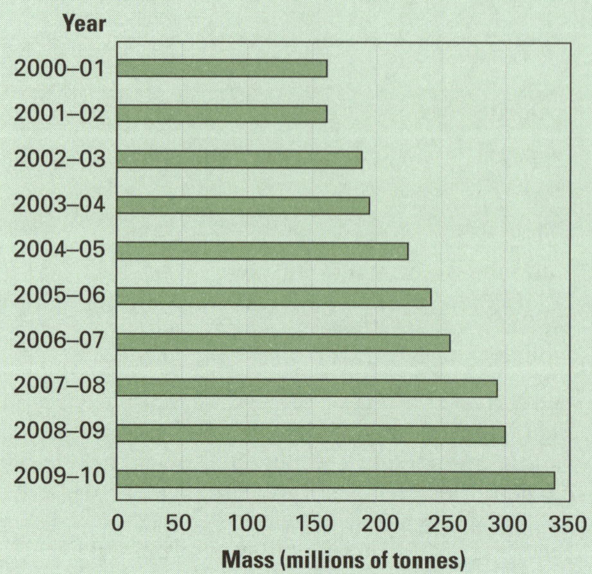

Figure 3.44 Iron ore exporting

a) How much iron ore was exported in the year 2003–04? *(1 mark)*

b) In which year did the exports first exceed 250 million tonnes? *(1 mark)*

c) Estimate what amount of iron ore was exported from Australia in 2010–11. What assumption are you making? *(2 marks)*

40. Figure 3.45 shows the type of waste recycled in Australian homes in 2006 and 2010.

a) Name four items that are most recycled. *(1 mark)*

b) Which items were less recycled in 2010 than they were four years earlier? *(1 mark)*

c) Which item is the least recycled? How can this situation be improved? *(2 marks)*

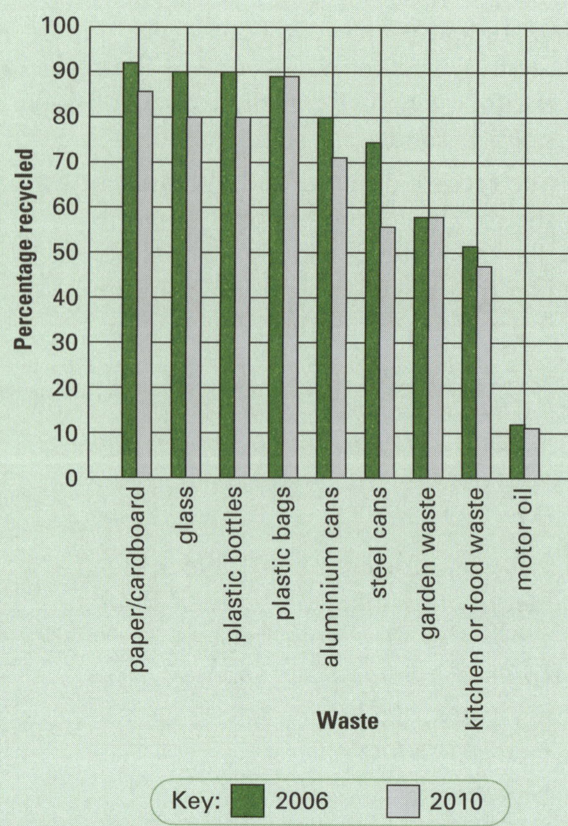

Figure 3.45 Percentage of waste recycled

Go to pp. 193–198 to check your answers.

CHAPTER 4

Energy

Overview

In this chapter you will learn about:

- kinetic energy
- potential energy—gravitational, chemical and elastic
- energy transformation
- heat energy—conduction, convection and radiation
- reducing energy consumption
- solar energy
- energy technologies.

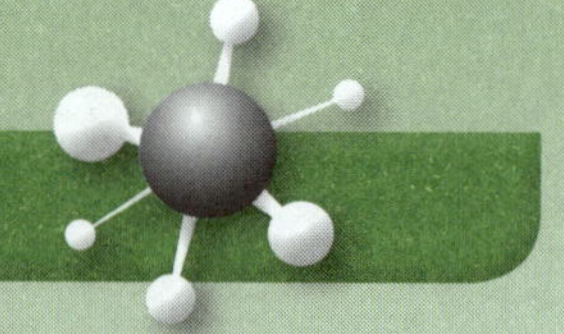

Glossary

Conduction of heat—heat energy is transferred through a solid as strongly vibrating particles collide with other particles in the solid

Conservation of energy—energy cannot be created or destroyed

Convection of heat—heat energy is transferred by fluid particles moving from one place to another

Kinetic energy—energy of moving bodies

Potential energy—stored energy (e.g. chemical potential energy, elastic potential energy, gravitational potential energy)

Radiation of heat—transferring heat energy in the form of electromagnetic waves (infra-red)

Solar energy—radiant energy from the Sun

Transfer of energy—movement of energy from one place to another; heat energy can be transferred by conduction, convection or radiation

Energy transformation—energy is converted from one form to another

4.1 Kinetic energy

We live in a world of energy. Energy causes change or has the potential to cause change. Some common forms of energy include heat energy, light energy, sound energy and electrical energy. In each of these examples, something is moving.

The particles of matter move faster when they are heated. When oxygen gas molecules are heated, their molecular speeds increase as shown in Figure 4.1.

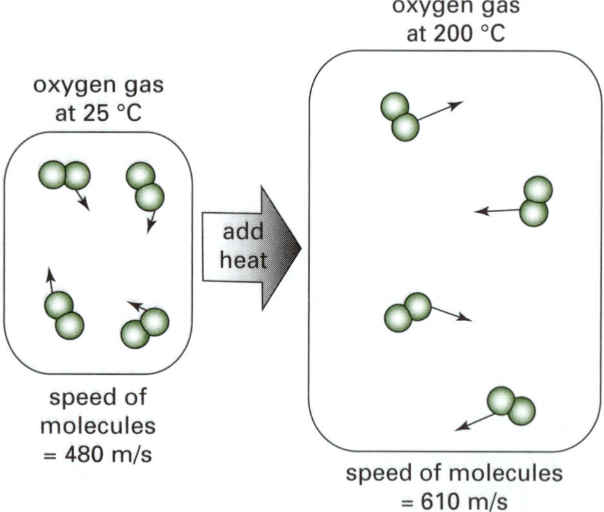

Figure 4.1 Speed of oxygen molecules increases when the gas is heated.

The particles of a medium move backwards and forwards when sound energy is transmitted. An example of this is seen in Figure 4.2, which shows sound waves being propagated towards the ear. The solid lines represent regions where the air particles are close together. Like all brass instruments, sound is produced by blowing air through closed lips. This produces a 'buzzing'-like sound in the mouthpiece and begins a standing wave vibrating in the air column inside the trumpet. The dimensions of the parts of the mouthpiece affect the timbre or quality of sound, the ease of playing and the trumpeter's comfort. The part behind the rim of the mouthpiece is the cup, which channels the air into a much smaller opening. Generally, the wider and deeper the cup, the darker the sound and timbre.

Figure 4.2 Sound from a trumpet starts when the player's lips vibrate against the mouthpiece.

Figure 4.3 shows a sequence (from a to d) of a sound wave moving through air from left to right. The particles of the air collide with particles ahead of them and transfer their kinetic energy forwards. They then rebound and absorb more energy from the sound source (e.g. sound waves moving through an organ pipe).

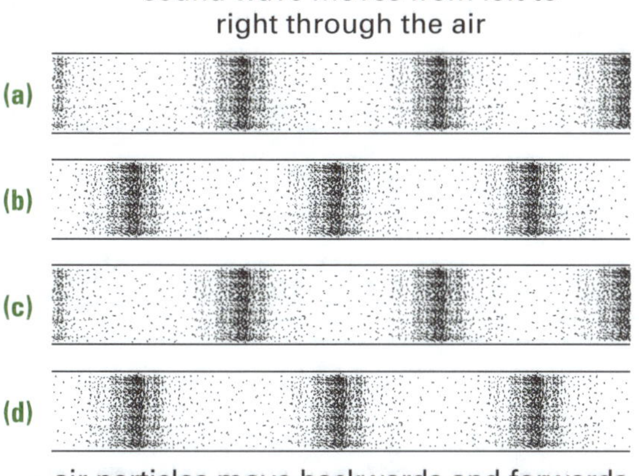

Figure 4.3 Sound energy transmitted through the air

Light energy comes from moving particles of light called photons. Moving electrons carry electrical energy.

Moving things possess kinetic energy. Speeding cars, rotating windmills, waves on the sea and strong winds all possess kinetic energy. A heavy truck travelling at 60 km/h has more kinetic energy than a car travelling at the same speed. We know this because it can cause more damage if it collides with another object. Increasing its speed can increase the kinetic energy of the car. Kinetic energy depends on both the mass and speed of a body.

Car crashes can lead to serious injury and death. Drivers are required to drive more slowly in the 40 km/hr speed zones around schools. If a driver is travelling at 50 km/hr or 60 km/hr, the stopping distance is greater when the brakes are applied. At 40 km/hr the stopping distance is considerably shorter and it gives time for the drivers to stop the car before a collision with a child crossing the road. As well, less damage is done to the pedestrian should they be hit.

4.2 Potential energy

Energy can be stored. This stored energy is called potential energy. Stored energy becomes available when it is changed into another form.

Let's consider three examples.

- Chemical potential energy can be stored in food and fuels. You eat food to obtain materials to power your body and for your cells to carry out their normal processes that keep you alive. Petrol stores energy which can be used to power your car. A battery stores energy that can then be used to power a light bulb or a mobile phone. In all these cases the chemical potential energy is stored in the chemical bonds of the materials used.
- Gravitational potential energy can be stored in an object held above the Earth's surface. Gravitational potential energy depends on the mass of the object as well as its height. Therefore, a heavy object has more gravitational potential energy than a lighter object at the same height. When an object falls down from a height, its gravitational potential energy is converted to kinetic energy. Its kinetic energy is at a maximum when the object is just about to hit the ground. In Figure 4.4 body A has a greater gravitational potential energy than body B because they have the same mass but A is higher off the ground. Bodies X and Y have the same gravitational potential energy as X is half as massive but it is at twice the height of Y.
- Elastic potential energy can be stored when a spring is compressed or twisted or a rubber band is stretched. The energy is released when the objects return to their normal shapes. Figure 4.5 shows how a compressed spring can store elastic potential energy and when released it can apply a force to a ball and push it away. The ball gains kinetic energy.

A trampoline stores elastic potential energy in its springs. When the jumper falls down onto the trampoline surface, the springs are stretched as shown in Figure 4.6. When they return to their normal length, the energy is converted back into the kinetic energy of the jumper. Trampoline springs store elastic potential energy when stretched. The twisted spring in a mousetrap stores plenty of potential energy.

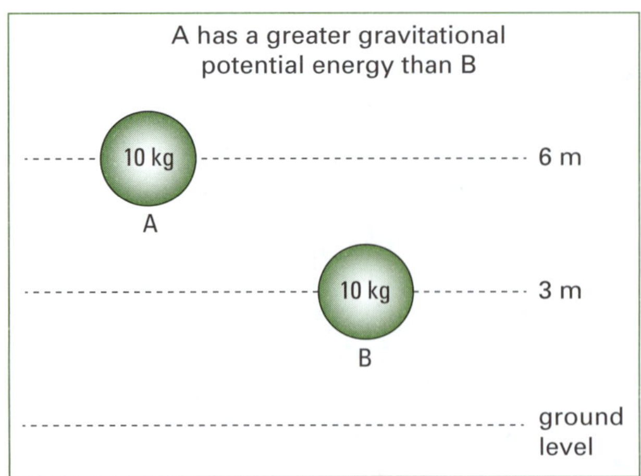

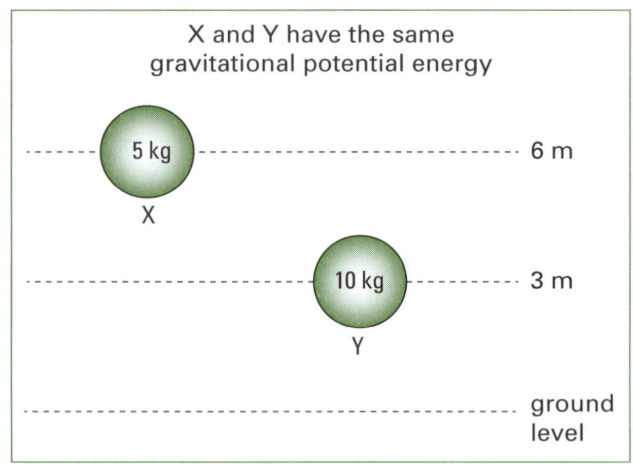

Figure 4.4 Gravitational potential energy

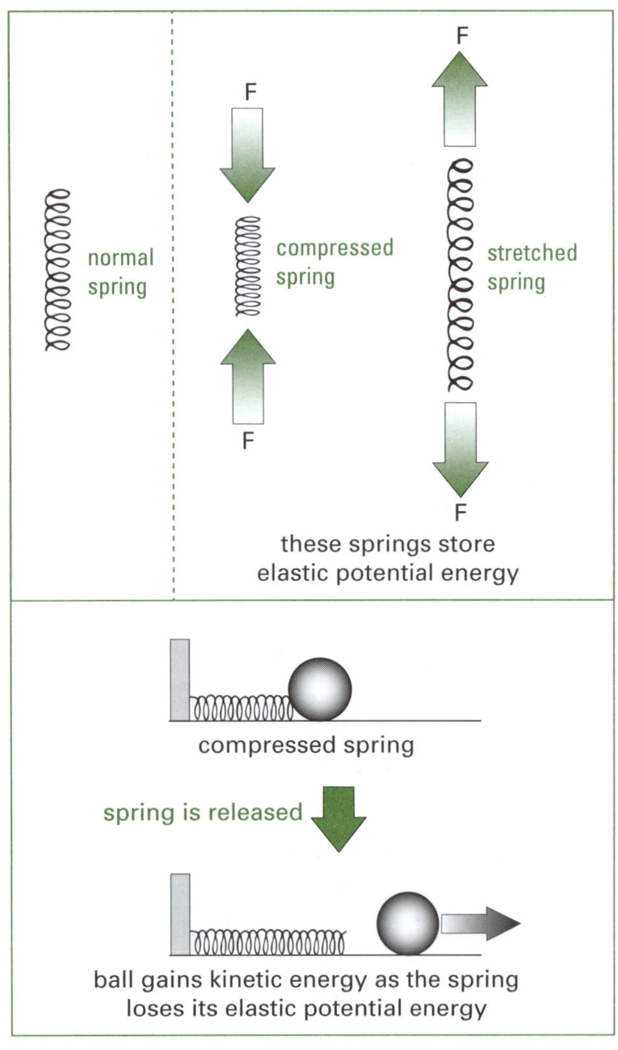

Figure 4.5 Stretched or compressed springs store energy.

Figure 4.6 Some examples of springs storing elastic potential energy

Machines such as bulldozers, cranes and train engines are well known in our modern society. The word *mechanical* comes from the Greek for 'machine'. Mechanical energy was therefore originally related to the ways in which machines worked. Today it has a wider meaning. Mechanical energy refers to the types of energy in which motion or the possibility of motion is present. Potential energy and kinetic energy are two types of mechanical energy.

4.3 Energy transformation

Energy can be changed from one form to another. Often more than one kind of energy is produced.

Examples

1. A light bulb transforms electrical energy into light energy and heat energy.
2. Most guns fire by the force of an explosive. Chemical energy from gunpowder is changed to heat energy, sound energy and some light energy. The gases produced in the explosion contain elastic potential energy because they are tightly compressed. The hot, expanding gases push the bullet out of the barrel at a high speed giving it kinetic energy.
3. Space vehicles usually have both liquid fuel and solid fuel rockets at launch. These fuels store chemical potential energy (see Figure 4.7). At the launch, the fuel is burnt to produce heat energy, sound energy and some light energy. The compressed combustion gases possess elastic potential energy. The force required to launch comes from the kinetic energy of these expanding hot gases as they escape from the rocket motors. As the rocket takes off and its altitude increases, it gains gravitational potential energy. At each conversion energy is wasted as heat and sound. The engines need to burn fuel to maintain sufficient kinetic energy to reach Earth orbit.
4. Coal is called a fossil fuel because it is the compressed remains of ancient plants. Coal contains chemical potential energy (see Figure 4.8). At the power station, the coal is burnt to produce heat energy. This heat is

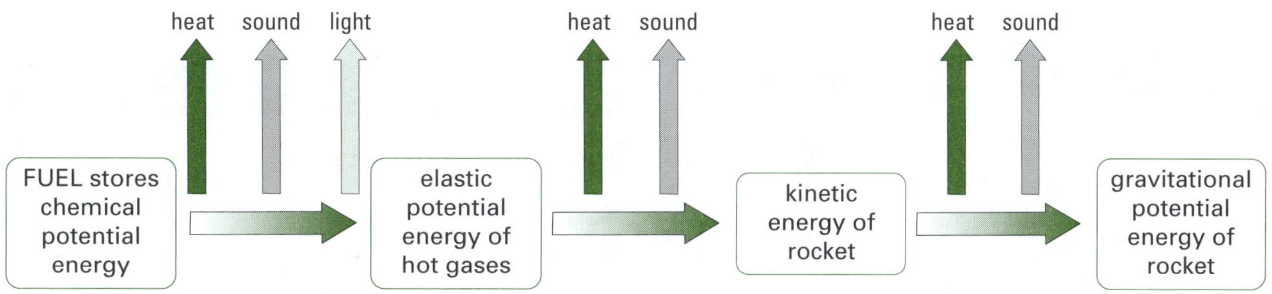

Figure 4.7 Rocket fuel stores chemical potential energy.

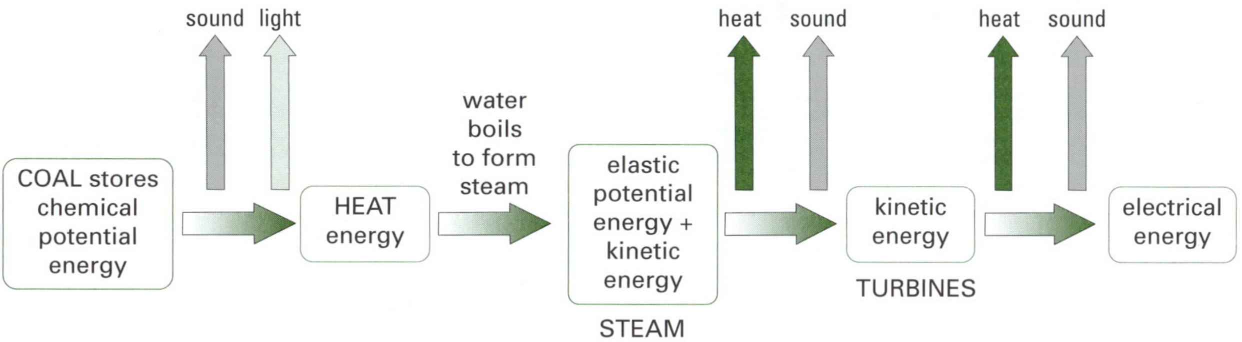

Figure 4.8 Coal stores chemical potential energy.

used to boil water into steam. The compressed steam contains elastic potential energy and kinetic energy. The compressed steam is used to turn the wheels of a turbine and energy is therefore transformed into kinetic energy. As the turbine rotates, it generates electrical energy in the copper wires. Some heat and sound energy are wasted forms of energy during the conversion of the energy of the steam into electrical energy.

5. Electric circuits are good examples of the transformation of electrical energy into various forms of energy such as heat and light (see Figure 4.9). A simple electric circuit consists of an energy source (e.g. battery), electrical wires (usually made of copper), a switch and a resistor (e.g. a heating coil or a light bulb). When the switch is closed, the chemical potential energy of the battery is converted to electrical energy. This energy is transmitted from the positive pole of the battery around the circuit in the form of an electric current. As the current passes through the resistor, it loses energy in the form of heat and/or light and then the low energy current returns to the battery to be re-energised.

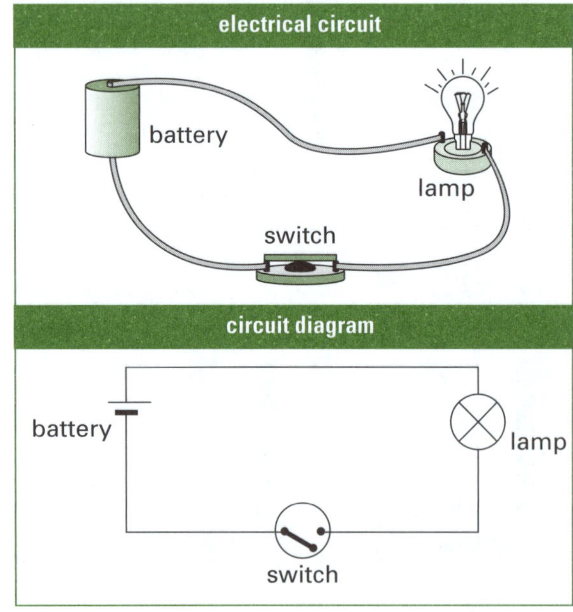

Figure 4.9 Electrical circuit

Electrical circuits are often designed to convert electrical energy into heat energy. Electric blankets and electric hot water jugs are examples of this. Figure 4.10 shows the result of a simple experiment in which two electrical immersion heaters are used to heat a given mass of water for a fixed time. In the first experiment the two coils are wired in a series circuit and in the second diagram they are wired in a parallel circuit. The experiment shows that the water gets much hotter when the immersion coils are wired in parallel. This tells us that a greater electric current has flowed through each coil in the parallel circuit.

The law of conservation of energy

Over 150 years ago scientists realised that energy can neither be created nor destroyed; it can only be converted from one form to another. This became known as the law of conservation of energy. This means that you can only get out as much energy from a machine as you put into it. Usually you don't even get that. This is because some of the energy is changed to a form we can't immediately use—it is wasted, usually as heat. However, the energy is not destroyed.

Experiment 1

Falling tennis balls

Aim

- To investigate the rebound height of a dropped tennis ball versus the dropped height

Method

1. Use a tennis ball for this experiment.
2. Drop it from different heights (H_1) onto a hard surface (e.g. a vinyl floor). Measure the height (H_2) to which it rebounds. (See Figure 4.11.)
3. Vary the drop height and repeat the experiment for six different drop heights.

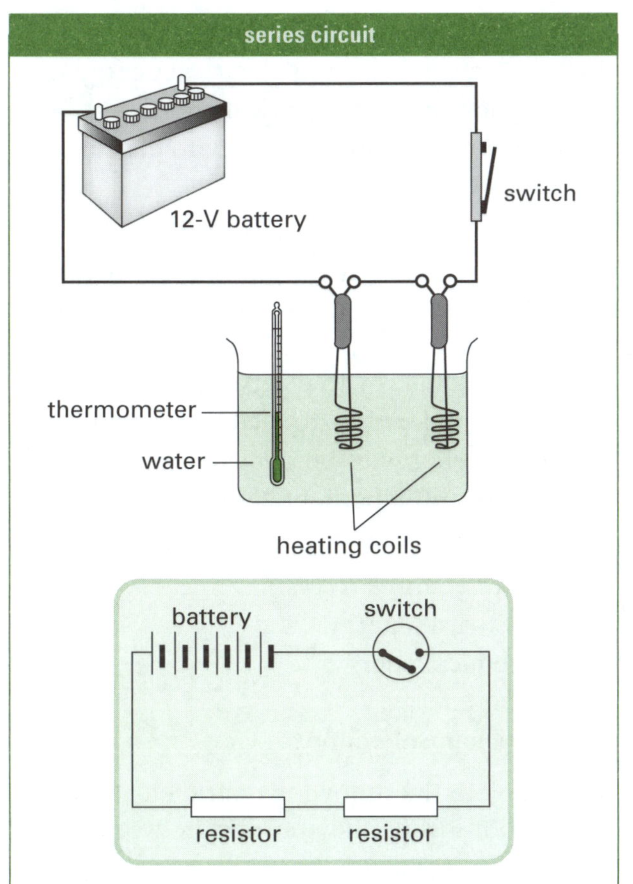

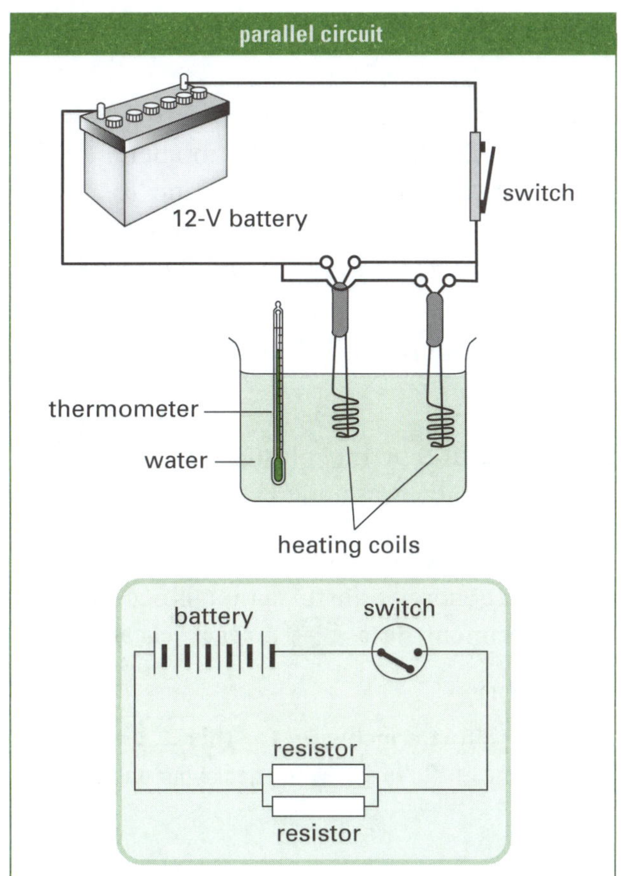

Figure 4.10 Series and parallel circuits with immersion heaters

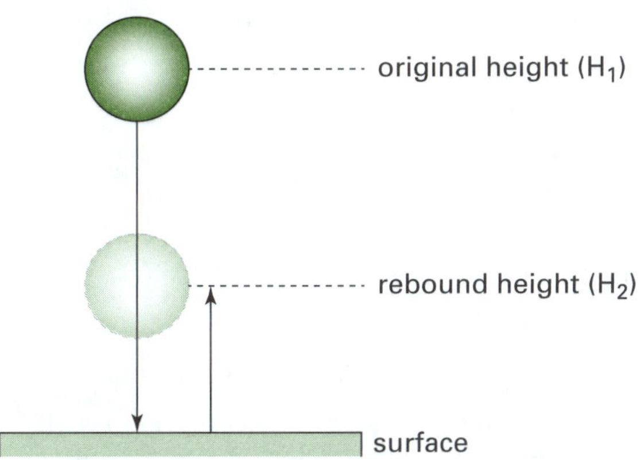

Figure 4.11 Rebound height

Results

Table 4.1 shows the results of the experiment.

Table 4.1 Results of ball-dropping experiment

H_1 (cm)	200	160	140	110	80	50
H_2 (cm)	110	90	79	59	42	24

Analysis

Graph this data and draw a line of best fit.

Discuss why the ball does not rebound to its original height. Go to p. 198 to check your answer.

Conclusion

Write a short conclusion for this experiment. Go to p. 198 to check your answer.

Experiment 2

Chemical energy

Aim

- To measure heat changes in a chemical reaction

Method

1. Place a level scoop of dry plaster of Paris in a dry test tube.

2. Measure the temperature of the dry powder. Record this temperature.
3. Using a water dropper, add one drop of water to the powder. Stir with the thermometer and record the temperature. Continue to add drops of water. Record the temperature after each drop of water is added and the mixture stirred.
4. Continue until no further change occurs.

Results

The temperature of the plaster increased as each drop of water was added.

Analysis

Explain what energy transformation is occurring in this experiment. Go to  p. 198 to check your answer.

Conclusion

Write a short conclusion for this experiment. Go to p. 199 to check your answer.

Test yourself 1

Part A: Knowledge

1. Which one of the following sentences could be used to define the term 'energy'? *(1 mark)*
   A Energy causes things to happen.
   B Energy is heat.
   C Energy is a type of force.
   D Adding energy makes things colder.

2. What type of energy is present in a stretched rubber band? *(1 mark)*
   A chemical energy
   B electrical energy
   C elastic potential energy
   D gravitational potential energy

3. The switch on a torch is pushed and the light bulb illuminates the path ahead. Identify an energy transformation that has occurred during these events. *(1 mark)*
   A sound energy to heat energy
   B chemical energy to electrical energy
   C light energy to electrical energy
   D chemical potential energy to elastic potential energy

4. Wind is used to turn windmill blades to produce electricity for a farm. Identify an energy transformation that has occurred. *(1 mark)*
   A kinetic energy to electrical energy
   B potential energy to electrical energy
   C solar energy to kinetic energy
   D gravitational potential energy to kinetic energy

5. A battery stores *(1 mark)*
   A electrical energy.
   B chemical energy.
   C heat energy.
   D mechanical energy.

6. Complete the following restricted-response questions using the appropriate word. *(1 mark for each part)*
   a) Light energy comes from moving particles of light called ....
   b) A heavy object has more gravitational potential energy than a lighter object at the same ....
   c) A trampoline stores .... potential energy in its springs.
   d) Potential energy and kinetic energy are two types of .... energy.
   e) Compressed steam contains elastic potential energy and .... energy.

7. Use the code letters to match the terms or phrases in each column. *(1 mark for each part)*

Column 1	Column 2
A heat	F gunpowder
B coal	G temperature
C sound	H chemical energy
D mass	I kinetic
E chemical energy	J gravity

Part B: Skills

8. The formula to calculate gravitational potential energy is:

$$GPE = 9.8mh$$

where m = mass of the body (kg)
h = height above the ground (m)

GPE is measured in joules (J).

Calculate the gravitational potential energy of a 10-kg mass held at a height of 10 m above the ground. *(1 mark)*

9. A body has a gravitational potential energy of 300 J when held at a fixed height above the Moon's surface. The body is now dropped. What will be the kinetic energy of the body when it is just about to hit the ground? Explain. *(2 marks)*

10. Sound can travel through water at a greater speed than it does in air.

a) What type of energy does sound carry? *(1 mark)*

b) Suggest a possible reason why sound travels faster in water than in air. *(1 mark)*

11. A volume of 3 mL of hydrochloric acid is placed in one test tube and 3 mL of sodium hydroxide is placed in a second test tube. The temperature of both solutions is 21 °C. The hydrochloric acid is now poured into the sodium hydroxide tube and a thermometer shows that the temperature of the mixture rises to 28 °C. Explain these observations in terms of energy transformations. *(2 marks)*

12. A 3-cm steel ball is dropped from a height onto a 5-cm thick piece of plasticine. The ball penetrates 1 cm into the plasticine.

a) What type of energy does the ball originally possess? *(1 mark)*

b) What type of energy does the ball possess as it is about to hit the plasticine? *(1 mark)*

c) Describe the energy transformations as the ball penetrates the plasticine. *(1 mark)*

13. When sweat evaporates from your skin, the skin becomes cooler. Analyse the energy changes in this process. *(2 marks)*

14. Figure 4.12 shows the transformation of energy. Code letters (A, B and C) are various types of energy. Identify each energy type from this list: kinetic energy; sound energy; heat energy; chemical potential energy; gravitational potential energy; light energy. *(3 marks)*

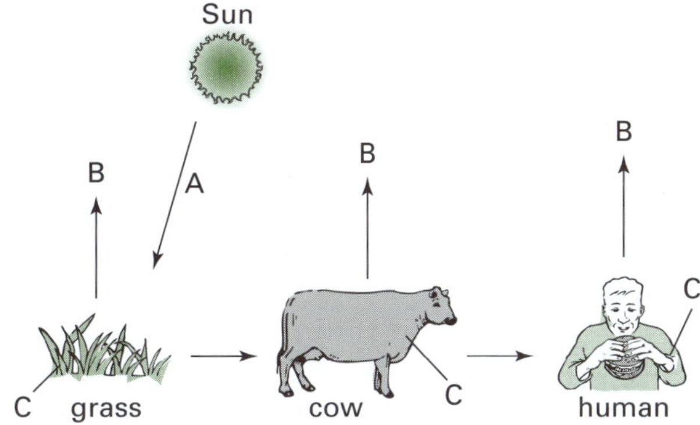

Figure 4.12 Energy transformation

Go to p. 199 to check your answers.

4.4 Heat energy

The history of heat

Before people recognised that heat was a form of energy, it was thought that the rise in temperature of a hot body was due to some invisible substance. This substance, which was called caloric, somehow flowed into the pores of the body and the body was heated. It flowed out when the body was cooled. Attempts were made to detect caloric by weighing the body when hot, and then when cold. But, of course, there was no change in weight. It was therefore assumed that this mysterious substance had no weight. The caloric theory broke down when it could not explain why things got hot for no apparent reason.

One of the experiments, which led to the caloric theory being rejected, was done by Benjamin Thompson (later known as Count Rumford, 1753–1814). While boring a cannon for the Bavarian government, he noticed that the drill generated a large amount of heat. He could place this metal in a bucket of water and get the water to boil. According to the caloric theory, if the cannon and drill got hot, then some other object must get cold and lose caloric. But nothing did. The supply of heat seemed inexhaustible! He reasoned that the only thing which could explain the continued rise in temperature was the increased motion of the particles. Here the kinetic energy in the drill was being changed to heat energy.

Temperature is a measure of how hot something is. We usually use an instrument called a thermometer to do this. The important difference between heat and temperature is that heat is a form of energy and temperature is a measure of the amount of that energy which is present in a body.

Transferring heat

Heat is a form of energy. Consider a simple electric radiator. The wires glow and you can feel the heat produced. Energy is being changed from one form to another. In a radiator, electrical energy is changed into heat and light. Heat energy is very important to us. We use it in the kitchen to cook, to boil water for our shower, to iron clothes, to warm us on winter nights and to help us cook meat on a barbecue. The amount of heat energy in a hot body is determined by the mass of the body, its temperature and the material that the body is composed of. Ten kilograms of copper at 70 °C has more heat content than 1 kg of copper at the same temperature.

The atoms or molecules of a hot object are vibrating or moving about rapidly. As the hot object cools down, its atoms or molecules slow down. The environment around the object gains the heat that is lost from the hot object. This transfer of energy to the environment can occur in three ways: conduction, convection and radiation.

Conduction of heat

When you hold the end of a steel rod in a fire, it will gradually become so hot that you will need to let go. In order to continue to hold the rod, you will need a non-conducting material as a handle (see Figure 4.13). Heat from the fire flows, or is conducted, through the rod towards the cooler end of the steel. The handle of the rod is made of a material that is a poor heat conductor or insulator. Heat moves from hotter objects to colder ones, never the other way around.

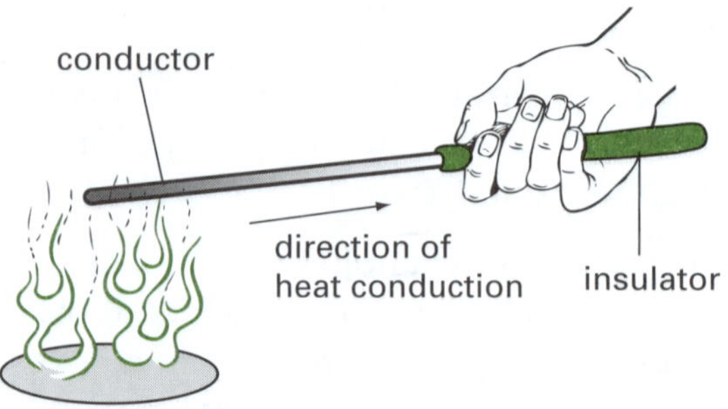

Figure 4.13 Heat conduction

Convection of heat

Heat moves in liquids and gases by convection. When air is heated, it expands. This makes it less dense, and it rises. Warm air rises and cooler (more dense) air moves in to take its place. In turn, this cooler air is warmed and it rises too. This process is shown in Figure 4.14.

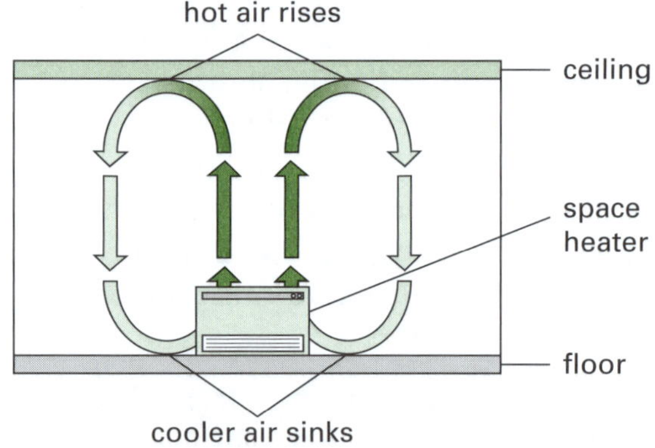

Figure 4.14 Convection currents heat the room

In conduction, the heated material doesn't move from one place to another. But in convection, the moving material moves the heat. Convection is responsible for forcing smoke up a chimney. It can also explain sea breezes and cloud formation (see Figure 4.15). Convection currents also occur through water. For example, warm ocean currents from the tropics move towards cold regions and, in turn, colder currents move back.

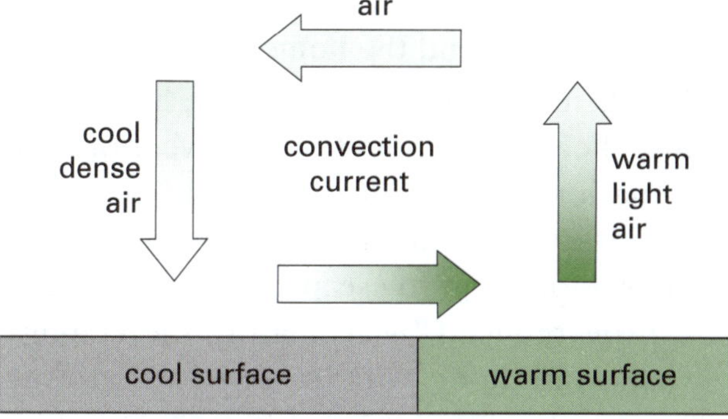

Figure 4.15 Convection currents result from the uneven heating of air by different surface temperatures of land or water. The cool, heavier air forces warmer air higher and establishes a convection cell.

Radiation of heat

In both conduction and convection, heat is moved by particles vibrating and passing on the heat, or by the particles themselves moving. But heat can also travel through empty space where there are no particles. Moving atoms or molecules create waves or radiant energy. One form of radiant energy is called infra-red rays. Infra-red rays are similar to visible light waves except they have a longer wavelength. Although you can't see infra-red waves, you can certainly feel them. The hotter the object, the more infra-red rays it gives off. When infra-red rays strike an object such as your skin, they speed up the particles in that object, thereby warming it. This is how the Earth gets its warmth from the Sun. Heat energy is transferred by radiation.

Experiment 3

Heat insulators

Aim

- To determine which polymer (plastotherm or polytherm) is a better heat insulator

Method

1. Set up the apparatus shown in Figure 4.16.
2. A temperature probe is inserted in a hole bored 5 mm below the top of each plastic board. The probe is connected via an interface to a computer that displays temperature changes with time.
3. One litre of 90 °C water was placed in turn on each plastic board and the temperature readings commenced.
4. The results for each experiment were then graphed.

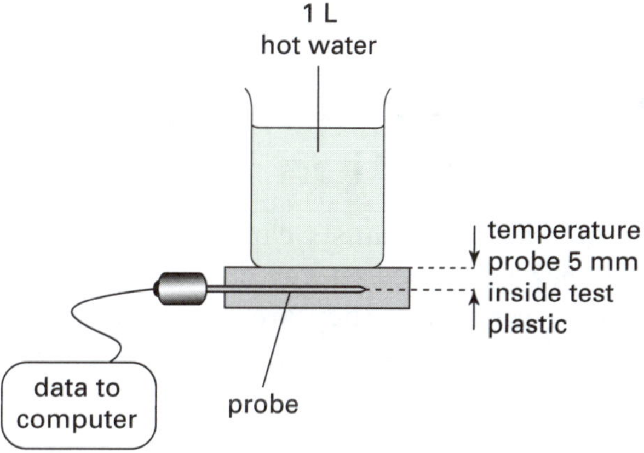

Figure 4.16 Testing plastics for heat transfer

Results

The results are shown in Figure 4.17.

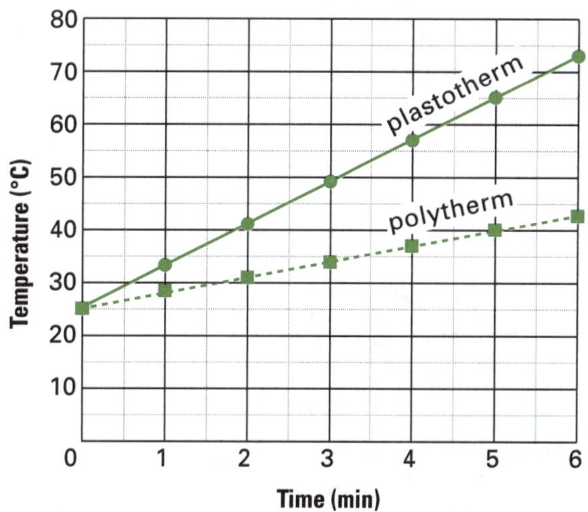

Figure 4.17 Heat transfer in two plastics

Analysis

Use the graphical data to answer these questions.

1. Which plastic is better at resisting the transfer of heat?
2. Is heat transferred by conduction, convection or radiation into the plastic?

3. Use the graph to predict the temperature of the polytherm after 210 seconds.
4. What variables were kept constant during the experiment? Go to p. 199 to check your answers.

Conclusion

Write a short conclusion for this experiment. Go to p. XXX to check your answer.

4.5 Reducing energy consumption

Engineers and scientists can modify existing machines to reduce energy consumption and improve efficiency. They are also involved in designing new machines that are more energy efficient.

Houses can be designed to reduce heat losses during winter and to prevent overheating during the summer. The use of reverse-cycle air conditioners to warm and cool the house requires a lot of energy and the burning of large amounts of fossil fuels. This also contributes to global warming due to the production of greenhouse gases.

Here are some ways to reduce energy consumption around the home.

- Home insulation. In the winter, homes can lose heat by conduction, convection and radiation. Heat is lost through the floor, ceilings, walls and windows. Large windows allow more heat to escape. These heat losses can be reduced by insulating the ceilings, walls and floor with materials that trap air, which is a poor conductor of heat (see Figure 4.18). Wall cavities can be filled with fibrous or foam materials. In the roof space, the ceilings can be insulated with fibreglass batts. Metal foils can also be used to reflect radiant energy back into the house. Heavy curtains can be drawn across windows to reduce heat losses through the glass. In the summer, the house is also kept cooler by the insulation. Metal foils in the roof space can also reflect radiant energy back to the outside of the house.
- Shading. The construction of awnings, pergolas and verandahs around the house also reduce direct sunlight on the walls and windows. This will keep the house cooler in summer and reduce the use of expensive air conditioning.

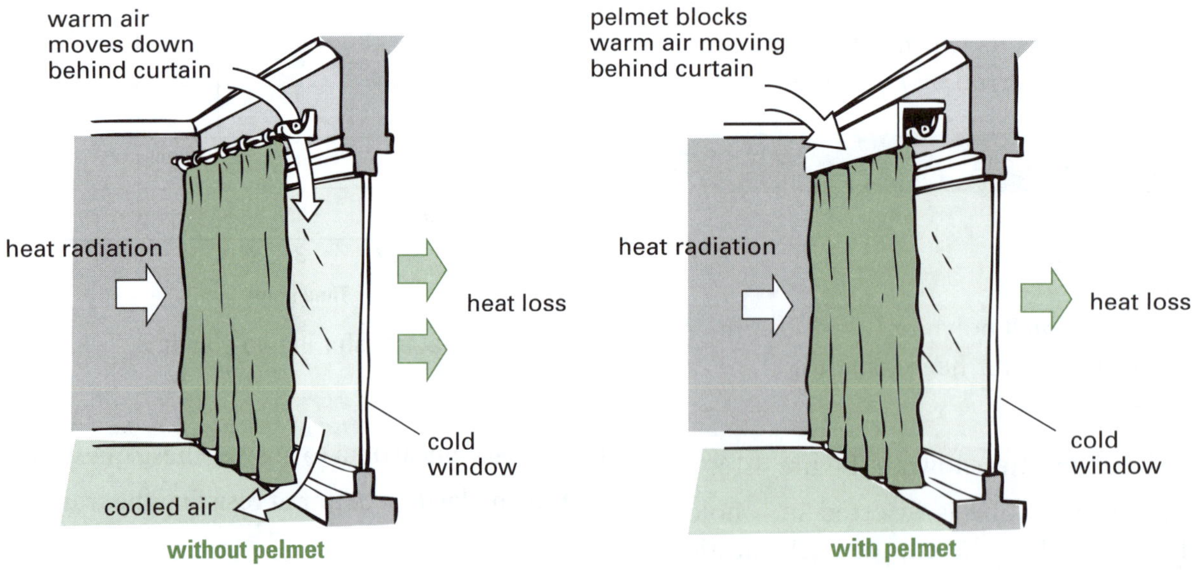

Figure 4.18 Curtains and blinds, together with pelmets, will help to keep heat inside the home by simply preventing warm air from coming in contact with the cold window glass. Less heat loss occurs.

- Water-efficient showerheads. Conventional showerheads deliver water at the rate of 25 L/minute. Energy-efficient showerheads deliver water at 9 L/min and therefore save both energy and water.
- Fluorescent and LED lights. Compact fluorescent and LED lights have much greater energy efficiencies than incandescent globes.
- Star ratings on appliances. Always buy the most energy-efficient appliances. These appliances have more stars on their energy stickers. Such appliances are more expensive to purchase but over the life of the appliance much more energy is saved and the total energy cost is lower.
- Stand-by mode. Many electrical appliances such as televisions and computers are left on stand-by mode when not in use. Under these conditions, some energy is still being consumed. In order to save energy, the appliances should be switched off at the power point when not in use.

Energy-efficiency improvements in cars

Humans pulled or pushed the first carts. When cattle and horses were domesticated, they drew the carts and wagons. In snow-covered countries dogs pull sleds and, of course, Santa has his reindeer. In the 14th century the first wind-powered vehicle was developed in Italy. Following the invention of the steam engine in the late 18th century, vehicles could be powered using steam. The first steam cars were so heavy that they needed to run on iron rails. Such vehicles eventually developed into trains. In 1860 the first practical gas-powered internal combustion engine was developed in Paris. In 1893 the first petrol-powered vehicle was developed in the United States.

Engineers have continued to make improvements to car designs to make them more energy efficient. Some of these improvements include:

- making cars more aerodynamic in shape in order to reduce drag
- improving engine valve timing so that fuel is not wasted
- reducing frictional energy losses in the engine
- designing low-resistance tyres
- improving fuel injection systems
- using lightweight materials to reduce the total weight of the car
- reducing use of energy during idling
- improving power steering
- more efficient alternator systems for powering air conditioning and battery recharge.

With increasing petrol costs, motorists are sensitive to the fuel efficiency of cars. So manufacturers are increasingly building vehicles which are more economical with fuel. This means cars need to be lighter, more aerodynamic and have more efficient engines. Lighter cars use less fuel because they need less energy to start and stop than heavier cars. CSIRO has developed high-performance magnesium and aluminium alloys to create strong, lightweight car parts that are not more expensive than similar structural components made from steel. The new magnesium alloy is 33% lighter than other aluminium alloys and stronger than the same weight of steel. It is claimed that using this alloy instead of iron in a car's engine block could result in greenhouse gas savings of 2.75 t of CO_2 equivalent over the car's life.

4.6 Solar energy and solar-powered vehicles

Electric cars are powered by an array of batteries. No chemical fuels are used. Electric cars were developed in the mid-to-late 19th century but fell out of favour due to advances in internal combustion engines powered by fuels such as petrol and diesel. In the latter part of the 20th century, new types of electric vehicles were being developed due to the impact of fossil fuel combustion on the environment.

The major drawback of electric cars is the high cost of the lithium-ion batteries that are used to power them. Other issues are the restricted range (~200 to 400 km) that these cars can travel without recharging. Currently there are very few recharging sites. Battery switch stations could be developed where a recharged battery can replace a depleted one but this may be impractical as it may leave businesses open to litigation. Alternatively, fast-charging stations are being established where motorists can get an ~80% recharge in 30 minutes. In Australia, fast-charging stations are being built in major cities and most should be in operation by 2013. Other fast-charging stations are progressively being built across the United States, China, France and Japan. Other nations are sure to follow.

Green groups welcome the use of electric cars but an important issue is that the charging of the batteries is performed using electricity derived from burning fossil fuels such as coal. Green groups are pushing for the use of non-fossil fuel generated electricity.

Solar energy could be used to directly power electric cars by collecting this energy via a photovoltaic (PV) array of solar cells. There are a number of problems in using PV arrays. These include the:

- high cost of PV cells
- efficiency of the PV array in converting radiant energy into electrical energy
- weight of the PV cells
- lack of sunlight on some days and absence of sunlight at night
- potential damage to solar panels due to continuous vibrations during driving
- area of the array required to absorb sufficient solar energy.

Solar-powered cars would also need to include rechargeable batteries to ensure sufficient driving range.

Currently, solar cars are only built for solar car races (see Figure 4.19). They have a single seat and do not carry loads other than the driver in order to save weight. Consequently, they are not suitable for the average driver or family.

Figure 4.19 Solar car racer

4.7 Energy technologies

Let's examine various technological applications of different energy forms.

Sensors

Sensors have a wide variety of everyday uses. Optical sensors transmit various forms of light energy (e.g. visible light, microwaves, infra-red waves) into the environment to detect changes in the environment. For example, many shops and buildings have motion detectors to open the front door to allow a customer to come in. The sensor emits infra-red waves which bounce off the person and return to the sensor unit. The pattern of reflected waves is altered by the presence of a person at the door. The sensor is activated and sends a signal to open the door.

Touch screens have become very common with tablet computers and mobile phones. Several methods are used. One method involves a special conductive layer being applied to the glass panel. This layer stores electric charge. When you touch the screen some charge is transferred to your finger as the layer loses charge at that spot. Electronic circuits measure this change in charge and transfer the information to the software driver.

A mercury switch is an interesting type of sensor (see Figure 4.20). Mercury is a liquid. Any sudden movements that cause the switch to tip sideways will cause the mercury to move and touch both electrical contacts. The circuit is then completed and an alarm is sounded.

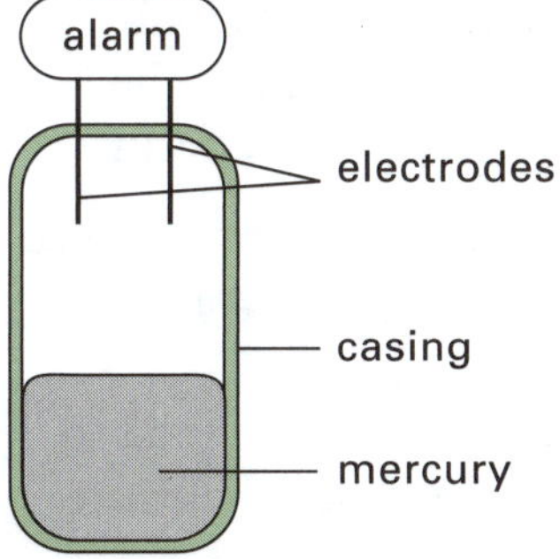

Figure 4.20 A mercury switch is a simple sensor. It can detect sudden movements or a change in orientation.

Lasers can be used in a wide variety of sensors that have many different applications.

- In the beverage industry, lasers can be used to determine whether the level of liquid in a bottle is correct.
- In vehicles, lasers can be used to detect the proximity of the vehicle to other objects and so avoid collisions (see Figure 4.21). They can also ensure that there is sufficient clearance of large vehicles as they pass under a bridge. One possible configuration is to have two front laser scanners covering a wide angle (around 220°) of road ahead, located inside the car just below the headlights. A rear laser scanner would cover 150° of the road behind and be situated centrally near the rear number plate.

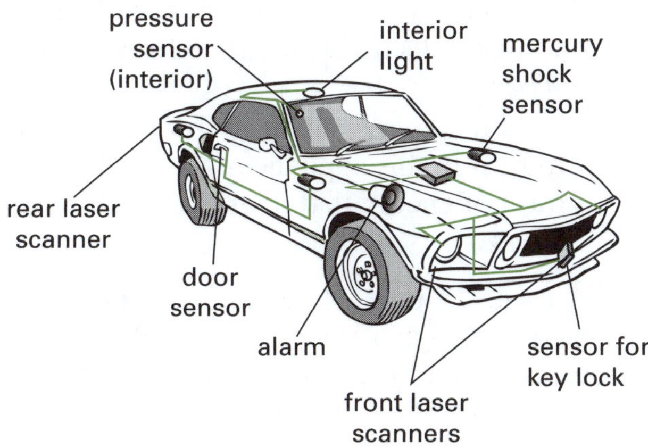

Figure 4.21 Modern cars have many sensors built into them.

- In digital cameras, laser sensors can ensure that automatic focusing is optimised.
- In shipping, the laser sensor can assist in docking to ensure collisions are avoided.
- Laser sensors can monitor the size of objects (e.g. vegetables) on a conveyor belt and detect undersized material.

Earthquake prediction

Earthquakes can produce great damage and often loss of life. Seismometers are the traditional method of measuring earthquakes. These devices monitor the time for each type of earthquake wave to arrive at the monitoring station. As Figure 4.22 shows, the first waves to arrive are called P waves and these are followed by S waves and then the destructive L waves. The greater the time difference between the arrival of the P and S waves, the further away is the epicentre of the quake. Time is the critical factor and scientists are working on developing early warning devices that may save more lives by giving residents more time to move to safer areas.

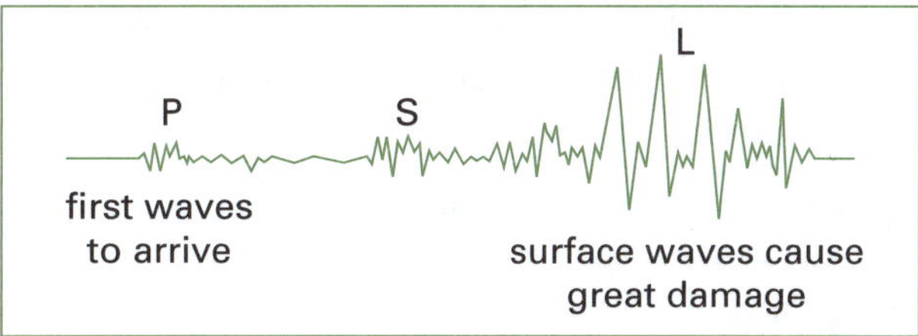

Figure 4.22 Seismometer recording of earthquake waves

One area of earthquake research involves magnetism. Prior to an earthquake, the crustal rocks experience great forces as they push against each other. This process produces changes in the crystalline structure which in turn generates small electrical currents called piezoelectricity. Accompanying this are changes in the magnetic field intensity. Sensors can be used to monitor these small changes in the magnetic field and provide warning of an imminent earthquake. The technology is called magnetic gradiometry and has been applied in Israel near the southern area of the Dead Sea.

Test yourself 2

Part A: Knowledge

1. The caloric theory was proposed to explain why objects got hot or cold. This theory has since been proven wrong. The caloric theory explained that objects got cold because *(1 mark)*
   - **A** they absorbed caloric from the environment.
   - **B** they lost caloric to the environment.
   - **C** some objects took longer than others to absorb caloric.
   - **D** their surfaces could not absorb caloric.

2. Select the true statement. *(1 mark)*
   - **A** Two kilograms of iron at 50 °C has the same heat content as 5 kg of iron at 50 °C.
   - **B** Thirty grams of water vapour at 100 °C has a greater heat content than 30 g of liquid water at 100 °C.
   - **C** Five grams of ice at 0 °C has the same heat content as 5 g of liquid water at 0 °C.
   - **D** One hundred grams of copper at 40 °C has the same heat content as 100 g of aluminium at 40 °C.

3. An example of a poor heat conductor is *(1 mark)*
   - **A** silver.
   - **B** gold.
   - **C** steel.
   - **D** glass.

4. Select the true statement. *(1 mark)*
   - **A** Warm air sinks and cooler air rises.
   - **B** Cool air is more dense than warm air.
   - **C** Conduction of heat is responsible for forcing smoke up a chimney.
   - **D** Infra-red rays have a shorter wavelength than visible light rays.

5. Select the true statement. *(1 mark)*
   - **A** Heavy curtains can reduce heat loss through windows during winter.
   - **B** Awnings and pergolas should be constructed on the east side of a house to keep it cooler during the summer.
   - **C** No energy is wasted when appliances are left on stand-by mode when not in use.
   - **D** Car bodies should be made of steel rather than aluminium to reduce the weight of the car and save on energy.

6. Complete the following restricted-response questions using the appropriate word. *(1 mark for each part)*
   - **a)** Solar-powered cars would also need to include .... batteries to ensure sufficient driving range.
   - **b)** Compact fluorescent and LED lights have much greater energy .... than incandescent globes.
   - **c)** In order to reduce these heat losses the ceilings, walls and floor should be ....
   - **d)** Temperature is a measure of how .... something is.
   - **e)** The atoms or molecules of a hot object are .... or moving about rapidly.

7. Use the code letters to match the terms or phrases in each column. *(1 mark for each part)*

Column 1	Column 2
A laser sensors	F electric charge storage
B touch screens	G radiation
C star ratings	H energy efficiency
D infra-red	I thermometer
E temperature	J camera focusing

Part B: Skills

8. **a)** Give an example of an appliance that changes electrical energy to heat energy. *(1 mark)*
   b) Is all the energy changed from electrical to heat in this appliance? Name another form of energy that might be produced. *(2 marks)*

9. Figure 4.23 shows one end of a copper rod being heated in a Bunsen burner flame. Select three of the following captions to replace the labels A, B and C. *(3 marks)*
   1. Vibrating particles bump into neighbouring particles causing them to vibrate.

2. Convection currents in the copper transfer heat away from the flame.
3. Heated atoms of copper vibrate more.
4. Heat is conducted along the rod as vibrating particles transfer energy.
5. Heat energy is radiated along the rod and causes the rod to heat up.

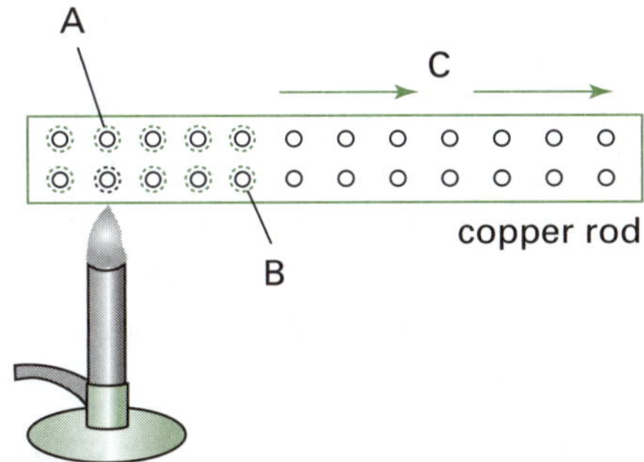

Figure 4.23 Heat conduction

10. Figure 4.24 shows a convection current producing a sea breeze.

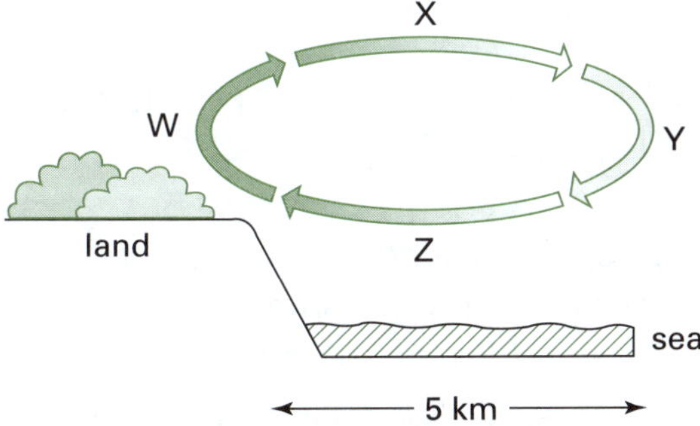

Figure 4.24 Convection current

Select four of the following captions to replace the labels W, X, Y and Z. *(4 marks)*

1. Cool air moves towards the land.
2. Warm land heats the air which then expands and rises.
3. Cool air is more dense and sinks.
4. Air cools and becomes more dense.
5. Warm rocks conduct heat to the air which then rises.

11. The heat from hot rocks can be extracted and converted to electrical energy. The flow chart in Figure 4.25 shows the stages (A to E) in this process. Match the following processes to the code letters in the flow chart. *(5 marks)*

1. Electricity is produced for the grid.
2. Water is pumped through pipes down to the hot rocks.
3. Steam cools back to liquid water.
4. Water boils to form steam.
5. Steam returns to the surface to turn turbines and produce electricity.

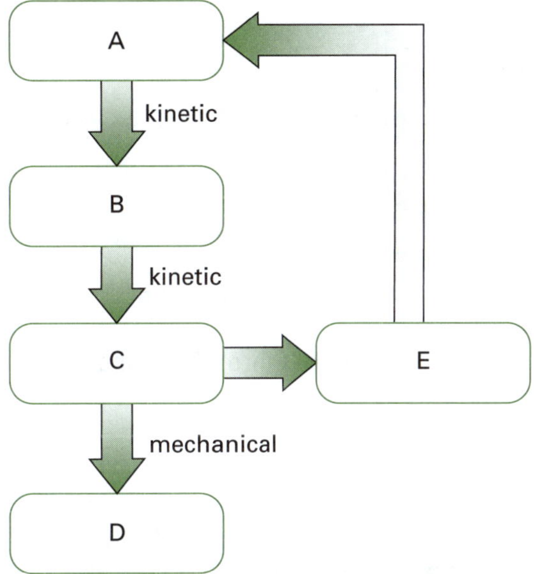

Figure 4.25 Extraction of energy from hot rocks

12. In winter it costs a lot of money to heat a home. Unless we stop the heat escaping too quickly, we will have wasted our money. Insulating your home will help reduce heat loss. Not all parts of a house lose the same amount of heat. Table 4.2 shows the ways in which heat can be lost from a typical house.

Table 4.2 Ways in which heat can be lost from a typical house

Places where heat is lost	% of heat escaping
walls	30
roof	25
floor	15
draughts	15
windows	15

a) Convert this information into a column graph. *(5 marks)*

b) Will this graph be the same for all houses? Why? *(2 marks)*

13. Lots of heat is lost through glass windows. Six joules of heat per second is lost through windows with an area of 1 sq. m when the outside temperature is 1 °C lower than the inside temperature. Consider the glass window in Figure 4.26.

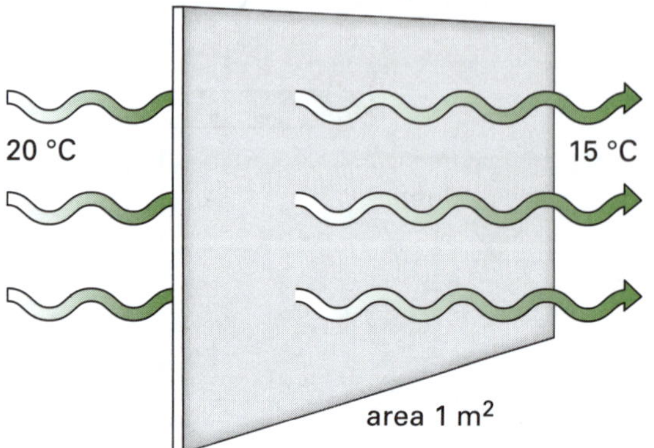

Figure 4.26 Heat loss through a window

a) Calculate the heat lost per second from this glass window. *(1 mark)*

b) Suggest a strategy to reduce this loss. *(1 mark)*

14. On a future space city on the Moon, solar cells absorb 100 units of energy into their storage batteries. This is used to produce 43 units of electricity for the city. By the time the electricity reaches the power points, there are 35 units of electricity available to charge the batteries of the electric cars, which run around the city on padded rubber roads. The cars only get 20 units of energy from their fully charged batteries to produce movement.

Draw a flow chart of these energy changes. *(7 marks)*

Go to pp. 199–200 to check your answers.

Summary

1. Moving things possess kinetic energy. Kinetic energy depends on both the mass and speed of a body.
2. Stored energy is called potential energy.
3. Energy can be stored as chemical, elastic and gravitational potential energy.
4. Chemical potential energy is stored in the chemical bonds of a substance.
5. Elastic potential energy is stored when materials are subjected to compression or stretching forces.
6. Gravitational potential energy is energy stored in a mass due to its height above the ground.
7. Energy can be transformed from one type to another.
8. In most transformations of energy some energy is wasted in the production of heat.
9. The law of energy conservation states that energy cannot be created or destroyed. Energy can only be converted from one form to another.
10. Heat energy always flows from a hot body to a cold body. The atoms or molecules of a hot object are vibrating or moving about rapidly.
11. Heat energy can be transferred by conduction, convection and radiation.
12. Poor heat conductors are called insulators.
13. Improvements in machines reduce energy consumption.
14. Energy consumption in the home can be reduced by installing insulation, turning off appliances when not in use and installing solar hot water and solar electricity panels.

Syllabus checklist

Are you able to answer every syllabus question in this chapter? Tick each question as you go through the list if you are able to answer it. If you cannot answer it, turn to the appropriate page in the guide as is listed in the column to find the answer.

For a complete understanding of this topic		Page no.	✓
1	Can I describe the type of energy possessed by moving things?	113	
2	Can I recall what stored energy is called?	113	
3	Can I explain the three ways energy can be stored?	113	
4	Can I recall where a substance's chemical potential energy is stored?	113	
5	Can I recall when elastic potential energy is stored?	113	
6	Can I describe gravitational potential energy?	113	
7	Can I recall examples of how energy can be transformed from one type to another?	115–116	
8	Can I explain why transformations of energy is never 100% efficient?	116	

For a complete understanding of this topic		Page no.	✓
9	Can I explain the law of energy conservation?	116	
10	Can I recall the flow of heat energy and what is happening to the atoms or molecules of a hot object?	120	
11	Can I recall the ways in which heat energy can be transferred?	120–121	
12	Can I recall what poor heat conductors are called?	120	
13	Can I recall ways in which energy consumption in the home can be reduced?	122–123	
14	Can I describe and give examples of improvements in machines that will reduce energy consumption?	123	

Chapter test

Go to p. v for *Tips for tests and examinations*

Part A: Multiple-choice questions

(1 mark for each)

1. Electric motors are able to change
   - **A** kinetic energy to electrical energy.
   - **B** electrical energy to kinetic energy.
   - **C** light to electrical energy.
   - **D** electricity to sound energy only.

2. In a gas oven, chemical potential energy is changed to
   - **A** heat and light energy.
   - **B** sound and light energy.
   - **C** heat energy only.
   - **D** electrical energy.

3. Which form of energy is different from the other three?
   - **A** a stretched spring in a trampoline
   - **B** compressed gas in a cylinder
   - **C** an electric current in a conductor
   - **D** a compressed mattress spring

4. Consider the following energy transformation:

   electrical energy → light energy + heat energy

   This transformation would best be shown in a
   - **A** bicycle.
   - **B** mobile phone.
   - **C** torch bulb.
   - **D** stereo speaker.

5. The energy stored in the wax of a candle is
   - A chemical energy.
   - B heat energy.
   - C light energy.
   - D kinetic energy.

6. The compressed air molecules in a car tyre possess
   - A elastic potential energy only.
   - B kinetic energy only.
   - C kinetic energy and elastic potential energy.
   - D kinetic energy and sound energy.

7. A ball is thrown up into the air until it reaches its highest point. Which statement is true?
   - A Its gravitational potential energy decreases as it rises.
   - B Its kinetic energy decreases as it rises.
   - C At the highest point its kinetic energy is at a maximum.
   - D Its elastic potential energy at the top of its flight is at a minimum.

8. Identify one major problem with the widespread use of photovoltaic devices such as solar panels in vehicles.
   - A high cost per unit
   - B high sunlight intensity damages the cells
   - C cells are very light and need to be strongly bolted down
   - D panels get covered readily with grease from the road

9. In order to increase fuel efficiency in cars, manufacturers build vehicles that
   - A are aerodynamic.
   - B are made from strong steel alloys.
   - C use renewable ethanol fuel.
   - D are heavier and can withstand vibrations better.

10. Placing a pelmet above the curtains on a window
   - A reduces cold draughts moving back into a heated room.
   - B increases the amount of solar energy that enters the room during winter.
   - C maximises convection currents in winter.
   - D prevents hot air accumulation between the window and the curtain.

Part B: Short-answer questions

11. Complete the following restricted-response questions using the appropriate word. *(1 mark for each part)*
   - a) Many cooking implements have handles that are made from wood or plastic or other poor conductors of heat. These materials can be called heat ....
   - b) Burning a piece of wood releases a lot of .... energy.
   - c) .... currents move smoke up chimneys, and help to explain why warm ocean currents move to colder regions.
   - d) Solar energy could be used to directly power electric cars using a .... array of solar cells.
   - e) Coal is called a .... fuel because it is the compressed remains of ancient plants.

12. Use the code letters to match the terms or phrases in each column. *(1 mark for each part)*

Column 1	Column 2
A radiant energy	F movement
B fibreglass	G infra-red
C kinetic	H elastic potential energy
D cross bow	I motion detector
E optical sensor	J insulation

13. If you hold onto an ice cube, it starts to melt in your hand. Why? *(1 mark)*

14. Lavinia set up some apparatus in a beaker of water as shown in Figure 4.27. The flask contains coloured water.
   - a) State one observation Lavinia will make as the beaker of water is heated. Explain. *(2 marks)*
   - b) How is heat transferred from the heated water to the coloured water in the flask? *(2 marks)*
   - c) What type of heat transfer ensures that all the water in the beaker will quickly

reach the same temperature even though only the base of the beaker is heated? *(1 mark)*

d) After the experiment, the electric heater was removed. Lavinia noticed that she could feel heat several centimetres from the sides of the beaker. Describe what method of heat transfer is happening in this case. *(2 marks)*

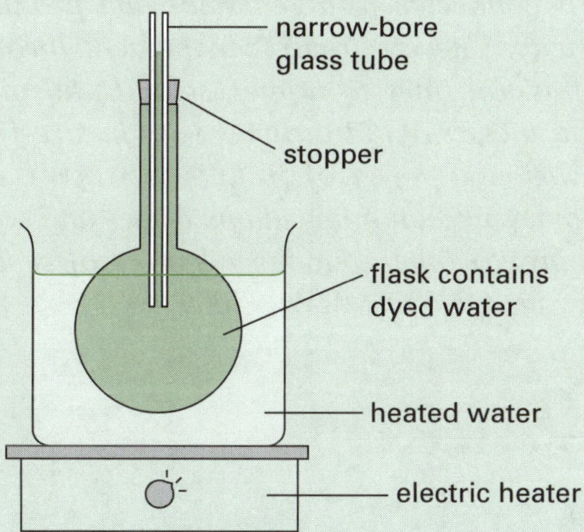

Figure 4.27 Heating coloured water

15. The thermos, or vacuum bottle, can keep liquids warm or cold for many hours. James Dewar invented the first such bottle, or Dewar flask, in 1885. A Dewar flask can keep liquids cold or hot for long periods of time. It does this by controlling the rate of heat transfer by conduction, convection and radiation. Use Figure 4.28 to explain which parts of the flask reduce heat transfer by each of these processes. *(4 marks)*

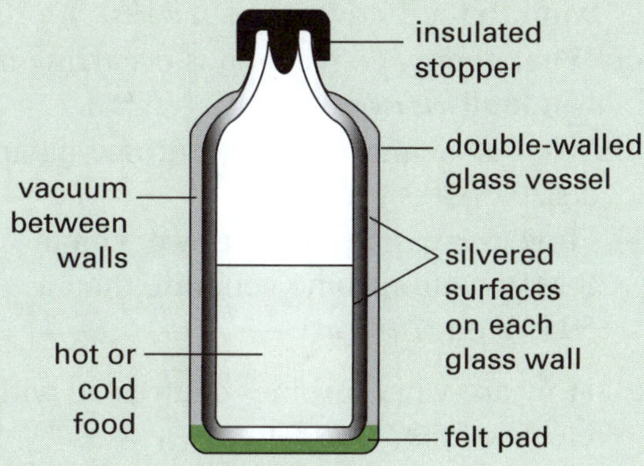

Figure 4.28 Thermos flask

16. For many years the tea in teapots was kept warm by a tea-cosy. A tea-cosy was a knitted cover that left openings for the spout and handle. Design an experiment to find out whether tea in a teapot can be kept warmer with a woollen or synthetic tea-cosy placed on the teapot.

a) What are the variables in this experiment? *(4 marks)*

b) What would the control be for such an experiment? *(1 mark)*

c) List the equipment required for this experiment. *(3 marks)*

d) Write a method for your experiment. *(4 marks)*

e) If the woollen tea-cosy is better, what results would you expect from your measurements? *(1 mark)*

f) How does a woollen tea-cosy slow down heat loss? *(1 mark)*

17. A black metal can and a silver metal can of the same size, metal and shape are each filled with hot water at 100 °C. A thermometer is placed in each can and the temperature of the hot water monitored over the next 20 minutes (see Figure 4.29). The air temperature during the experiment was 28 °C.

The temperature data collected are shown in Table 4.3.

Table 4.3 Temperature data for experiment

Time (min)	Temperature of silvery can (°C)	Temperature of black can (°C)
0	100	100
2	96	94
4	92	88
6	88	83
8	83	77
10	80	72
12	77	68
14	73	63
16	69	58
18	66	53
20	62	49

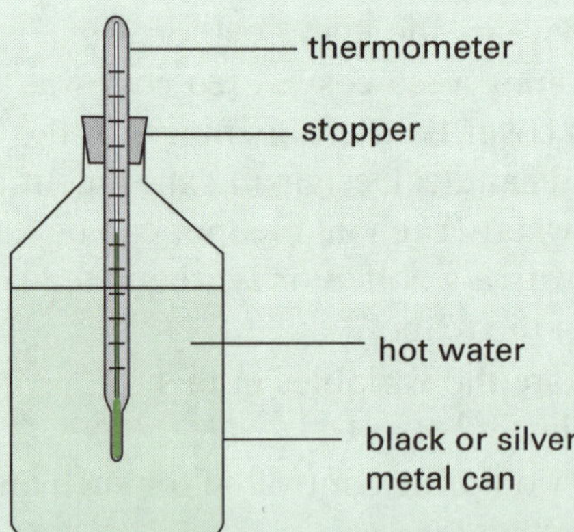

Figure 4.29 Experiment with hot water and metal cans

a) On one sheet of grid paper draw a line graph of temperature versus time for each can. Use a different colour for the temperatures of each can. Draw the lines of best fit. *(5 marks)*

b) Which can loses heat at a faster rate? *(1 mark)*

c) Given sufficient time, what will be the final temperature of each can, assuming the laboratory remains at a constant temperature? *(1 mark)*

18. Heat may be transferred by conduction, convection and radiation. For each of the following examples, state the way(s) that heat energy is transferred.
    a) Heat energy from the Sun travels to the Earth. *(1 mark)*
    b) An aluminium saucepan gets hot when placed on an electric hotplate. *(1 mark)*
    c) Smoke from a bushfire rises up into the sky. *(1 mark)*
    d) A boy feels warm as he sits 2 m in front of an electric radiator. *(1 mark)*

19. Wetsuits worn by divers and surfers consist of foam rubber with bubbles of air trapped in them. How can such suits keep the divers warm? *(1 mark)*

Read the following information about wind energy then answer Questions 20 and 21.

The uneven heating of our atmosphere by the Sun and the unevenness of the Earth and its rotation cause winds. The terrain over which winds move, including bodies of water and plant life, further modifies them. Wind energy or wind power describes the process by which wind can be used to generate mechanical power or electricity. Wind turbines such as windmills can convert the kinetic energy of the wind into kinetic energy for the mill which is used to pump water, grind grain or generate electricity. A wind turbine works somewhat like a fan in reverse (see Figure 4.30). The wind turns the blades which spins a shaft. This is connected to a generator and produces electricity. The electricity can be sent through transmission lines to homes, schools, hospitals and businesses. Wind turbines vary in size from those below a power of 50 kilowatts (kW) used for homes, telecommunications dishes and water pumping, to others having a power of several megawatts (MW). (1 MW = 1000 kW)

Figure 4.30 Wind turbines

20. a) What causes winds on Earth? *(1 mark)*
    b) Give two examples of how people use wind to their advantage. *(2 marks)*
    c) What energy conversion is occurring in a windmill? *(5 marks)*
    d) What is meant by a wind turbine having a size of 50 kW? *(1 mark)*
    e) How many times more power can a 3-MW wind turbine generate than a 50 kW one? *(1 mark)*

21. a) How are wind turbines contrasted with electric fans? *(2 marks)*
    b) Small turbines are sometimes used in connection with batteries and diesel generators.

i) What would be an advantage to such a connection? *(3 marks)*

ii) Where might these hybrid systems be of most use? *(2 marks)*

c) A controller in the wind turbine starts the generator to produce electricity when the wind speeds reach 30 km/h, and shut it off at about 110 km/h.

i) Suggest a reason for the generator not being used with wind speeds lower than 30 km/h. *(1 mark)*

ii) Suggest a reason for the generator not being used with wind speeds greater than 110 km/h. *(1 mark)*

22. Just as there are energy sources, there are also energy receivers. The ocean and the atmosphere are energy receivers because they warm up when they receive energy from the Sun (an energy source). For each of the following sources of energy, name one energy receiver.

a) electric guitar *(1 mark)*

b) gas oven *(1 mark)*

c) cyclone *(1 mark)*

23. Figure 4.31 shows a boy using a pogo stick.

Figure 4.31 Pogo stick

a) Explain how a pogo stick works. *(1 mark)*

b) When does the pogo stick have the maximum elastic potential energy? *(1 mark)*

c) When does the pogo stick have the maximum gravitational potential energy? *(1 mark)*

d) When does the boy have the greatest kinetic energy? *(1 mark)*

e) When does the boy have the least kinetic energy? *(1 mark)*

24. The temperature of the Earth's crust increases with depth. This temperature increase varies from place to place, but in mines near Broken Hill in New South Wales, the rise in temperature is approximately 1° for every 60-m increase in depth.

a) If the surface temperature is 20 °C, calculate the approximate temperature at the bottom of a 1200-m mine shaft. *(1 mark)*

b) Heat from hot rocks in the Earth's crust is seen as a possible future energy source. In some regions where hot rocks are found, the temperature of the crust is much hotter than the normal average. Some temperature–depth data has been collected in a region such as South Australia where hot rocks exist below the surface at a depth of about 5 km. Two values in Table 4.4 are missing.

Table 4.4 Temperature–depth data

Depth below surface (km)	Temperature (°C)
0	20
1	50
2	100
3	
4	200
	250

i) Is there a pattern in this temperature–depth data? *(1 mark)*

ii) Predict the value of the temperature at a depth of 3 km. *(1 mark)*

iii) Predict the value of the depth when the temperature is 250 °C. *(1 mark)*

c) In order to extract the heat energy from these hot rocks, the zones containing the hot rocks need to be located. The graph in Figure 4.32 shows different locations where the temperature has been tested. Each line represents a region of the same temperature. Between which surface markers should the Hot Rock

Power Company build its power station? Why? *(2 marks)*

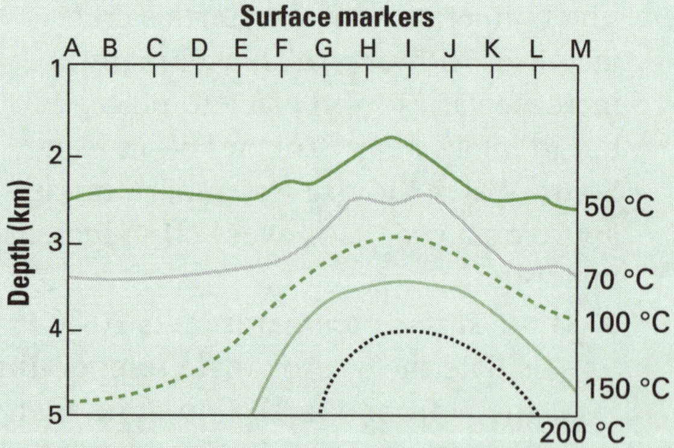

Figure 4.32 Heat from hot rocks

25. A student constructed a steam generator. She used simple laboratory equipment as shown in Figure 4.33.

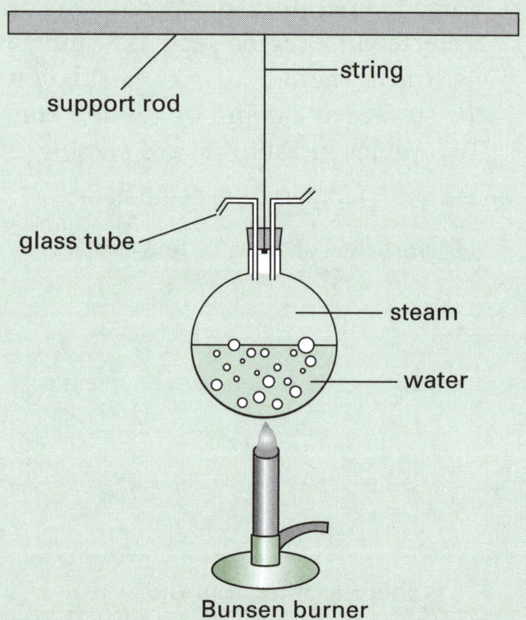

Figure 4.33 Steam generator

a) Predict what will happen as the Bunsen burner continues to heat the water. *(1 mark)*

b) Describe all the energy transformations that will occur as the steam generator operates. *(4 marks)*

c) Propose one design change that will increase the speed of steam production. *(1 mark)*

26. Two students were talking about fluorescent lights and incandescent lights.

Tim said: 'Because a fluorescent light produces less light energy than the electrical energy put in, this must mean that the law of energy conservation does not apply to fluorescent light'.

Fred said: 'The fluorescent light does obey the law of energy conservation. Some of the energy is wasted in other forms such as heat'.

Which student is correct? Why? *(2 marks)*

27. Ball A and ball B have the same mass and are rolling along a low-friction surface as shown in Figure 4.34. Which ball has the greater kinetic energy? *(1 mark)*

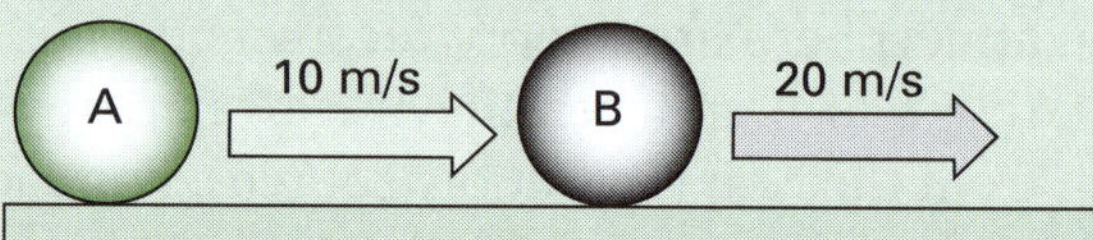

Figure 4.34 Balls A and B rolling along a low-friction surface

28. Ball X and ball Y are rolling along a low-friction surface with the same speed as shown in Figure 4.35. Which ball has the lower kinetic energy? *(1 mark)*

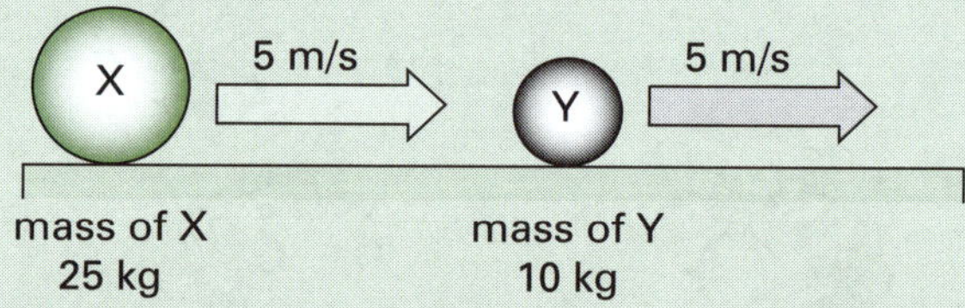

Figure 4.35 Balls X and Y rolling along a low-friction surface

29. A car company is testing two remotely controlled vehicles. Each vehicle is allowed to crash into a solid concrete barrier at the end of the track. A 1.5-t truck and a 2.5-t truck are both travelling at 25 m/s at the moment of the collision. Which truck will receive the greater damage? Explain. *(2 marks)*

30. A 2-kg mass (A) is held at a position of 3 m above the floor. A second 2-kg mass (B) is held at 1.5 m above the floor.

a) Which mass has the greater gravitational potential energy? *(1 mark)*

b) Both masses are now released and allowed to fall under gravity onto a force meter on the floor. Which body will register the greater impact force? *(1 mark)*

31. Figure 4.36 shows an archer about to fire an arrow from his bow. Is the elastic potential energy stored in the string or the bow? Explain. *(2 marks)*

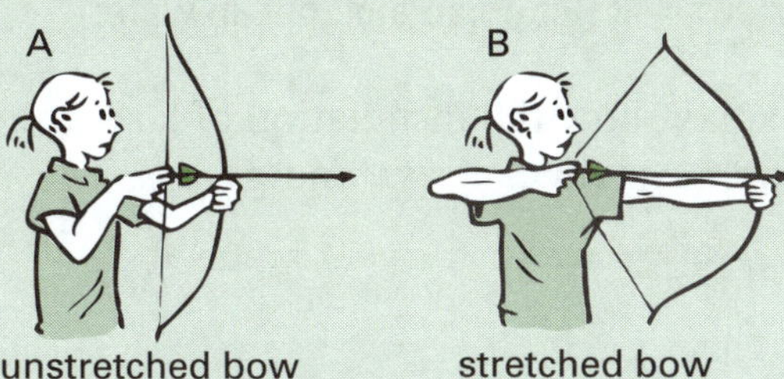

Figure 4.36 Bow and arrow

32. Less than 50% of the energy from petrol goes into moving a car forward. Suggest some ways the remaining energy is wasted. *(2 marks)*

33. The United States uses a different temperature scale to most other countries of the world. This is the Fahrenheit scale. In Australia we use the Celsius scale. The following formula allows us to convert temperatures between the two scales.

$$5(F - 32) = 9C$$

where F = Fahrenheit temperature
C = Celsius temperature.

Calculate the temperature in Fahrenheit when the noonday temperature in summer is 40 °C. *(2 marks)*

34. Use the formula in the previous question to complete Table 4.5 by calculating the Fahrenheit temperatures. *(2 marks)*

Table 4.5 Temperatures at which water freezes and boils

	Temperature (°C)	Temperature (°F)
Water freezes	0	a)
Water boils	100	b)

35. Car brakes get very hot as they are applied to bring the vehicle to a stop. The faster the car was travelling, the hotter the brakes get to stop the car. Explain. *(2 marks)*

36. Figure 4.37 shows energy transformations as an arrow is fired into a target. Complete the energy chain by replacing the code letters with appropriate energy forms. *(3 marks)*

37. Consider the experiment shown in Figure 4.38. A student placed a metal gauze on a tripod just above a Bunsen burner. She then turned the gas on and brought a candle flame to a position above the gauze and lit the gas passing through it. She saw

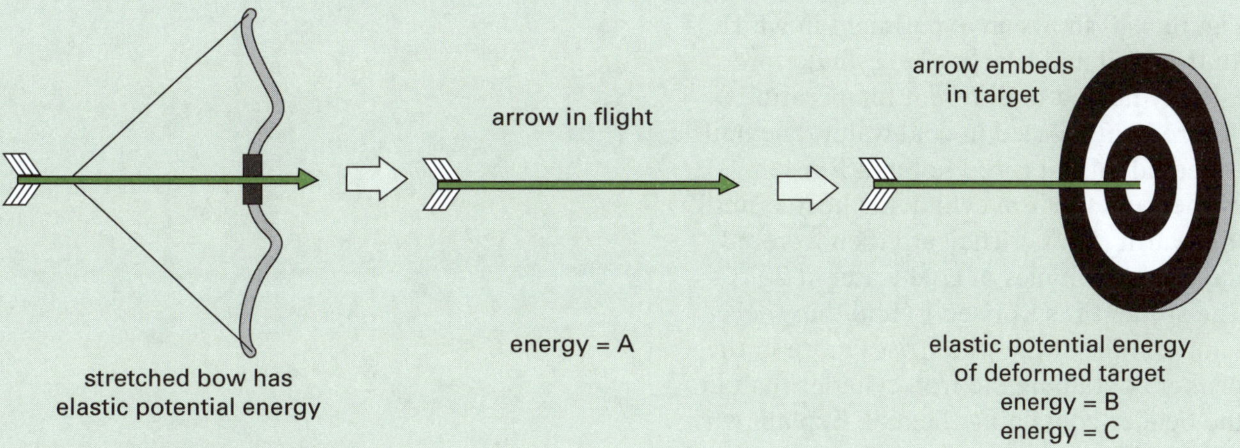

Figure 4.37 Arrow fired into target

the gas burn above the gauze, but not below it. Explain why this happens. *(2 marks)*

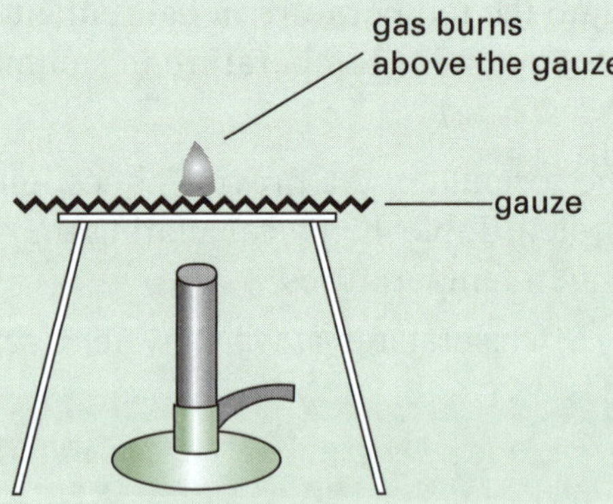

Figure 4.38 Gas burns above gauze

38. Consider the experiment shown in Figure 4.39. A tungsten wire is supported on a glass stand in a glass bulb. Inside the bulb there is a vacuum. The wire is heated electrically until it is red hot. A hand placed 10 cm from the apparatus gradually gets hot. Explain how heat is transferred in this experiment. *(2 marks)*

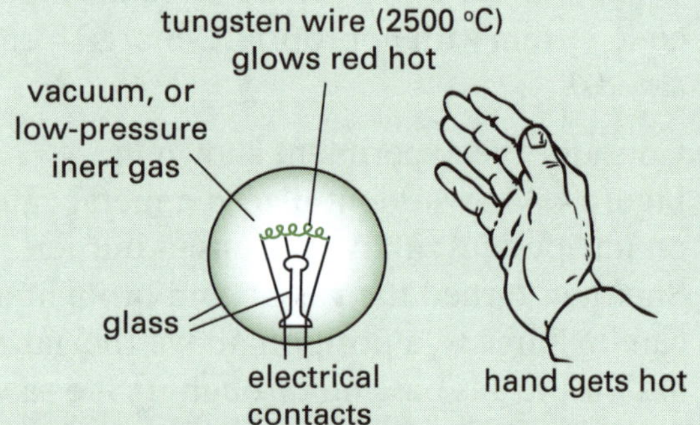

Figure 4.39 Heating a hand

39. Figure 4.40 shows an experiment in which an iron nail and a large iron cylinder are initially heated to different temperatures before being placed in cold water. The nail is heated until it is red hot in a Bunsen burner and the iron cylinder is heated until it is warm (50 °C). They are then lowered into equal volumes of cold water at 20 °C. The student is surprised to find that the temperature of the water rises more in the beaker containing the iron cylinder than in the beaker containing the nail. Explain why this has happened. *(2 marks)*

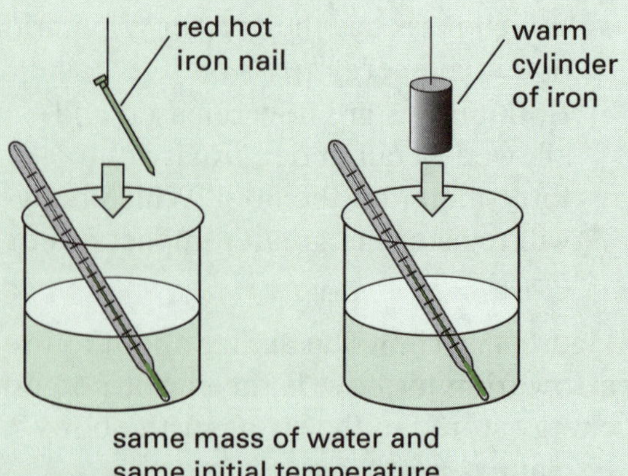

Figure 4.40 Hot objects placed in water

40. Explain how the installation of a solar hot water system can save energy. *(2 marks)*

Go to pp. 200–203 to check your answers.

CHAPTER 5

Investigations and problem solving

Overview

In this chapter you will learn about:

- working as an individual and working in teams
- conducting background research
- making observations
- variables and fair testing
- making and testing a hypothesis
- safety during a first-hand investigation
- choosing equipment
- accuracy and reliability
- writing scientific reports
- graphing and drawing diagrams
- making conclusions.

Glossary

Constant—a factor that is kept the same during an investigation

Dependent variable—a measurement that changes due to the independent variable

First-hand investigation—an investigation where original material is collected by the investigator

Generalisation—a statement that is true in the majority of cases

Hypothesis—a broad, general explanation of a collection of observations that predicts a result and suggests a way of testing it

Independent variable—a factor which you change on purpose

Inference—a possible and reasonable explanation of an observation

Parallax error—an error in reading an instrument due to the observer's eye and the pointer not being in a line perpendicular to the plane of the scale

Second-hand investigation—an investigation where the original material sourced from the work of other people is used

5.1 Working individually and working collaboratively

Students need to be able to work individually on some occasions as well as collaboratively in teams on others. There are advantages to both ways of working. There are also responsibilities that must be considered.

Working individually requires students to:

- appreciate the importance of taking personal responsibility in planning and performing a task
- work to realistic timelines and goals
- persevere with an activity to achieve a reasonable end point
- accept personal responsibility for a safe working environment for themselves and others in the laboratory, and
- evaluate the effectiveness of their work in completing the task.

Working collaboratively in teams requires students to:

- identify the different roles of each team member
- negotiate and allocate responsibility to each team member
- accept and perform the roles decided by the team
- collaborate with others in setting and working to agreed timelines and goals
- accept personal responsibility for maintaining a safe working environment for all team members
- work cooperatively to monitor progress of the investigation, and
- evaluate the process used by the team and the effectiveness of the team in completing the task.

5.2 Conducting background research using secondary sources

The following points should be noted when doing background research and gathering information from secondary sources:

- Use a wide range of resources to gather secondary information:
  - library books
  - internet
  - technical magazines
  - CDs/DVDs.

  Ensure you reference sources from which you have collected information.

- Use key words to locate information or to find the meanings of words and terms. You could look in:
  - the index of a book
  - the table of contents at the front of the book
  - a glossary to help you understand difficult technical words.
- Collate and summarise the collected information. You must not plagiarise (or copy) the work collected. You must read and make notes into a scaffold and then use these note summaries to construct your own text.
- Extract data. Information from secondary sources may contain various types of graphs (e.g. histograms, sector graphs, line graphs and divided bar graphs), flow charts and diagrams. You can use this type of information to extract data.
- Is information relevant or irrelevant? You must decide this as you gather second-hand data. For example, if you are gathering information on water movement in flowering plants then any data on non-flowering plants would be irrelevant.
- Not all the gathered data will be reliable. You need to check its reliability by comparing it with other sources. Science articles that appear in respected journals and that have been peer reviewed are more likely to be reliable than an article that is self-published.
  - Articles written by creation scientists are not reliable, as they do not follow the scientific method.
  - Many articles from companies, rather than educational or research sites, may be biased.
  - Scientific product claims may be written by commercial companies. In advertising their product, various companies use scientific terms and data to back up the claim that their product is superior. For example, 'Scientific tests show that Sorbo paper towel is more water absorbent than all other brands'. These statements cannot be accepted as reliable without independent testing.
- Organise your data. Your collected data needs to be organised for easy analysis. Tables, graphs, scaffolds and diagrams are commonly used to organise information. Computer databases and spreadsheets are becoming increasingly important as ways of organising collected data (see Figure 5.1).

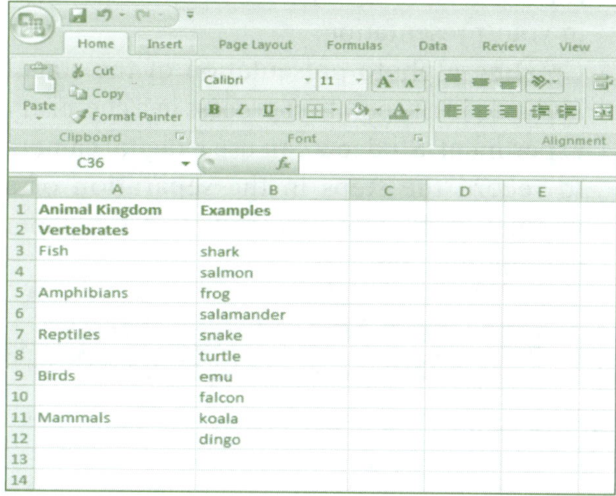

	A	B	C	D	E
1	**Animal Kingdom**	**Examples**			
2	**Vertebrates**				
3	Fish	shark			
4		salmon			
5	Amphibians	frog			
6		salamander			
7	Reptiles	snake			
8		turtle			
9	Birds	emu			
10		falcon			
11	Mammals	koala			
12		dingo			
13					
14					

Figure 5.1 Using a spreadsheet to organise information

5.3 Investigating a problem: observing

The first step of a scientific investigation takes place when you make an observation. This might be an observation of an event such as a pinecone falling from a tree. It might be an observation of some feature of an animal such as the presence of webbing between the toes of a bird. These observations may then lead you to think about the cause or reason behind the observation.

Observations can be made using various technologies. Digital cameras and video cameras are ideal to make a recording of your observations. The digital files can be published or sent via the internet through social media applications to other students for review and comment.

The following are examples of experiments using digital technology.

1. Observing the metamorphosis of a caterpillar into a butterfly (see Figure 5.2). Digital photos can be taken each day of the change in the caterpillar as the metamorphosis occurs. These photos can be combined to produce a digital presentation. If you have trouble photographing some parts of your presentation, you can access photo libraries on the internet. Only copyright-free images can be used. You can also include drawings in your presentation.
2. Creating a photo presentation of the stages involved in a separation process in a chemistry experiment. Students can use a digital camera to record the steps in the separation of a mixture such as plant pigments using paper chromatography.

Experiment 1

Observing *Eucalyptus* nuts, flower buds and leaves

Aim

- To observe diagrams of a variety of nuts and flower buds and leaf dimensions from different *Eucalyptus* trees and to record their similarities and differences

Method

1. Students are provided with five labelled diagrams of gumnuts and flower buds, and numerical data on leaves from different species of *Eucalyptus* (see Figure 5.3).
2. The similarities and differences between them are recorded.

Analysis

1. Examine the provided diagrams and construct a table of similarities and differences.
2. Draw a diagrammatic key that can be used to classify these species. Go to p. 203 to check your answer.

Conclusion

Write a suitable conclusion for this experiment. Go to p. 203 to check your answer.

5.4 First-hand investigations: fair testing and variables

Some problems will be investigated by conducting research in the laboratory or in the field. When you actually research and come up

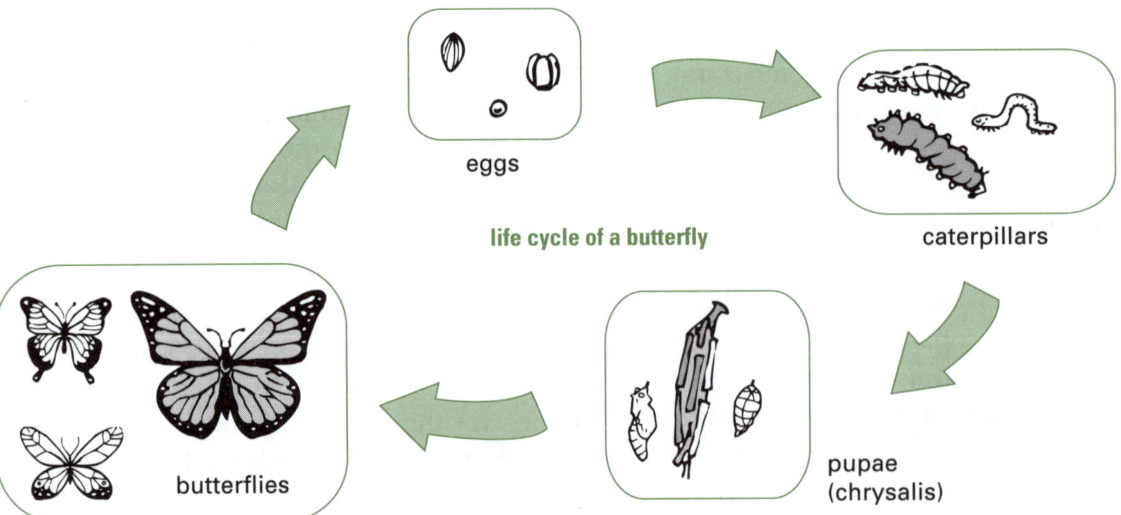

Figure 5.2 Using digital technologies to create a presentation

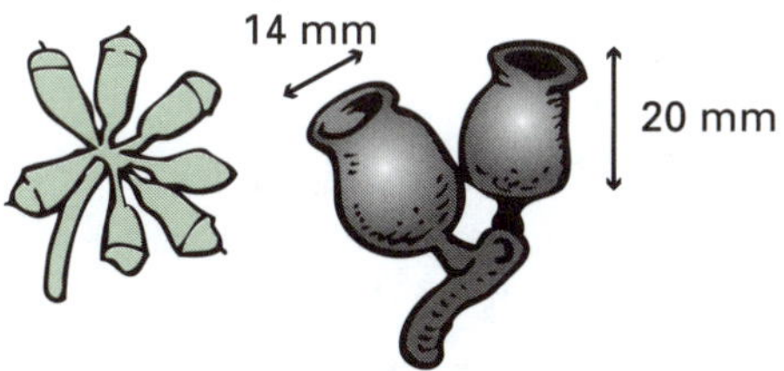

A *Eucalyptus gummifera* (red bloodwood)
Urn-shaped nuts
Leaves: 10–16 cm long
20–50 mm wide
Flowers: 7 per cluster
Woody nuts

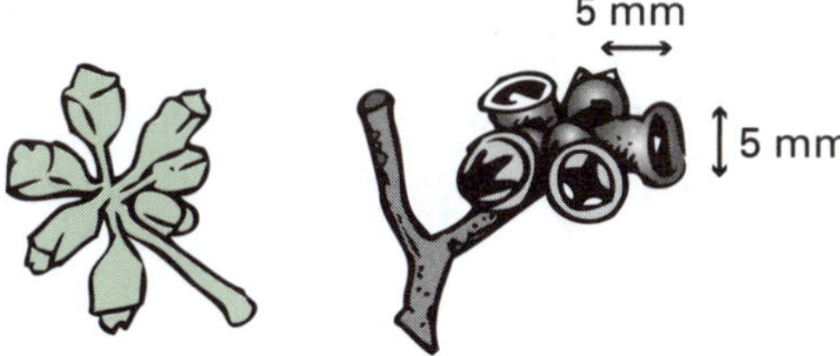

B *Eucalyptus saligna* (Sydney blue gum)
Bell-shaped nuts
Leaves: 10–20 cm long
15–30 mm wide
Flowers: 7–11 per cluster
Woody nuts

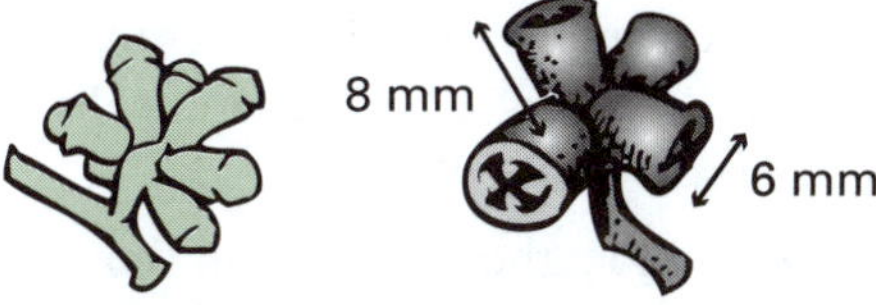

C *Eucalyptus botryoides* (bangalay)
Barrel-shaped nuts
Leaves: 10–14 cm long
30–60 mm wide
Flowers: 7 per cluster
Woody nuts

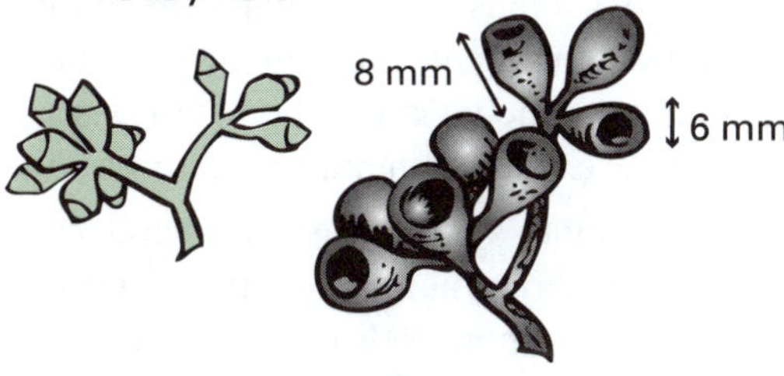

D *Eucalyptus paniculata* (grey ironbark)
Pear- or cone-shaped nuts
Leaves: 7–13 cm long
15–30 mm wide
Flowers: 7 per cluster
Woody nuts

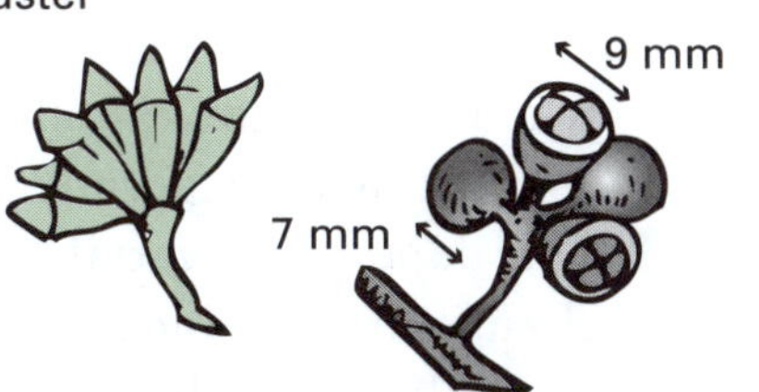

E *Eucalyptus punctata* (grey gum)
Globular-shaped nuts
Leaves: 6–10 cm long
20–30 mm wide
Flowers: 7–11 per cluster
Woody nuts

Figure 5.3 *Eucalyptus* flower buds, nuts and leaves

with original data, this is usually called a first-hand investigation. Some problems will be researched by second-hand investigation. This involves using information (data) developed by other persons and working with it. Table 5.1 shows some of the steps in a first-hand investigation.

Making a hypothesis

The next step in the scientific method is to generate a hypothesis. What is a hypothesis?

- It is a general explanation of the observation(s) or solution(s) to a problem being investigated.
- A hypothesis is *not* a wild guess. You might say that it is an educated guess.
- Hypotheses are temporary statements that lead towards the development of scientific theories.

Hypotheses are only made after looking at all the available evidence and information that has been collected. This is why background reading of identified data sources is so important. The work of others may help you to produce a good hypothesis. Hypotheses are the result of a scientist making inferences, making predictions and developing generalisations.

Table 5.1 Steps and examples of the typical scientific method

Some of the steps of the typical scientific method	Simple example
Observations—observed or measured results	'I have observed various cells under the microscope and so far all have cell membranes and only the leaf cells had a cell wall. Animal cells had no cell wall.'
Inferences—following an observation, a scientist may make an inference; this is a reasonable explanation which could explain the observation. An inference may ultimately turn out to be incorrect.	'Perhaps plant cells need cell walls for support.'
Predictions—the next step is to test the inference by making a prediction and performing an experiment to test the prediction. These experiments may support the inference or they may not. A new explanation may need to be sought.	'I'll have to check this out. So I'll examine 20 more different plant cells; I predict that the thickest cell walls will be found in cells from the stems (trunks) of tall trees.'
Generalisations—as more supporting data is collected, the scientist may see a pattern emerging. At this point he or she may be able to make a generalisation about the topic under investigation. A generalisation is a statement that is true in the great majority of cases.	'I can now conclude: the cell walls in the trunks of woody trees are much thicker than other plant cells.'

The hypothesis must be able to:

- predict the results of future investigations about the problem, and
- suggest ways in which it can be tested to determine whether or not it is true.

Hypotheses are not always correct. If the data collected during the investigation does not support the hypothesis, then a new hypothesis must be developed.

Fair testing and variables

Many factors can influence the results of an experiment. These factors are called variables. In order to investigate a problem experimentally, you must keep all, but one, variable constant so that only the variable being investigated can affect the result. This is known as fair testing.

- Experimental results are only reliable if the tests are fair.
- During an investigation, the experimenter tests what happens when he or she alters one variable. The effect of a change in this independent variable on another variable (the dependent variable) is investigated.
- The independent variable is the one that the investigator systematically allows to change. It is the change you allow to happen when conducting an experiment.
- The dependent variable is affected by the change in the independent variable. The controlled variables are the other factors that can alter the result if they are not held constant.

Further examples

Example 1: Crystal concentration

- **Hypothesis:** the surface texture has an effect on growing crystals
- **Independent variable:** surface texture
- **Dependent variable:** how much of the growing surface is covered with crystals
- **Variables to be kept constant:** nature of solute; solute concentration; amount of solution; type of container; temperature, humidity, light and other environmental factors

Example 2: Energy changes

- **Hypothesis:** a ball dropped onto the ground from a certain height will bounce back to that height
- **Independent variable:** height from which the ball is dropped

- **Dependent variable:** height to which the ball returns
- **Other variables to be controlled:** size and type of ball; surface to be bounced from; dropping rather than throwing; wind direction and speed (preferably none)

Control

Measurements may only be reliable in some experiments if a control is used. This is particularly true in biological or biochemical systems where many factors can alter a result. A control is an experiment that is performed as a comparison with those in which the independent variable is allowed to change.

Example 3: Metals and acids

- **Hypothesis:** the more acidic a solution, the faster a metal in it will corrode (see Figure 5.4)
- **Control:** test tube containing water and the metal cube, but no acid added
- **Independent variable:** volume of acid
- **Dependent variable:** amount of metal corrosion
- **Controlled variables:** same size test tube; same metal (e.g. Zn or Fe); identically sized metal cube; same concentration of acid; same volume of water or water–acid mix; same time allowed for the reaction to occur; same temperature

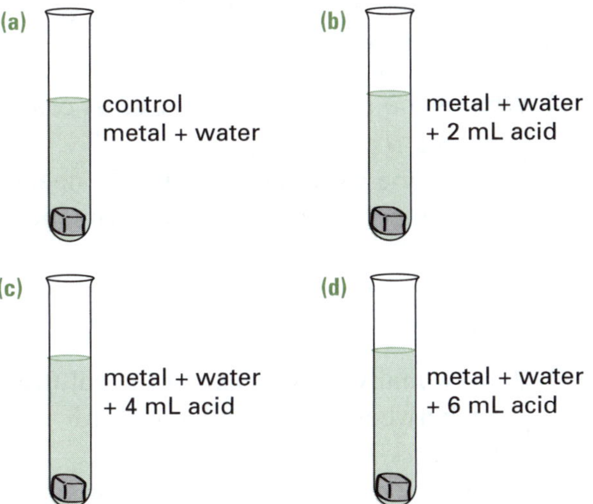

Figure 5.4 A controlled experiment testing the effect of the amount of acid on a metal cube

Test yourself 1

Part A: Knowledge

1. Which of the following is *not* a responsibility of a student working in a team? *(1 mark)*
   - **A** negotiate various responsibilities
   - **B** work cooperatively
   - **C** participate in group discussions
   - **D** personally decide on a timeline for the investigation

2. A student gathered second-hand data on the movement of water in the stems of flowering plants. Which one of the pieces of data is not relevant? *(1 mark)*
   - **A** an internet article on the transpiration of water from roots to leaves in *Eucalyptus* plants
   - **B** a senior textbook describing the exclusion of salt from algae growing in an estuary
   - **C** a university paper describing the use of radioactive water molecules in the investigation of water movement in the xylem vessels of *Acacias*
   - **D** a DVD showing how dyed water moves in the cut stems of celery plants

3. Consider the following statement made by a parent to her child: 'The blood on your shirt suggests that you were fighting'. This statement is an example of *(1 mark)*
   - **A** an inference.
   - **B** an observation.
   - **C** a generalisation.
   - **D** a prediction.

4. A student plans and performs an investigation to determine whether the amount of fertiliser added to soil will affect the height of bean seedlings. In this investigation the control would be *(1 mark)*
   - **A** the height of the plant.
   - **B** the type of soil.
   - **C** soil without added fertiliser.
   - **D** different brands of fertiliser.

5. A student planned an investigation to determine whether 1 g of magnesium will dissolve slower at 20 °C if the hydrochloric acid is diluted. The independent variable is the *(1 mark)*
   A mass of magnesium.
   B temperature of the acid.
   C concentration of the acid.
   D rate of the reaction.

6. Complete the following restricted-response questions using the appropriate word. *(1 mark for each part)*
   a) The .... variable is a factor which you change on purpose.
   b) In order to avoid .... you must read and make notes into a scaffold and then use these note summaries to construct your own text.
   c) In .... testing you must keep all but one variable constant so that only the variable being investigated can affect the results.
   d) The controlled variables are the other factors that can alter the result if they are not held ....
   e) Measurements may only be .... in some experiments if a control is used.

7. Use the code letters to match the terms or phrases in each column. *(1 mark for each part)*

Column 1	Column 2
A spreadsheet	F liquid surface
B observations	G technical words
C glossary	H digital technology
D teams	I tables and graphs
E meniscus	J collaboration

Part B: Skills

8. **a)** Bud wanted to determine whether a new brand of fertiliser containing higher amounts of trace metals was better than an existing brand for his pot plants. Suggest a hypothesis he can test in an experiment. *(1 mark)*
   b) Name four variables Bud needs to control. *(4 marks)*

9. Rose noticed that a pot plant in her home always faced towards the window. Her mother would regularly turn the plant so that its flowers would face into the room, but a few days later the plant would again be facing the window. Rose thought that somehow the plant was affected by the light coming through the window. Her younger sister, Violet, was convinced the plant was pining for its freedom.
   Which idea can be tested through the methods of science? Why? *(2 marks)*

The following information is required for Questions 10 and 11.

In an experiment, the numbers of yeast cells observed in a culture over 10 days were as shown in Table 5.2.

Table 5.2 Number of yeast cells observed in a culture

Day	Number of yeast cells
0	18
1	27
2	39
3	45
4	56
5	64
6	70
7	70
8	60
9	48
10	32

10. **a)** What is the initial number of yeast cells counted? *(1 mark)*
    b) Suggest one variable which would need to be controlled in such an experiment. *(1 mark)*
    c) Construct a line graph of these data. *(5 marks)*

11. **a)** By how many yeast cells did the culture grow between the second and third days? *(1 mark)*
    b) After how many days did the number of yeast cells in the culture start to decrease? *(1 mark)*

c) On the sixth and seventh days, the number of yeast cells remained the same. Suggest a possible reason for this. *(1 mark)*

12. Read the following account of an experiment conducted by Charles Darwin and his son Francis in the late 19th century. Answer the questions which follow.

Darwin and his son had noticed that the growing tips of grass seedlings were influenced by light (see Figure 5.5).

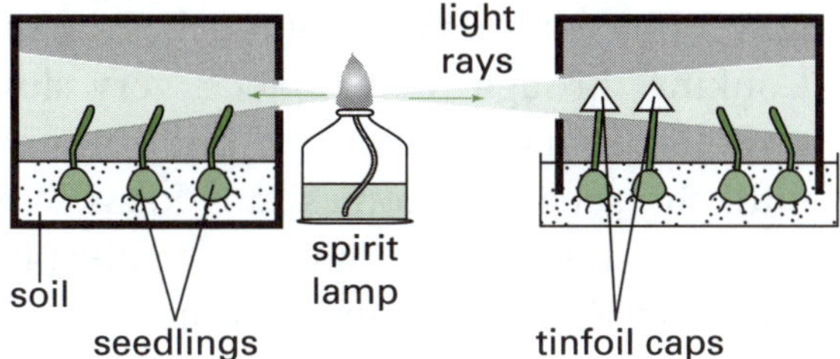

Figure 5.5 Darwin's experiment

The tips of the seedlings always grew towards the light. The actual bending and curvature of the stem began several millimetres below the tip and extended a distance of up to 20 mm down the stem.

Darwin came up with the idea that the light might be causing the production of a chemical in the tip of the seedling. Perhaps, thought Darwin, this chemical was moving down the stem and causing it to bend.

Francis and Charles tried cutting off portions of the tip of the plants before exposing them to light. They found that the removal of a few millimetres of tip failed to stop the bending. However, removing 4 mm of the tip prevented the bending and the plants grew straight up. They also found that placing tinfoil caps over the seedling tips prevented the bending process in the presence of light.

a) What were Darwin's initial observations? *(1 mark)*

b) What was Darwin's hypothesis? *(1 mark)*

c) How did Darwin test his hypothesis? What variables would he need to consider? *(2 marks)*

d) Did the results of Darwin's experiment support his hypothesis? Explain. *(2 marks)*

e) Did Darwin prove that his hypothesis was correct? Explain. *(2 marks)*

13. The flow chart in Figure 5.6 shows the steps in developing and testing a hypothesis. Select three terms from the following list to replace the code letters A, B and C. *(3 marks)*

1. Develop a generalisation.
2. Reject all conclusions.
3. Review all available observations and evidence.
4. Test the hypothesis.
5. Write a conclusion.

14. A student wrote down the following list of variables for his investigation:

- length of the steel, brass and copper rods
- diameter of each rod used
- type of heat source applied to one end of each rod
- temperature of the environment during the experiment (must be kept constant during the experiments)
- initial temperature of each rod.

Suggest a hypothesis that the student may be testing. Explain your reasoning. *(2 marks)*

Go to pp. 204–205 to check your answers.

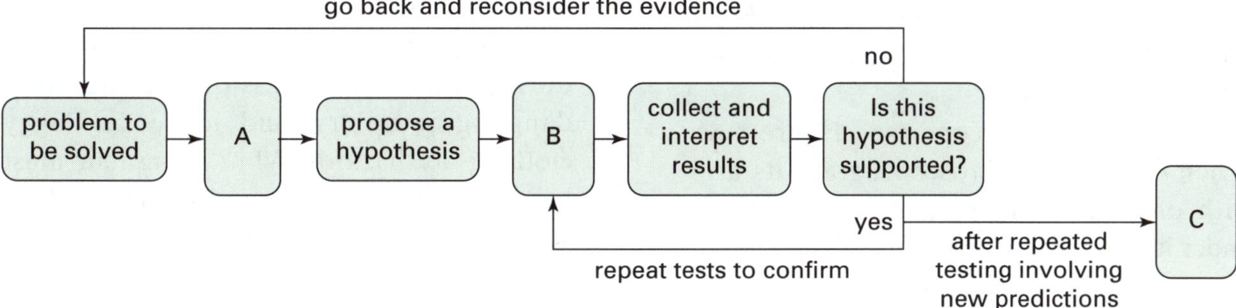

Figure 5.6 Developing and testing a hypothesis

5.5 Safety in a first-hand investigation

Experiments should always be done in a safe manner to avoid injuries. You must always listen to and follow the teacher's instructions. Your teacher will inform you of the safety issues for each experiment.

Safety signs such as that shown in Figure 5.7 can be found in the laboratory. Chemicals have safety signs on their labels. Read the label carefully before using the substance.

Figure 5.7 Poisonous substances have this safety sign on their label.

Here are some examples of safety issues.

1. Using a Bunsen burner. When using a Bunsen burner to heat a beaker of water supported on a tripod and gauze, ensure that a blue flame is used (the inlet hole is open) and that the Bunsen burner is not placed under the gauze until all the apparatus has been assembled. After the heating is completed, the Bunsen should either be turned off at the tap or placed to the back of the bench with the hole closed to produce a yellow safety flame. These yellow flames are easy to see and so you are less likely to accidentally burn yourself. Do not touch the hot equipment until it has cooled down. Make sure your safety glasses are on.
2. Using a light microscope (see Figure 5.8). When carrying a microscope, grasp its arm with one hand and place the other hand under its base. There is then less likelihood of dropping it (expensive to replace), especially on your foot. (It is called a light microscope, but it's heavy.) Adjust the microscope by turning the coarse adjustment knob to raise the body tube. Using a low power objective lens, adjust the iris diaphragm. While looking through the eyepiece, adjust the mirror until you see a bright white circle of light. Place a slide on the stage and centre the specimen over the opening. Use the stage clips to hold the slide in place. Look at the stage from the side and carefully turn the coarse adjustment knob to lower the body tube until the low power objective lens almost touches the slide. Looking through the eyepiece, very slowly turn the coarse adjustment knob upwards until the specimen comes into focus. When it is almost in focus, turn the fine adjustment knob until the specimen appears sharp. Change to the high power objective lens by looking at the microscope from the side and carefully revolving the nosepiece until the high power objective lens clicks into place. Make sure the lens does not hit the slide.

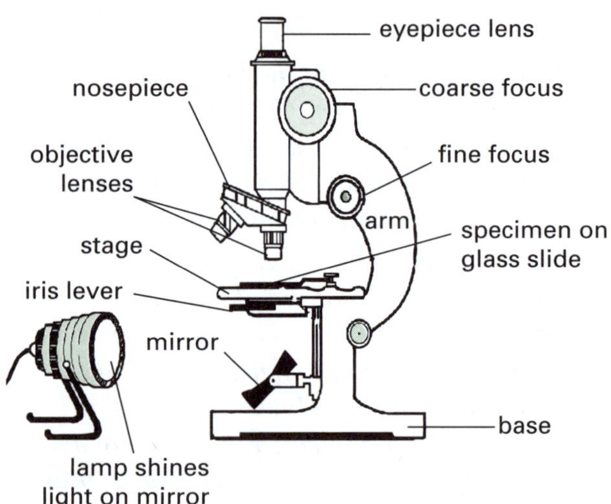

Figure 5.8 The light microscope

3. Dressing for experiments. Dress properly during laboratory sessions. Long hair, dangling jewellery, and loose or baggy clothing are hazards. Why? Long hair must be tied back, and baggy clothing must be secured. Shoes must completely cover the foot and not be slippery. Often chemicals may be inadvertently spilled and may land

on your feet, so it's important to have them covered. Any time chemicals, heat or glassware are used, students will need to wear safety goggles. Protective fire-resistant gloves should also be used where appropriate. Protective gear needs to be comfortable but, at the same time, protect you (see Figure 5.9).

Never handle broken glass with your bare hands. Use a brush and dustpan to clean up glass. Place broken glass in a designated glass disposal container.

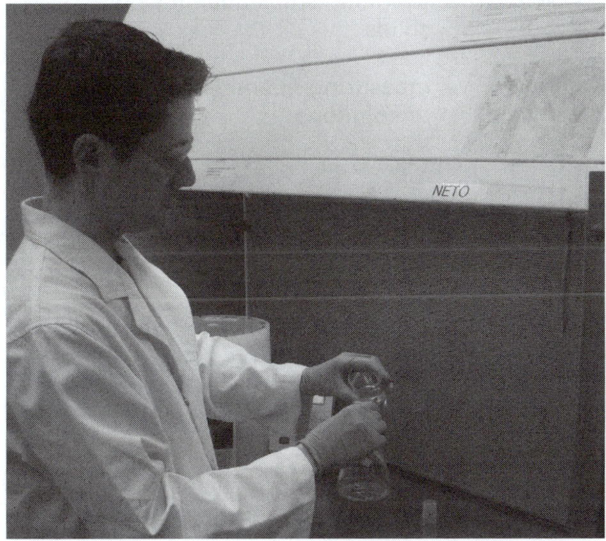

Figure 5.9 Scientists who work with chemicals must wear a lab coat, safety glasses and rubber gloves.

5.6 Choosing equipment and increasing accuracy and reliability

The selected equipment needs to be assembled to produce an apparatus that will perform the required task. Assembling pieces of equipment and using the apparatus requires good manipulation skills.

Accuracy

Making accurate measurements is important when conducting an investigation. This involves choosing the most appropriate measuring devices available. Reducing errors and choosing equipment that can measure in a more precise way can improve accuracy.

One important error is called parallax error. This is an error that can be avoided. For example, when measuring the volume of a liquid on a scale, your eye must be placed directly over the scale. If you look at the scale from an angle, the reading will be incorrect, as Figure 5.10 shows.

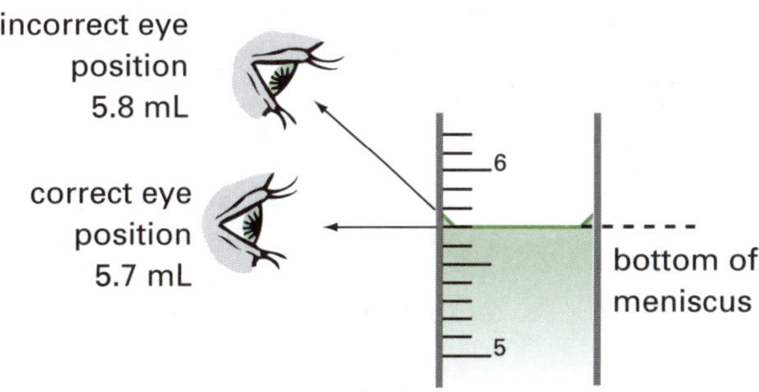

Figure 5.10 Avoiding parallax error in reading a scale

Figure 5.11 provides some examples of common laboratory equipment and how they can be used.

Reliability

Reliability can be improved by making repeated measurements over a number of trials. The number of trials that are performed depends on the type of experiment. For example:

1. Testing a new drug for migraine headaches. In this case the drug will need to be tested on thousands of volunteers to determine whether there are adverse side effects to the users and whether the drug helps sufferers compared with those who are not.
2. Testing a sample of sea water for its salt content. In this case a minimum of five repeats is all that is needed. The five repeats should give answers that cluster closely. An average of the five results can then be calculated.

Using alternative technologies in experiments

Figure 5.12 shows two methods that can be used to record motion. In (a), a stopwatch is used to record the time for a vehicle to move certain distances along a marked track. In (b), a motion sensor beam is used to monitor the movement of a person walking towards the sensor. The

Equipment	What it is used for
test tube 	• observing reactions when chemicals are mixed • heating solids, liquids and solutions over a Bunsen burner, using a wooden test-tube holder to hold the test tube
beaker 	• temporary storage of liquids and solutions • observing reactions involving larger quantities of liquids than in test tubes • heating solutions
conical flask 	• observing reactions in solutions • often used when the contents need to be kept well mixed by swirling • temporary storage of liquids and solutions • boiling solutions
tripod 	• support for gauze and equipment during experiments
gauze 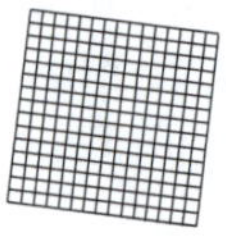	• support for beakers, conical flasks or evaporating basins during heating experiments
mortar and pestle 	• grinding materials to fine powders or to extract materials into liquids
evaporating basin 	• evaporation of solutions to dryness or partially to allow crystallisation
watch glass 	• natural evaporation of small quantities of solutions • weighing small quantities of solids
Petri dish 	• temporary storage of specimens • growing cultures of microorganisms • weighing small quantities of solids
measuring cylinder 	• measuring quantities of liquids or solutions
filter funnel 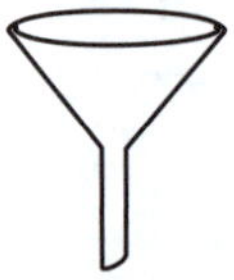	• support for filter paper during filtration
Bunsen burner 	• heating materials and equipment during experiments

Figure 5.11 Apparatus and its uses

computer program calculates the position of the person with time and uses a graphing application to plot this data on a graph.

The motion sensor in (b) produces a more accurate result than the method used in (a). Greater errors occur in experiment (a) because it relies on the hand–eye coordination of the experimenter. There is a delay between observing the vehicle cross a line on the track and the depression of the stop button on the stopwatch. However, measurements like these will allow you to calculate kinetic energy.

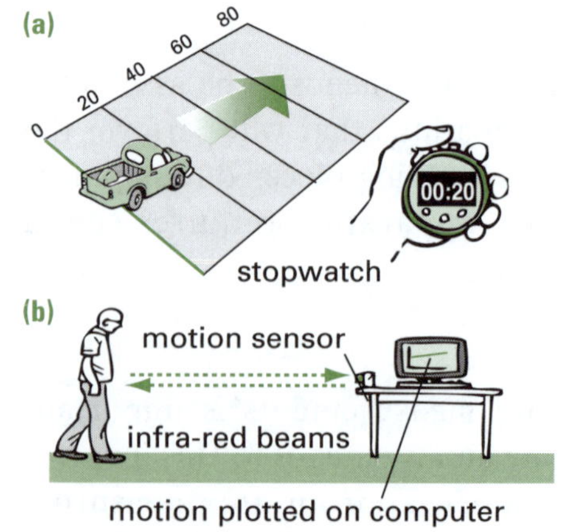

Figure 5.12 Alternative methods to measure motion

A simpler apparatus is shown in Figure 5.13. Additional masses can be added to the cart to increase its mass. A string is attached to the cart. The string passes over a pulley and weights are added to the other end of the string. As the weights fall, the trolley accelerates. Using a video camera with stop framing, a video can be taken of this motion. When replayed one frame at a time, the instantaneous speed can be determined and its kinetic energy can be calculated.

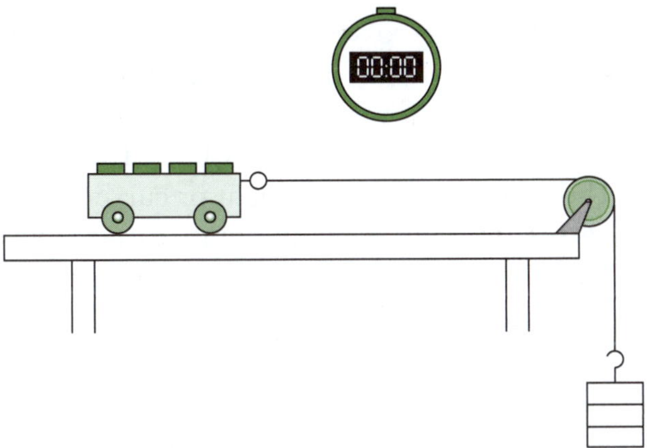

Figure 5.13 A cart accelerating from one end of the track is being pulled along by a falling weight. Experiments like this allow you to calculate the kinetic energy of the cart.

5.7 Scientific reports

Recording and presenting data

Scientific reports can be presented in a variety of ways including a written report, photos, sketches, videos, computer slide show, debates or orally. Presenting a written experimental report is different to the way you would present an exposition. The following are some different methods you can use to present information.

- Procedure. This text type is used to describe how an experiment is to be performed. Present tense is used. The steps of the procedure are usually numbered or presented as sequential dot-points.
- Experimental record. This extends the procedure to include results and conclusions. Results are written in past tense, while the conclusion can be written in the present tense, past tense or both.
- Procedural recount. This text type is used to record the procedure of the experiment that has been performed. Past tense is used in a recount.
- Report. This type of text is used to present information on a particular topic (e.g. the different types of frogs in Australia) that has been investigated using first-hand and/or second-hand data. Factual and descriptive information is presented.
- Discussion. This text type identifies issues and presents points for and against. In the end you decide on a viewpoint and justify it.

At the end of your report you must acknowledge all sources of information in your bibliography. Use an accepted method of citing references. This will include the author, name of the article/book, chapter or page reference, publisher and date of publication. Also, for sources on the internet you should state as much of this information as you can (not just the web address).

Processing collected data

Tables, graphs and drawings are common methods of recording and presenting data in a scientific report.

Tables

Scientists often record their information in tables and as graphs. Consider the experiment shown in Figure 5.14. A student is slowly heating some ice in a beaker and measuring the temperature every two minutes. Notice the bulb of the thermometer needs to be near the middle of the ice or water being heated. If it were allowed to just sit in the beaker, the bulb touching the bottom of the glass would measure the heat coming through the glass instead of the heat in the substance.

Table 5.3 shows the temperature of ice and water every two minutes as it is being heated steadily. The table has the following features.

- There are two columns. (Other tables may have more columns for data.)

- Each column has a clear heading.
- The units (minutes and degrees Celsius) are shown in the column headings and are not repeated with each measurement.
- The table is enclosed by lines.

Table 5.3 Temperature versus time

Time (minutes)	Temperature (°C)
0	0
2	0
4	1
6	8
8	22
10	40
12	58
14	73
16	91
18	99
20	100
22	100
24	100

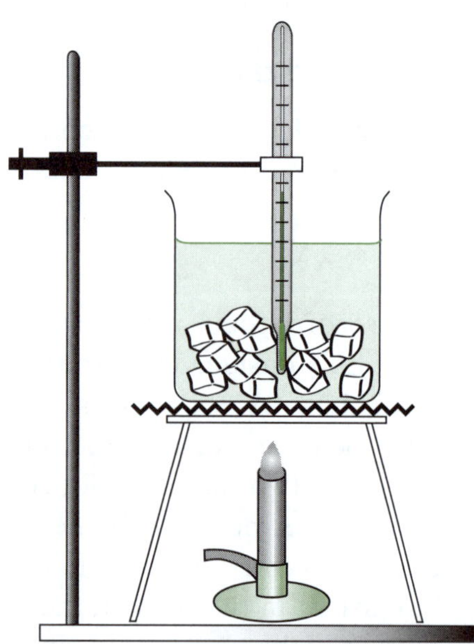

Figure 5.14 Apparatus for heating ice

Graphs

The information in Table 5.3 can be shown as a line graph. Here are the steps you would take to draw the graph.

1. Draw your axes, and correctly label them, including appropriate units. Time (the independent variable) is usually shown on the horizontal axis, so temperature (the dependent variable) would be on the vertical axis. There is no reason the vertical axis should start at 0. Make your graph reasonably large using grid paper. About half a page is a good size for a graph.
2. Mark the points on your graph with a small cross (×) or a large dot in pencil. A pencil allows you to easily erase points if you make a mistake.
3. Join the small crosses with a smooth curve or straight line. If the line is straight, use your ruler. If the plotted points on a line graph seem not to line up, draw a line of best fit. This is a smooth line, straight or curved, that passes through as many points as possible. The points it does not pass through should be scattered equally on both sides of the line. The graph shows a line of best fit.

Remember to always give your graph a title. Figure 5.15 shows the completed graph of the data in Table 5.3.

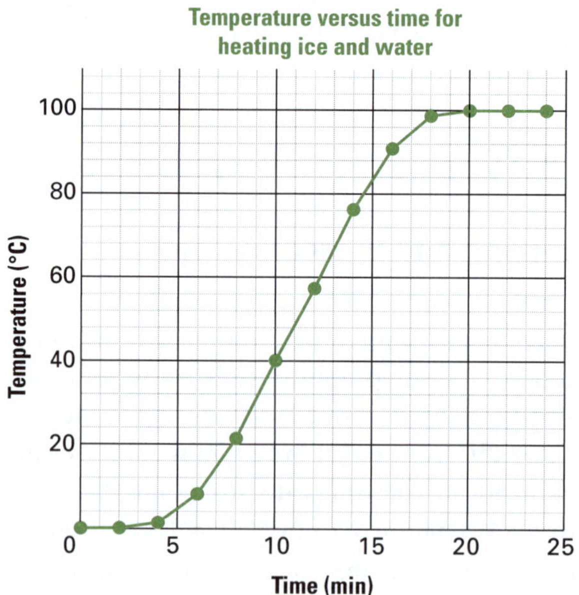

Figure 5.15 Line graph showing line of best fit

Other types of graphs are commonly used in science to display and analyse data. Some of these other types are shown in Figure 5.16.

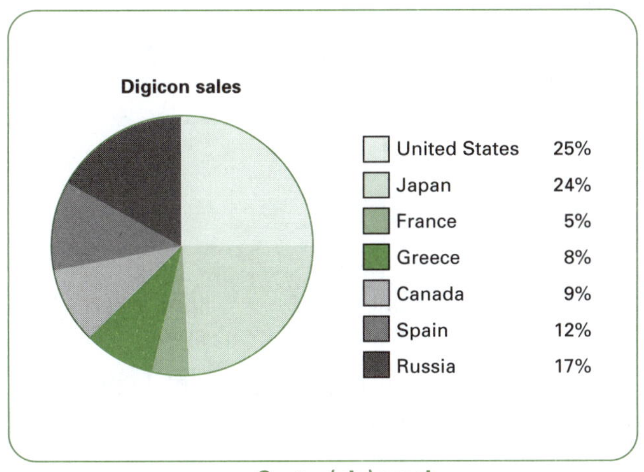

Sector (pie) graph

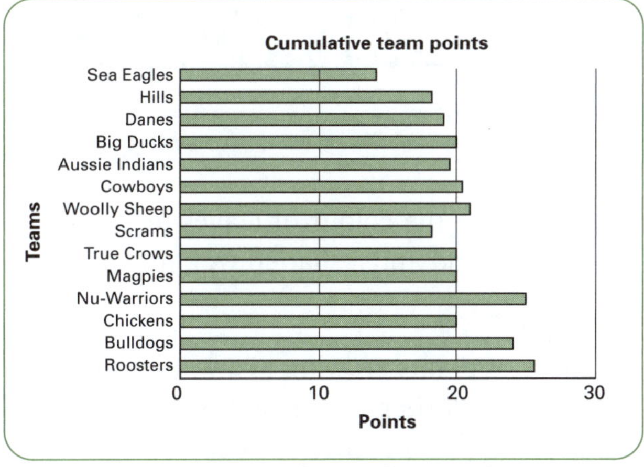

Bar graph

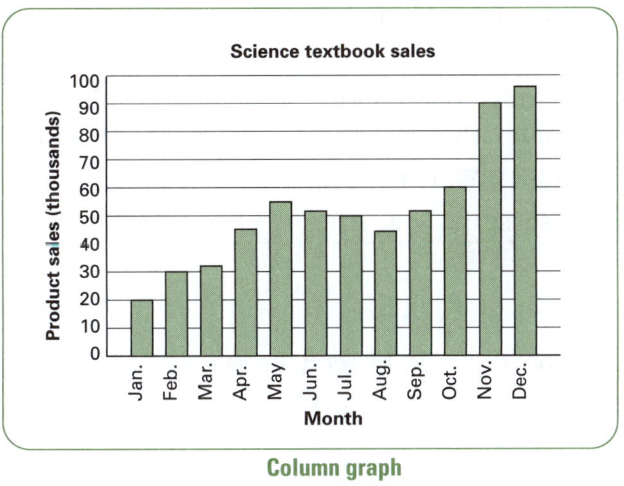

Column graph

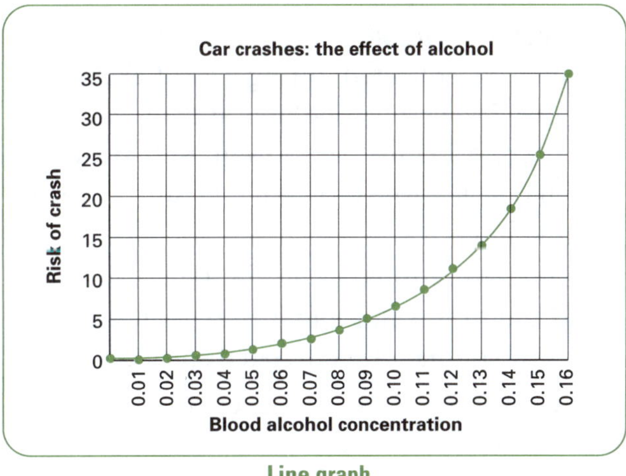

Line graph

Figure 5.16 Types of graphs

Drawing diagrams

Scientific drawings summarise a lot of important information about an experiment. Here are some important rules to follow.

- Use a pencil, rubber and ruler. A pencil and rubber allow you to make corrections.
- Your diagram should be at least one-third of a page (10 to 12 cm) in size.
- Draw the apparatus using a two-dimensional perspective.
- Remember to use your ruler wherever straight lines are needed.
- Use printing to label the parts of your diagram.
- Connect the labels to the diagram using ruled straight lines. No arrow heads are needed at the end of lines that have labels.

Figure 5.17 shows an example of a basic scientific drawing showing the apparatus used in an experiment.

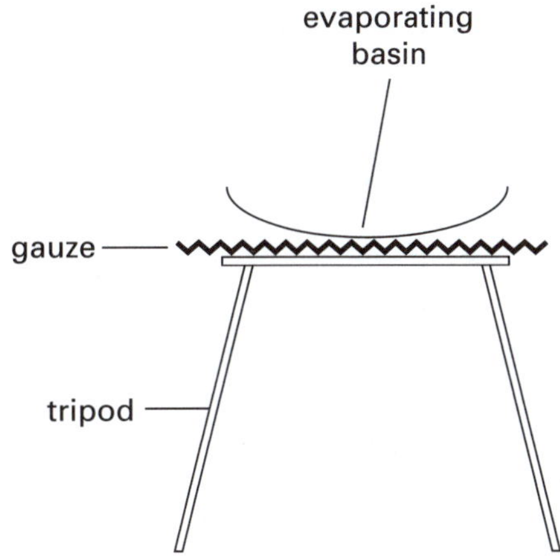

Figure 5.17 Science drawing

Experiment 2

Leaf colour

Aim

- To compare the colour of leaves in an open, dry woodland and a shaded wet, forest

Hypothesis

- Leaves of plants growing in a wet eucalyptus forest are darker green than similar plants growing in an open, dry eucalyptus woodland

Method

1. Numerous samples of leaves from different plants are collected during a field excursion to the dry, open woodland and the shaded, wet forest. Plants that grow in both locations should be a priority. The samples are returned to the lab.
2. Each leaf sample is ground in a mortar and then the chlorophyll is extracted into a known volume of methylated spirits.
3. The green intensity of each extract is compared using a scale from 1 to 10.

Results

The relative green colour of the extracts are plotted in Figure 5.18 as a function of tree position in the woodland and the forest. The green colour intensity of three samples (A, B and C) is shown in Figure 5.19.

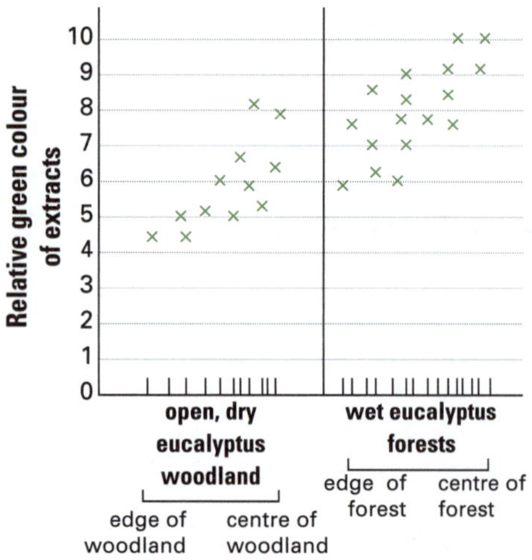

Figure 5.18 Relative green colour of leaves in a woodland and a forest

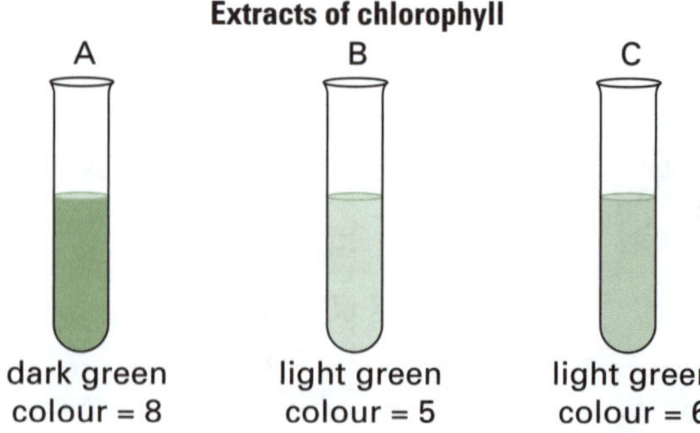

Figure 5.19 Relative green colour of leaves in a woodland and a forest

Analysis

Write an analysis of these results. In your analysis:

1. identify where in the woodland or forest the leaves that produced extracts A, B and C were obtained
2. discuss the wide scatter of points in the graph.

Go to  p. 204 to check your answer.

Conclusion

Write a suitable conclusion.

Go to  p. 204 to check your answer.

Experiment 3

Soil type and plant growth

Aim

- To determine if soil type affects seed growth

Method

1. Fill four pots with loamy soil. Label the pots L1, L2, L3 and L4.
2. Fill four pots with clayey soil. Label the pots C1, C2, C3 and C4.
3. Fill four pots with sandy soil. Label the pots S1, S2, S3 and S4.
4. Follow the instructions on the liquid fertiliser. Fertilise the soils with the same amount of fertiliser.
5. Take three seeds and plant them in one pot to a depth of about 8 mm. Space the seeds around the pot equally.

6. Repeat, planting the seeds in the other pots.
7. Place all the pots together in the same place where they will get the correct sunlight conditions and proper drainage.
8. Water each pot each day with 100 mL of water.
9. Observe the plants after about 6 weeks. Record the heights of the plants after this time.

Results

The height of each plant was recorded after 6 weeks (see Table 5.4).

Table 5.4 Height of each plant after 6 weeks

Pot	Plant height (cm)			Average height (cm)
L1	30	35	34	
L2	29	33	36	
L3	28	30	35	
L4	29	34	36	
C1	24	26	28	
C2	27	25	29	
C3	26	25	27	
C4	28	25	22	
S1	20	20	23	
S2	17	19	21	
S3	20	18	22	
S4	21	23	22	

Notes:

- Plants in loamy (L) soils appear healthy with large, very green leaves and strong stems.
- Plants in clayey (C) soils appear fairly healthy; some plants have few leaves and weak stems.
- Plants in sandy (S) soils appear pale, with few leaves and weak stems.

Analysis

1. Calculate the average height of the plants in each pot and record the results in your notebook copy of the table.
2. Calculate the average heights of plants in each of the three soil types.

Analyse these results. Go to p. 204 to check your answer.

Conclusion

Write a suitable conclusion for this experiment. Go to p. 204 to check your answer.

5.8 Conclusions, discussions and evaluations

Once you have processed your results, you will need to draw valid conclusions. The conclusions must relate directly to the experimental results. You cannot make a conclusion about information that you have not gathered or measured. Your conclusion must be brief and answer the aim of the experiment.

As part of the scientific method you will then provide plausible explanations of the phenomena being investigated. These possible explanations will be written in the discussion section of the report. There is also an opportunity at this point of justifying any inferences you have made about the collected data. In the discussion, you may make some predictions that could lead to further experimentation.

As more experiments are completed, you may be able to evaluate how successful or otherwise your experiments have been. Discuss what other members of your group think.

5.9 First-hand investigation

See Case study 1: Distance travelled by a metal ball on the following page.

Case study 1: Distance travelled by a metal ball

Hypothesis

- The greater the compression in the spring, the greater the distance that the ball lands away from its original position

Aim

- To measure the horizontal distance a metal ball lands from the foot of the table due to the compression of a spring, and to determine whether there is a linear relationship between the compression of the spring and the distance away the ball lands

Equipment

- long compressible spring
- tape measure
- ruler
- heavy weight
- gate preventing ball movement until required
- sensor pad

Method

1. Set up the apparatus as shown in Figure 5.20.
2. Place a ruler on the table to measure the compression of the spring in centimetres. The zero point of the ruler is where the heavy weight just touches the spring.
3. Move the weight to compress the spring by 2 cm.
4. The metal ball should now be compressed between the spring and the release gate. Place an electronic sensor pad on the floor. This will indicate the position where the ball lands.
5. Now release the gate. The ball should fly off, landing on the electronic sensor pad placed on the floor.
6. Measure the distance, in centimetres, from the foot of the table to the point on the sensor pad where the ball landed.
7. Repeat the procedure five times at selected higher spring compressions.

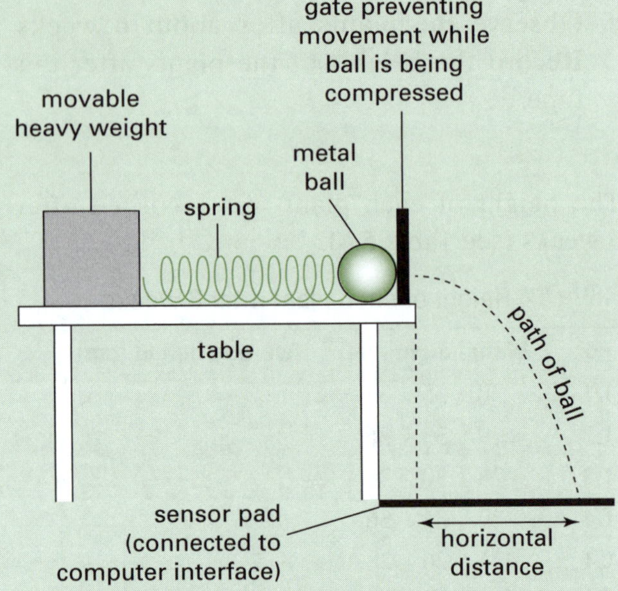

Figure 5.20 Experimental apparatus to determine the distances away a metal ball lands due to different spring compressions

- Variables to be kept constant: mass and size of marble; type and stiffness of spring; height of table above floor; position of release gate.
- Independent variable: spring compression.
- Dependent variable: distance metal ball lands from table.
- Accuracy: The accuracy of the measurements depends on the equipment selected and how carefully the experiment is conducted. The accuracy in determining the distance relies on two measurements, the compression of the spring and the distance from the table where the ball lands. Make sure these are measured precisely. Also, make sure the spring is not damaged in the process, such as over-stretching or over-compressing.
- Reliability: The measurements are reliable if repeated experiments (five or more)

Case study 1: Distance travelled by a metal ball cont.

lead to consistent results. Sometimes in a large set of repeated measurements, an individual measurement may differ widely from the other measurements. You may be justified in ignoring this measurement before calculating an average. Measurements cannot be reliable if they are not performed with high precision and accuracy.

Typical results

Table 5.5 shows the final average times for a fixed mass of sugar to dissolve at five temperatures.

Table 5.5 Final average times for a fixed mass of sugar to dissolve at five temperatures

Spring compression (cm)	Landing distance from table (cm)
2.0	12.5
4.0	24.5
6.0	38.1
8.0	49.9
10.0	61.5
12.0	72.8

Graphical results

The following points should be noted about Figure 5.21.

- The axes have titles and units.
- The graph has a title.
- The scales on each axis allow the data to fill most of the grid space.
- A curved or straight line of best fit is drawn through or near the data points.
- The graph is drawn on grid paper (for accuracy).

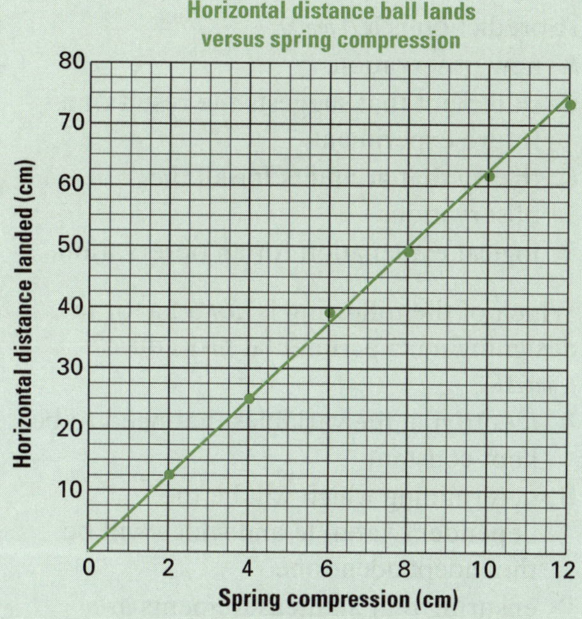

Figure 5.21 Graphical results

Conclusion

As the spring is compressed more, the distance from the foot of the table where the ball lands increases. A linear relationship is seen between these two variables.

Test yourself 2

Part A: Knowledge

1. An inference is a *(1 mark)*
   - **A** statement that is true in the majority of cases.
   - **B** statement that predicts what will happen in the next set of experiments.
   - **C** theory.
   - **D** logical explanation of an observation.
2. The independent variable in an investigation is *(1 mark)*
   - **A** the one that is allowed to change systematically.
   - **B** affected by the changes made during the experiment.
   - **C** the control.
   - **D** controlled so that its value does not change.

3. Which of the following statements is not a scientific hypothesis? *(1 mark)*
   A Plant cells are nicer to look at than animal cells.
   B Rocks with large crystals must have cooled slowly.
   C The more you chew your food, the easier it is to digest.
   D The energy of a moving object depends on its mass.

4. A prediction is a *(1 mark)*
   A new observation.
   B statement that suggests the result of a future experiment.
   C theory that accounts for all the observations.
   D logical explanation for an observation.

5. Which of the following is *not* a factor in designing an experimental procedure? *(1 mark)*
   A identifying the variables that need to be kept constant
   B determining which will be the dependent variable and which will be the independent one
   C ensuring that all measurements are reliable and accurate
   D collecting only those results that will support your hypothesis

6. Complete the following restricted-response questions using the appropriate word. *(1 mark for each part)*
   a) Further experiments may .... an inference or they may not.
   b) Generalisations are statements that are .... in the majority of cases.
   c) Hypotheses are .... statements that lead eventually to the development of scientific theories.
   d) The factors that influence the results of experiments are called ....
   e) The dependent variable is affected by a change in the .... variable.

7. Use the code letters to match the terms or phrases in each column. *(1 mark for each part)*

Column 1	Column 2
A control	F predictive explanation
B reliability	G allows comparison
C parallax	H repetition
D hypothesis	I error
E inference	J logical explanation

Part B: Skills

8. Suggest a logical inference that could be made for each of the following observations.
   a) Arthur noticed that heavy rubble was deposited on the inside bend of a river. *(1 mark)*
   b) Bees are hovering around a particular flower. *(1 mark)*
   c) A lump of solid carbon dioxide bubbled when placed in a beaker of water. *(1 mark)*
   d) Oil floats on water. *(1 mark)*
   e) Different-sized objects all landed at the same time when dropped from the same height. *(1 mark)*

9. The data in Table 5.6 shows the time it takes for starch, in the presence of an enzyme, to completely break down at different temperatures. Use it to predict the temperature for the fastest rate of the reaction. *(1 mark)*

Table 5.6 Breakdown of starch as a function of time

Temperature (°C)	Time (min)
10	40
20	20
30	10
40	8
50	24

10. Billy observed that the metal handle of a ladle gets hot when he is stirring a pot of hot soup (see Figure 5.22). He thought that how hot the handle got depended on the

quantity of soup in the pot. He made the following hypothesis which he then tested:

The temperature of the metal handle of the ladle depends on how much hot soup is in the pot.

Figure 5.22 Holding a soup ladle

a) Name the independent variable in Billy's experiment. *(1 mark)*

b) Name the dependent variable. *(1 mark)*

c) Name the variables that need to be controlled. *(5 marks)*

11. A student wanted to compare the extent to which different metal wires stretch when weights are attached to them (see Figure 5.23). The wires are iron, copper, steel and brass.

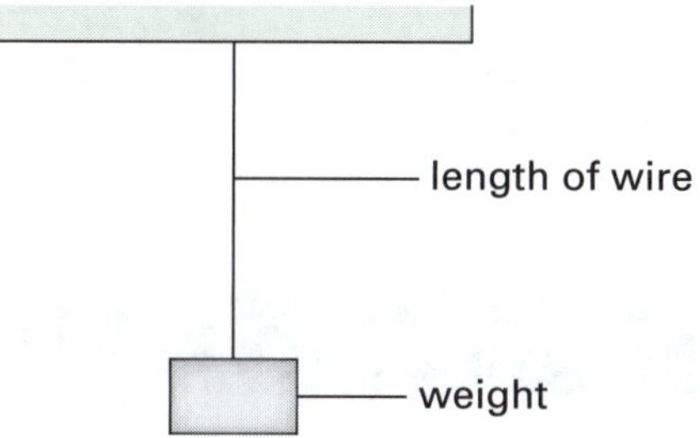

Figure 5.23 Hanging weight

a) List at least four pieces of equipment the student may use. *(4 marks)*

b) What are three variables the student must keep constant? *(3 marks)*

c) Why is it important that the student start off with the same wire length in each case? *(2 marks)*

d) Describe a procedure that allows a student to experimentally determine which wire stretches the most. *(6 marks)*

12. Suggest two advantages of working in teams. *(2 marks)*

13. Make a list of two variables that could affect each of the following measurements.

a) the time for a weather balloon to reach its target height *(2 marks)*

b) the stopping distance of a car when the brakes are applied *(2 marks)*

c) the distance that light can penetrate into a pool of water *(2 marks)*

d) the quantity of metal present in an ore *(2 marks)*

e) how long it takes for a person to digest a meal *(2 marks)*

14. Mary investigated how quickly a blue solution was decoloured by a mild bleaching agent. She conducted her investigation at different temperatures. The results of her experiment are shown in Table 5.7.

Table 5.7 Time solution takes to be decoloured at different temperatures

Temperature (°C)	Time to decolourise (s)
20	80
30	40
40	19
50	11
60	5

a) What variables did Mary need to control in her experiments? *(3 marks)*

b) Construct a line graph of Mary's data. Plot the temperature on the horizontal axis. *(5 marks)*

c) How did the speed of the reaction change as the temperature increased? *(1 mark)*

d) Do Mary's results support the following hypothesis? Explain. *(2 marks)*

The rate of a reaction doubles for each 10-degree rise in temperature.

Go to pp. 205–207 to check your answers.

Summary

1. The scientific method involves: observing; inferring; producing; experimenting; formulating hypotheses; testing and modifying hypotheses; making conclusions; generalising; and developing theories and laws.
2. Fair testing involves the control of variables. The independent variable is allowed to change and the dependent variable is affected by this change.
3. Experiments should be carefully designed to ensure that they are safe.
4. There are advantages and disadvantages in working individually or in teams.
5. Data can be gathered by first-hand experimentation and/or by second-hand data research.
6. Repeated measurements improve reliability.
7. Data can be presented in a variety of ways, including graphs and tables.

Syllabus checklist

Are you able to answer every syllabus question in this chapter? Tick each question as you go through the list if you are able to answer it. If you cannot answer it, turn to the appropriate page in the guide as is listed in the column to find the answer.

	For a complete understanding of this topic	Page no.	✓
1	Can I list the advantages and disadvantages of working both individually and in teams?	138	
2	Can I recall what fair testing involves?	142	
3	Can I recall what the scientific method involves?	142	
4	Can I recall ways to ensure experiments are safe and accurate?	146–147	
5	Can I explain the difference between collecting data by first-hand and second-hand methods?	140–142	
6	Can I explain why repeated measurements improve reliability?	147	
7	Can is state some ways in which data can be presented?	149–151	

Chapter test

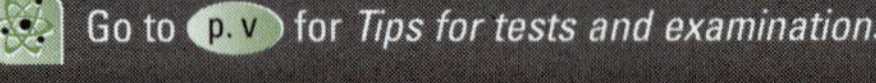
Go to p. v for *Tips for tests and examinations*

80 MIN

Part A: Multiple-choice questions

(1 mark for each)

1. In a scientific report, the correct sequence of recording is
   - **A** Aim, Method, Results, Conclusion.
   - **B** Aim, Method, Conclusion, Results.
   - **C** Conclusion, Results, Method, Aim.
   - **D** Method, Aim, Results, Conclusion.
2. An educated guess written as a tentative explanation is called a
   - **A** theory.
   - **B** variable.
   - **C** hypothesis.
   - **D** prediction.
3. The scientific method
   - **A** can test most questions.
   - **B** doesn't involve much experimenting.

C solves problems randomly.
D involves deduction and logic.

4. A scientific investigation usually begins with
   A a question.
   B an experimental idea.
   C observations of specific events.
   D a formulation of a hypothesis.

5. Which one of these words would you expect to cause the greatest problems in science?
   A where
   B what
   C when
   D why

6. A researcher is testing a new drug. Doctors give the drug to one group of patients and placebos (pills with no active ingredient) to another group. The group receiving the placebos
   A is in case some of the people being given the drug change their minds.
   B represents a control group.
   C allows the test to be repeated many times.
   D makes up the experimental group.

7. A hypothesis that is repeatedly verified may eventually become
   A a theory.
   B an educated guess.
   C a law.
   D a conclusion.

8. The step-by-step directions used in performing an experiment are the
   A aim.
   B procedure.
   C hypothesis.
   D problem.

9. Scientific laws
   A can be changed.
   B result from accumulated evidence and observations.
   C originate from hypotheses and theories.
   D can be described by all of the above.

10. The most difficult part of experimenting in science is
   A generating the problem.
   B forming a theory.
   C designing the experiment.
   D making observations.

Part B: Short-answer questions

11. The following is a list of steps in the scientific process. Complete the statements by placing each of the following words into the correct space: conclusion; controlled; information; record; variable; independent; control; problem; hypothesis; data; experiment; dependent. *(12 marks)*
   a) State the .... This is the question to which you are trying to find an answer.
   b) Gather .... about the problem from a variety of sources, including libraries, experts and the internet.
   c) Form a .... This is an educated guess of the answer based on the information that you gathered.
   d) Perform an .... to find the answer to a problem. This may involve the laboratory or some other convenient location.
   e) The .... is the thing that is changed in an experiment.
   f) The .... variable is the variable you investigate to determine its effect.
   g) The .... variable is the variable you measure, which should change in response to what you changed.
   h) The .... variables in an experiment do not change. They are kept the same in all steps of the experiment.
   i) The .... is the group in the experiment that does not have the variable.
   j) You should .... the results of your experiment.
   k) .... is defined as recorded observations and measurements.

l) You state a .... based on the results of the data gathered about the experiment.

12. Explain why each of the following is an important safety issue in a science laboratory. *(6 marks)*
    a) Any spill on the floor can cause an accident, so always clean it up immediately.
    b) When you work in the laboratory, long hair must be tied back.
    c) You should wear closed, sturdy leather shoes when in the laboratory.
    d) Never run in a laboratory.
    e) In order to prevent accidents during laboratory activities, you should always follow directions.
    f) Make sure all equipment is clean before and after you use it.

13. State four things the student in the experiment shown in Figure 5.24 is doing incorrectly. *(5 marks)*

Figure 5.24 Performing an experiment incorrectly

14. Figure 5.25 shows a student is performing an experiment at home.

Figure 5.25 Experimenting at home

Name two things she is doing correctly and two things she is doing incorrectly. *(4 marks)*

15. Explain the meaning of the safety signs shown in Figure 5.26. *(2 marks)*

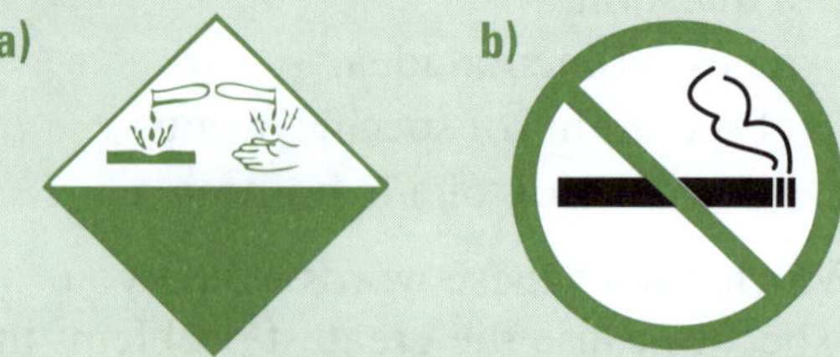

Figure 5.26 Safety signs

16. Name the pieces of laboratory equipment shown in Figure 5.27 (not drawn to scale). *(18 marks)*

Figure 5.27 Lab equipment

17. Most test tube racks are made from wood or plastic. Suggest a possible reason for racks not being made from metal. *(2 marks)*

18. a) Name the process occurring using the equipment shown in Figure 5.28. *(1 mark)*
    b) Replace the letters on Figure 5.28 with labels. *(9 marks)*
    c) How is the vapour entering the condenser cooled? *(2 marks)*
    d) Explain the arrows on the diagram. Why is it not good practice for them to be the other way around? *(4 marks)*

e) Boiling chips are usually placed in the flask being heated. What is the purpose of this? *(1 mark)*

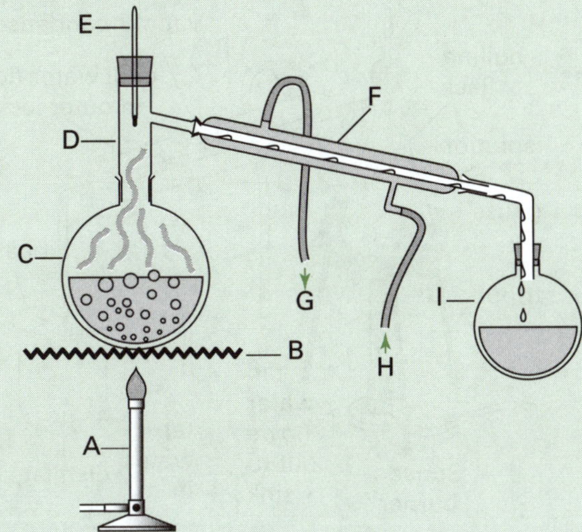

Figure 5.28 Physical separation process

19. An early form of distillation involved using the equipment shown in Figure 5.29.

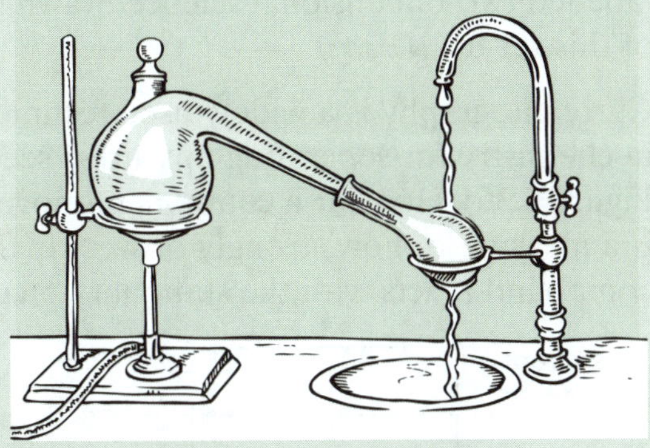

Figure 5.29 Early type of distillation

It was not as efficient as the apparatus shown in the previous question. Why was this? Use diagrams to help you answer this. *(6 marks)*

20. Comment on each of these statements made by commercial companies.

a) Nine out of 10 dentists recommend Dento toothpaste. *(2 marks)*

b) University tests show that Wizzo powder washes 22% whiter. *(2 marks)*

c) Our products are available from all leading retailers. *(2 marks)*

21. Read each of the following statements and decide whether they are examples of observations, inferences, predictions or generalisations.

a) A virus caused the cough. *(1 mark)*

b) Most compounds of copper are blue. *(1 mark)*

c) The minerals in the rock were a lighter colour to those in the rock from the bottom of the cliff. *(1 mark)*

d) If a more concentrated acid is added to zinc, the zinc will dissolve faster. *(1 mark)*

22. Read each of the following and decide whether or not each statement is a hypothesis. Justify your answer.

a) Cells divide when they reach a certain size. *(2 marks)*

b) Aboriginal peoples first learned to create fire by rubbing sticks together. *(2 marks)*

c) Igneous rocks are nicer to look at than sedimentary rocks. *(2 marks)*

23. A student grew some alum crystals in a beaker as shown in Figure 5.30 and measured their average sizes over a period of time.

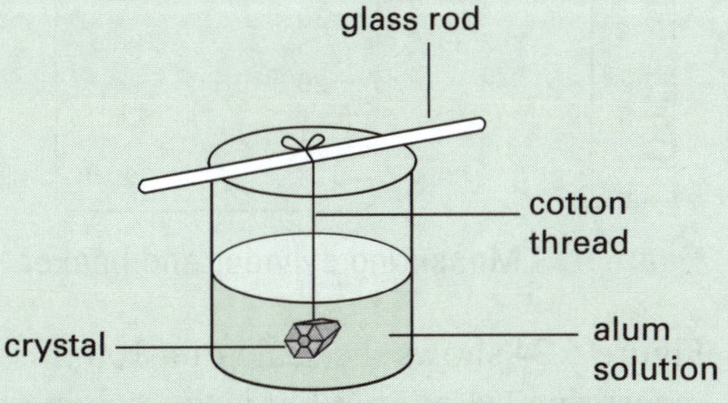

Figure 5.30 Alum crystals

Which is the dependent variable? Which is the independent variable? *(2 marks)*

24. What is the meaning of the signs shown in Figure 5.31? *(1 mark for each)*

Figure 5.31 Signs

25. Figure 5.32 shows two students (A and B) measuring the length of a glass prism using a millimetre ruler. Which student will obtain a more accurate reading? Explain. *(2 marks)*

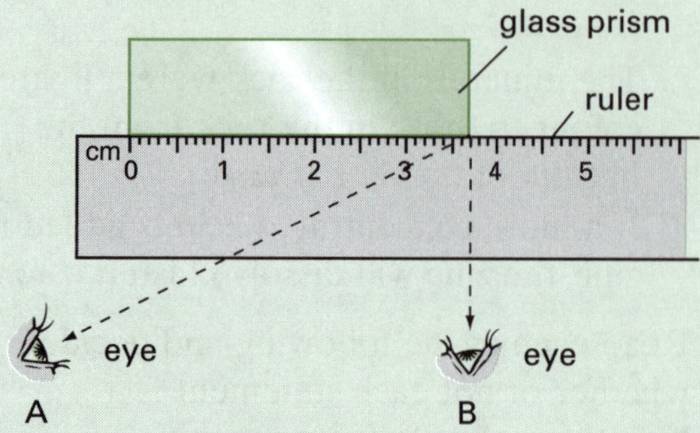

Figure 5.32 Parallax error

26. Figure 5.33 shows a measuring cylinder and a beaker. They are both used to measure the volumes of liquids. Which is more accurate? *(1 mark)*

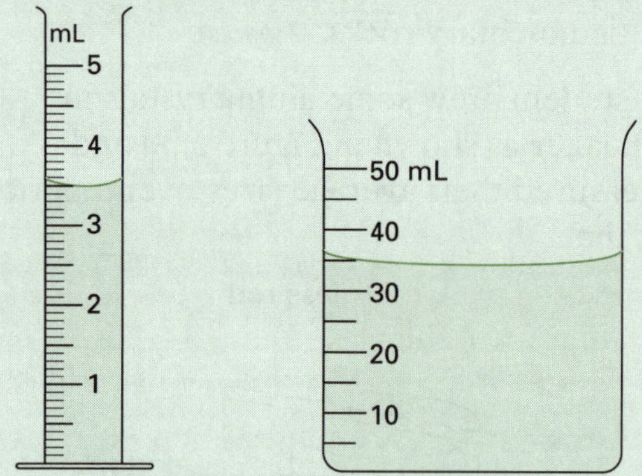

Figure 5.33 Measuring cylinder and beaker

27. Figure 5.34 shows the scale of a 100-mL measuring cylinder. What is the volume of liquid in the cylinder? *(1 mark)*

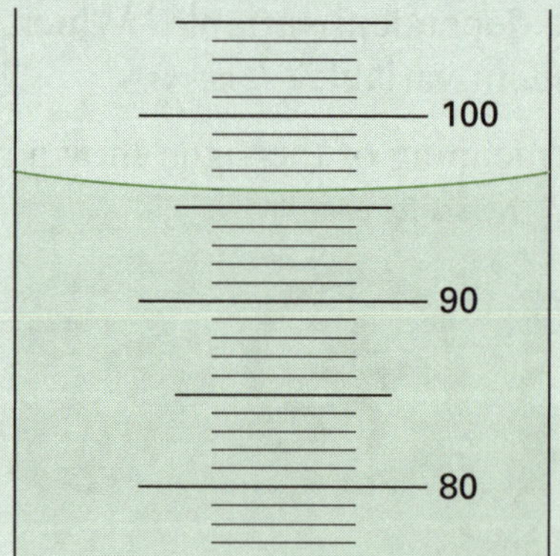

Figure 5.34 Measuring cylinder scale

28. Figure 5.35 shows apparatus set up for simple distillation.

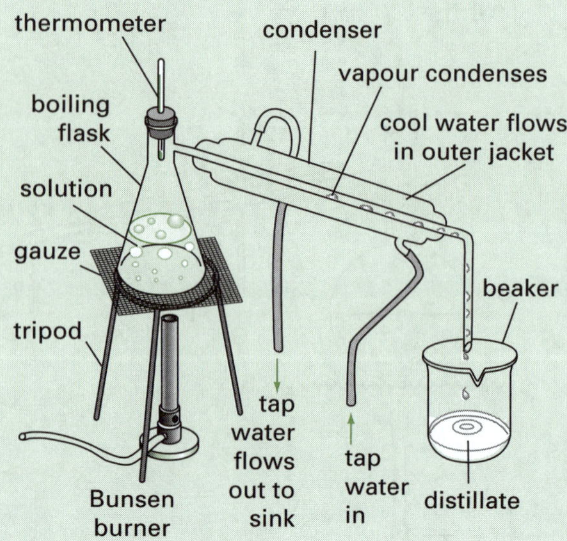

Figure 5.35 Distillation

This diagram shows a three-dimensional view. But in a science diagram all this information is not required. Draw a labelled two-dimensional science drawing of this set-up. *(4 marks)*

29. Chromatography is a widely used technique in chemistry for separating mixtures (see Figure 5.36). How far a component moves is a measure of how strongly or weakly the compound reacts with the stationary phase.

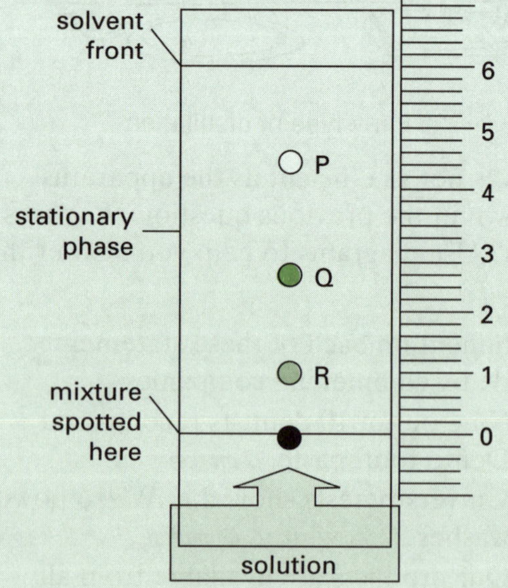

Figure 5.36 Chromatography

a) Which substance (P, Q or R) interacts more strongly and prefers to stay associated with the stationary phase? *(1 mark)*

b) If R has moved 1.0 cm, estimate the distance P has moved in this time. *(1 mark)*

c) The retention factor (*Rf*) is the ratio of the distance travelled by the compound up the chromatography paper versus the distance travelled by the solvent front. Calculate the *Rf* for component Q. *(2 marks)*

30. A piece of chromatography paper spotted with different substances is placed in a container with a shallow layer of a suitable solvent in it (see Figure 5.37).

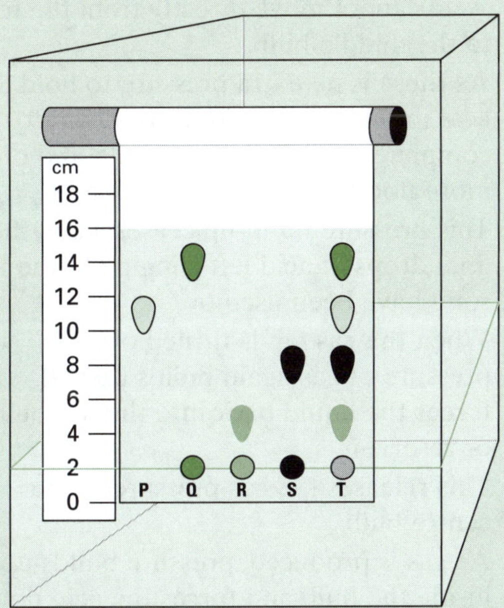

Figure 5.37 Paper chromatography

a) Which of the substances shows it is a mixture? *(1 mark)*

b) Explain why a lid is placed on the container. *(2 marks)*

c) A scale has been placed next to the diagram. In calculating *Rf* (see definition in the previous question), from what point on the scale should measurements be taken? Why? *(2 marks)*

d) Which component has an *Rf* of 0.68? *(2 marks)*

31. One way to collect gas is over water, as shown in Figure 5.38.

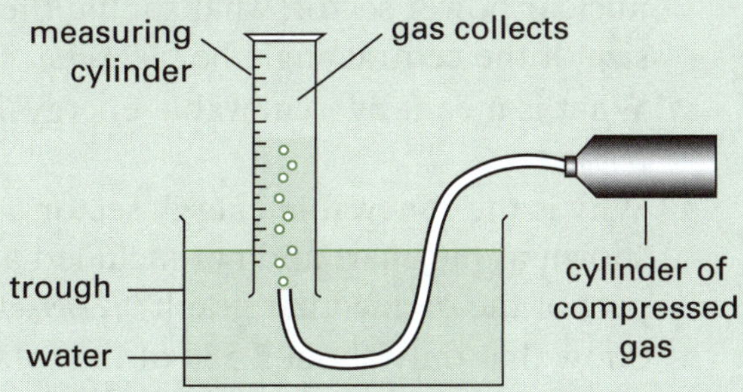

Figure 5.38 Collecting gas over water

a) What are three features of this gas that allow it to be collected in this way? *(3 marks)*

b) If the gas is flammable, would this present a problem to collecting it in this manner? *(1 mark)*

32. The divided bar graph and pie chart in Figure 5.39 set out energy consumption of a major country by primary sources.

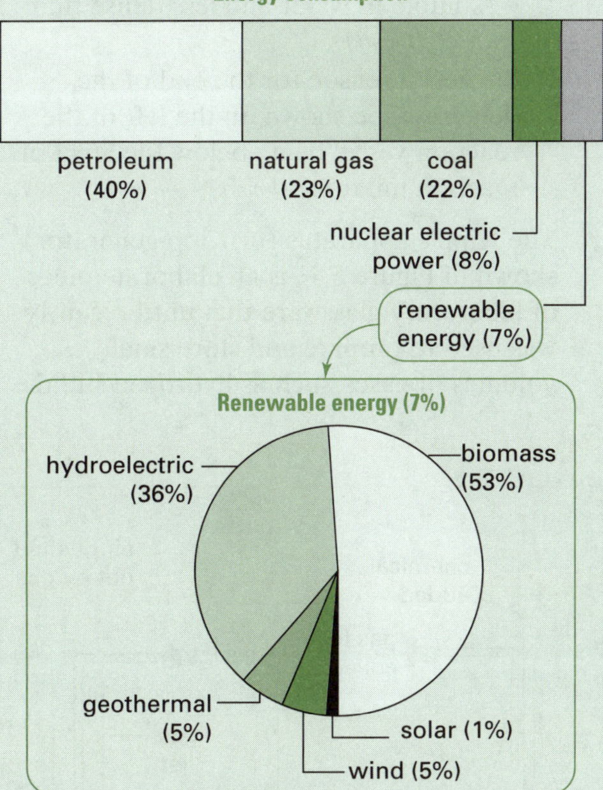

Figure 5.39 Energy consumption

a) What percentage of energy derives from fossil fuels? *(1 mark)*

b) Suppose the divided bar graph was drawn as a sector graph. In drawing the nuclear power sector, what should the size of the central angle be? *(2 marks)*

c) What is meant by 'renewable energy'? *(1 mark)*

d) Why is the renewable energy sector shown as pie chart and not included as part of the divided bar graph? *(2 marks)*

e) Show that only about 2.5% of the total energy comes from hydroelectric power. *(1 mark)*

f) What percentage of the total energy comes from biomass? *(1 mark)*

33. Figure 5.40 shows how gases can be prepared and three ways of collecting gases produced in different chemical reactions.

a) Which gas would be

i) insoluble in water, and does not react with it? *(1 mark)*

ii) soluble in water and more dense than air? *(1 mark)*

iii) soluble in water and less dense than air? *(1 mark)*

b) Suggest a reason for the end of the delivery tube shown on the left of the reaction vessel being below the level of reaction mixture? *(1 mark)*

34. The Kipp's apparatus (or Kipp generator) shown in Figure 5.41 is an elaborate piece of laboratory glassware that until recently was used to prepare and store small volumes of gases, such as hydrogen sulfide, carbon dioxide or hydrogen. It is named after its inventor, the Dutch pharmacist Petrus Johannes Kipp (1808–1864).

Hydrogen sulfide can be prepared by the reaction:

iron sulfide (solid pellets)	+	hydrochloric acid	→	iron chloride (soluble)	+	hydrogen sulfide (gas)

The following is an explanation of how the apparatus works, but the steps are out of sequence. Identify the correct order for an explanation of how the Kipp generator works. *(12 marks)*

A. The Kipp's generator consists of three glass bulbs connected one above the other.

B. Acid cannot move directly from the top to the middle bulb.

C. As there is no extra pressure to hold the acid in the top bulb, it drops down to completely fill the bottom bulb and once more flood the solid.

D. This pressure build-up ceases when the final drops of acid left clinging to the solid have been used up.

E. When the gas tap is turned off, the pressure of gas again builds up and forces the liquid back into the top bulb or reservoir.

F. This releases the gas pressure in the centre bulb.

G. As gas is produced, pressure builds up inside the bulb and forces the acid down

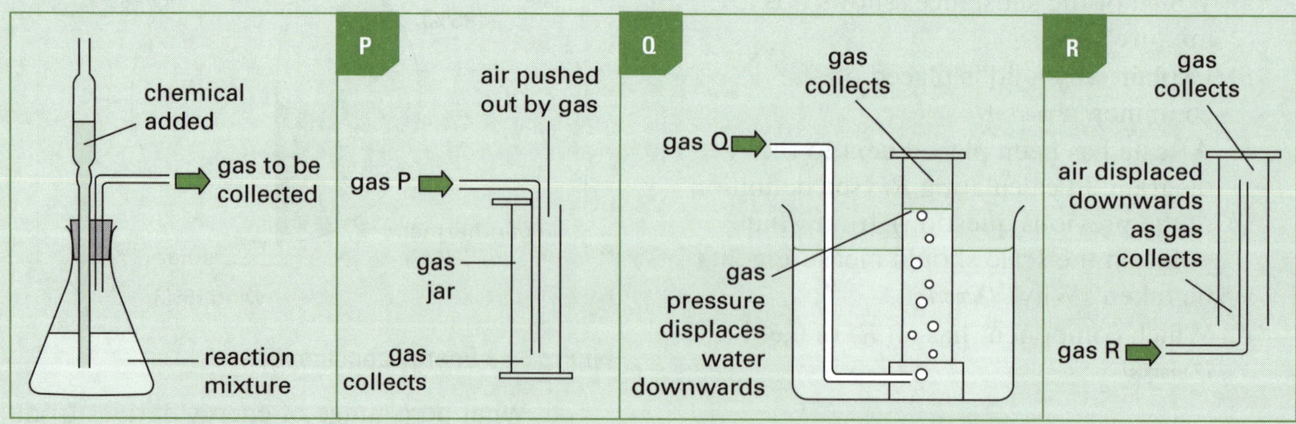

Figure 5.40 Collecting gases

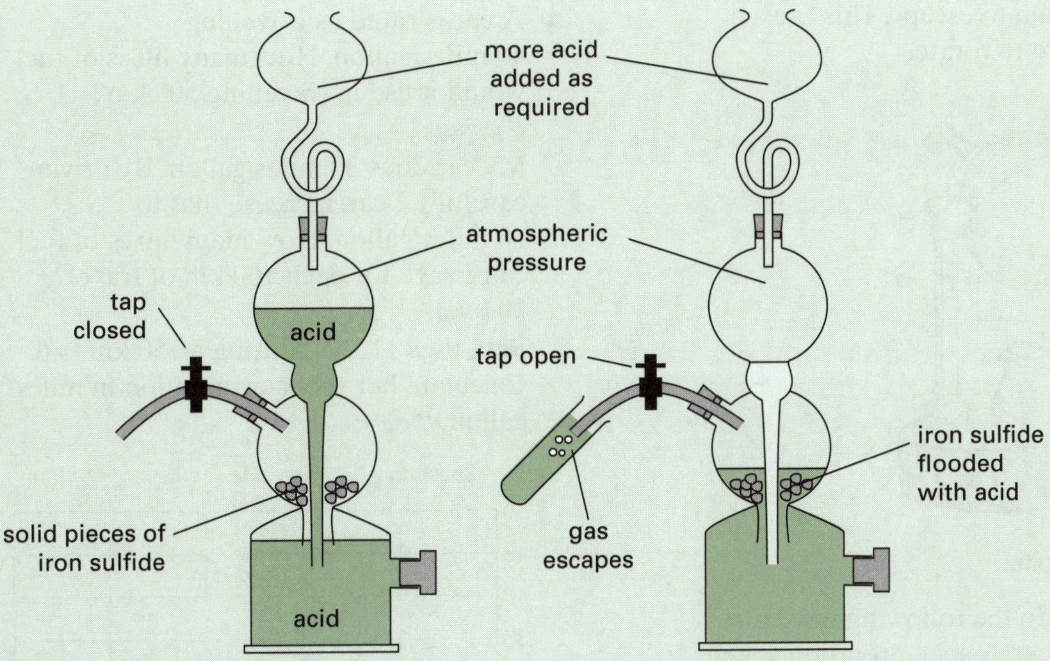

Figure 5.41 Kipp's generator

into the bottom bulb and up into the top bulb.

H. Enough acid is poured in at the top to fill the bottom section and flood the solid in the centre bulb.

I. When the acid is forced out of the centre bulb, gas generation stops.

J. The solid iron sulfide needed to make the gas is placed in the central bulb.

K. A glass fitting stops the solid iron sulfide from dropping into the bottom bulb.

L. The tap is turned on when gas is required.

35. A student set up the experiment shown in Figure 5.42 to examine the bands of colours that appear on a film of dilute detergent solution suspended on a circular copper frame. She used this film as a mirror to reflect light from the sky to obtain various patterns.

The bands appeared to change colour and position regardless of whether she used set-up P, Q or R. The film thins as it drains, and coloured bands appear.

In which situation would the film last the longest? Explain. *(2 marks)*

36. Heron lived during the first century AD in Alexandria. He invented the aeolipile (see Figure 5.43).

In the aeolipile, a sealed pot filled with water over a fire, supplied steam. Two tubes came up from the pot, releasing the steam into a spherical hollow metal ball. This sphere had two curved outlet tubes, which

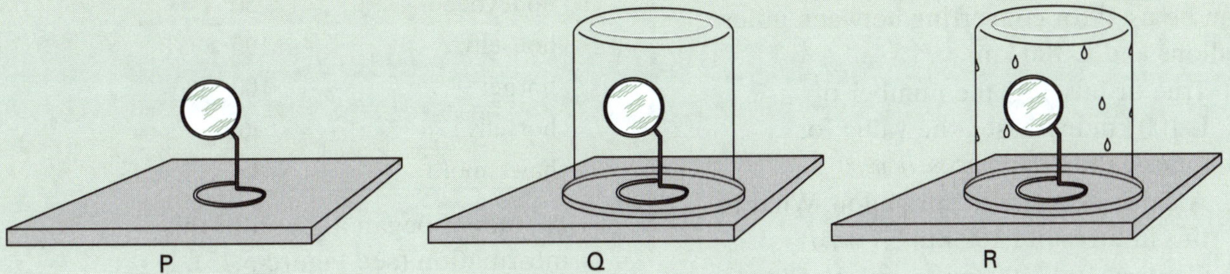

Figure 5.42 Bands of colour (P = stand with film; Q = stand with film inside a dry overturned beaker; R = stand with film inside a moist overturned beaker)

allowed the steam to escape. This caused the metal sphere to rotate.

Figure 5.43 Aeolipile

a) What effect do the following have on the metal sphere's speed rotation speed?
   i) adding more wood to the fire *(2 marks)*
   ii) adding more boiling water to the pot *(2 marks)*
   iii) making the ball larger *(2 marks)*
   iv) making the outlet tubes narrower *(2 marks)*

b) Suppose one of the outlet tubes was blocked. Would the sphere still rotate? *(1 mark)*

c) Suppose the outlet tubes were turned to face the same direction. Would the sphere still rotate? *(1 mark)*

d) Suppose the spherical ball was replaced with one shaped like a football. Would the sphere still rotate? *(1 mark)*

37. Fuel consumption in cars is now measured in litres of fuel used for each 100 km travelled (L/100 km). An older unit, still used in the United States, is miles per gallon (miles/gallon or mpg). Figure 5.44 can be used for converting between miles/gallons and L/100 km.

a) True or false: As the number of L/100 km increases, the value for miles/gallon decreases. *(1 mark)*

b) A car uses 25 miles per gallon. What is this in litres per 100 km? *(1 mark)*

c) Convert 19 L/100 km to miles/gallon. *(1 mark)*

d) A car is rated as travelling 40 miles/gallon. How many litres of fuel would it use in travelling 500 km? *(2 marks)*

e) My car does 20 miles/gallon. By driving carefully I can increase that to 30 miles/gallon. How many litres of fuel can I save for each 100 km of travel? *(3 marks)*

f) Peta uses 112.5 L during a 750 km trip. Calculate her fuel consumption in miles/gallon. *(2 marks)*

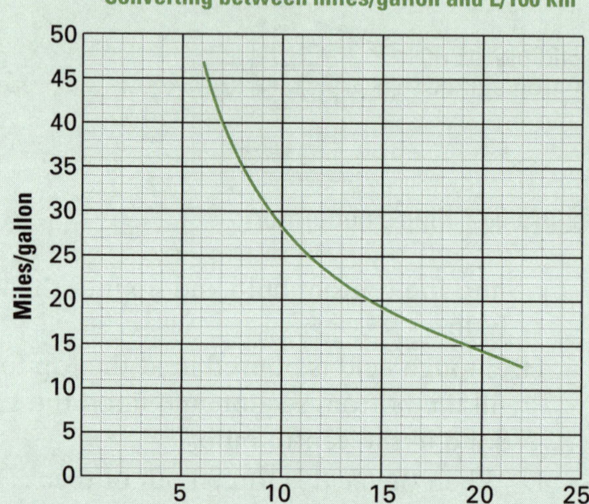

Figure 5.44 Unit conversion chart

38. Laboratory experiments show that not only does the wing beat vary in different species of insect but it can also vary in any one individual at different times. Table 5.8 gives the average wing beats each second for some insects.

Table 5.8 Average wing beats per second

Insect	Beats per second
honeybee	250
housefly	190
hornet	100
horsefly	96
hawk moth	85

A student began a graph of this information (see Figure 5.45).

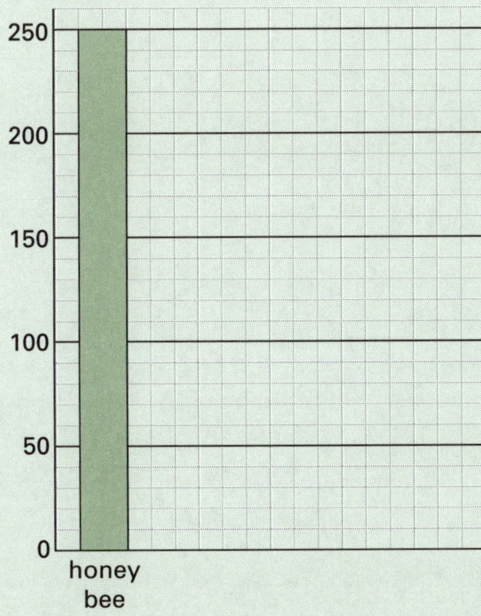

Figure 5.45 Insect wing speed

a) Complete the graph, labelling the axes and giving the graph a title. *(7 marks)*

b) What type of graph is this? *(1 mark)*

c) Would a line graph be more suitable in showing this information? Explain. *(2 marks)*

39. Table 5.9 gives the wing beats per second and speed under controlled circumstances for several insects.

Table 5.9 Wing beats per second and speed under controlled circumstances

Insect	Wing beats/ second	Speed (km/h)
P	250	9.2
Q	190	7.1
R	130	10.3
S	120	12.6
T	100	20.6
U	96	14.2
V	85	17.9
W	38	25.1
X	28	1.8
Y	16	5.3

George drew the scatter plot shown in Figure 5.46.

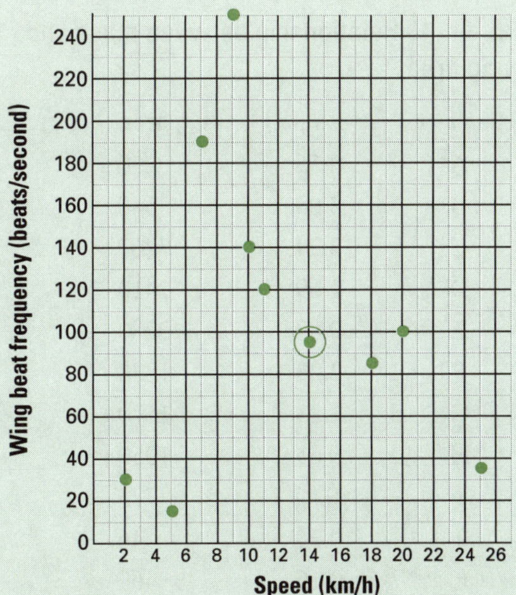

Figure 5.46 Frequency versus speed

a) Name the insect represented by the dot that is circled. *(1 mark)*

b) A student made the hypothesis that the greater the wing beat frequency, then the greater is the speed of the insect. Is this hypothesis supported by this data? Explain. *(2 marks)*

c) Suggest one thing the student could do to explore this hypothesis further. *(1 mark)*

d) Suggest three other factors besides wing beat frequency that could influence the speed of flight of an insect. *(3 marks)*

40. Dimitra was trying to determine whether there is any relationship between the mass and wing area of a number of South Atlantic seabirds. She produced Table 5.10.

a) This table is ordered on a particular variable. What is this variable? *(1 mark)*

b) Draw a scatter plot for this table. Place mass on the horizontal axis. *(6 marks)*

c) One of these results would be termed an 'outlier'. Which is it? *(1 mark)*

d) She made the hypothesis that the larger the wing area, then the larger is the mass of the bird. Is this hypothesis supported by this data? Explain. *(2 marks)*

Table 5.10 Relationship between mass and wing area

Sea bird	Mass (kg)	Wing area (cm^2)
A	8.55	5830
B	3.08	3540
C	3.24	3260
D	1.69	2410
E	0.89	2280
F	2.23	1830
G	1.23	1670
H	0.61	1050
I	0.42	773
J	0.44	646
K	0.16	469
L	0.13	221
M	0.04	215
N	0.12	197

Go to pp. 207–211 to check your answers.

Course test 1

This test covers material from all the chapters in this book. It consists of the following.

- Section A: 25 multiple-choice questions *(1 mark for each)*
- Section B: 15 restricted-response questions *(1 mark for each)*
- Section C: 11 knowledge, skill and processing data questions *(total 60 marks)*

Total marks for this test: 100

Time: 2 hours

Part A: Multiple-choice questions

(25 marks—1 mark for each question)

1. Which of the following is *not* an example of a fungus?
   - **A** yeast
   - **B** mould
   - **C** mushroom
   - **D** bacteria
2. The jelly-like part of the cell where chemical reactions take place and nutrients are used is the
   - **A** cell membrane.
   - **B** cytoplasm.
   - **C** nucleus.
   - **D** mitochondrion.
3. All of the following are examples of asexual reproduction, except
   - **A** a new breed of dog that is developed by crossing existing breeds.
   - **B** a new strawberry shoot that sprouts in a garden patch where strawberries are growing.
   - **C** cells from a single corn plant that are used to grow disease-resistant corn.
   - **D** mould that grows on jam left out of a fridge.
4. Figure T1.1 shows sexual reproduction in chickens.

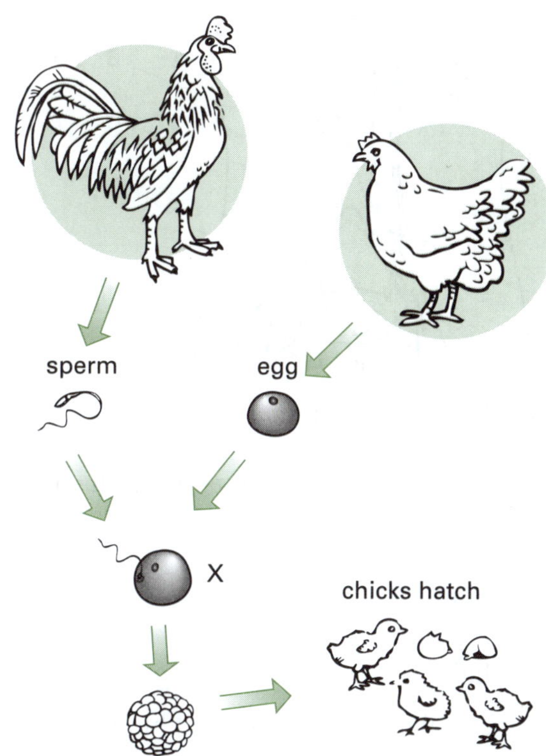

Figure T1.1 Sexual reproduction in chickens

The process marked X in the diagram is
   - **A** gestation.
   - **B** pollination.
   - **C** mitosis.
   - **D** fertilisation.
5. The purpose of therapeutic cloning is to
   - **A** insert healthy genes into cells.
   - **B** destroy unwanted pest species.
   - **C** create specialised tissues or organs for transplant.
   - **D** make cells grow faster and larger.
6. Which of the following are part of the respiratory system?
   - **A** oesophagus, ascending colon, duodenum
   - **B** aorta, pulmonary vein, ventricle
   - **C** tendon, biceps, gluteus maximus
   - **D** trachea, bronchi, alveoli

7. Which of the parts in Figure T1.2 (A–D) is *not* associated with the urinary system?

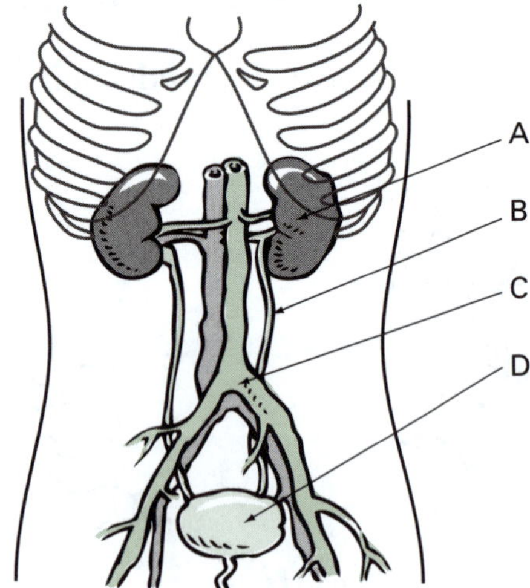

Figure T1.2 Urinary system

8. The proteins, actin and myosin, which are able to contract, are found in
   A muscle.
   B bone.
   C blood.
   D hair.

9. The skeletons of sharks and human embryos are composed of
   A bone.
   B cartilage.
   C collagen.
   D keratin.

10. The system that removes gaseous waste (such as carbon dioxide) from the bloodstream is the
    A urinary system.
    B circulatory system.
    C respiratory system.
    D digestive system.

11. The smallest unit of an element that has all the basic properties of the element is called the
    A atom.
    B molecule.
    C nucleus.
    D proton.

12. Matter that has a definite shape and is made up of tightly packed particles is classified as a
    A solid.
    B liquid.
    C gas.
    D plasma.

13. An example of a physical property is
    A flammability.
    B chemical composition.
    C weight.
    D reactivity.

14. Which of the following could indicate that a chemical change has occurred?
    A colour change
    B solution becomes warmer
    C a gas is formed
    D all of the above

15. The letters in Figure T1.3 represent a form of energy. They can be replaced with which of the following?

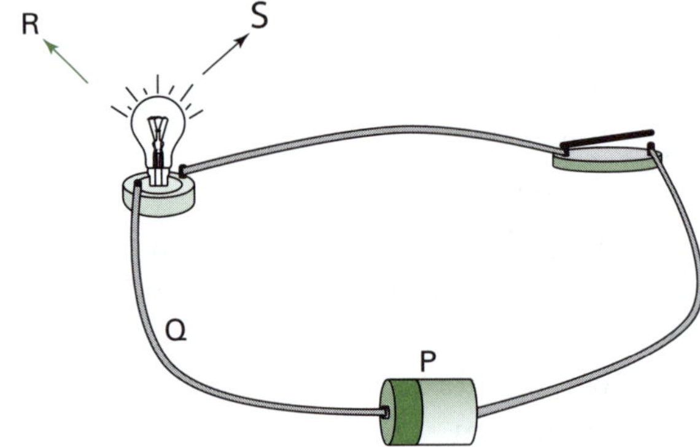

Figure T1.3 Circuit

	P	Q	R	S
A	potential	heat	light	kinetic
B	chemical	electrical	heat	light
C	kinetic	potential	heat	light
D	chemical	light	heat	electrical

16. When a test tube full of hydrogen is lit with a match, a sharp 'pop' is heard. This indicates that a reaction has occurred and the product is
    A carbon dioxide.
    B oxygen.

C water.
D hydrogen peroxide.

17. Atoms of different elements can be distinguished by their number of
A atoms.
B electrons.
C neutrons.
D protons.

18. Butane, $CH_3(CH_2)_2CH_3$, is a highly flammable, colourless gas that is sold bottled as a fuel for cooking and camping. The number of atoms in one molecule of butane is
A 2.
B 6.
C 11.
D 14.

19. Kinetic energy is energy that is
A due to motion.
B stored.
C yet to be produced.
D already used.

20. Which of these statements is *incorrect*?
A The cell is the basic unit of life.
B All organisms are made up of cells.
C Most cells are microscopic in size.
D Cells are covered by a cell membrane or a cell wall.

21. Figure T1.4 shows the reaction between methane (CH_4) and oxygen (O_2) to form carbon dioxide (CO_2) and water (H_2O).

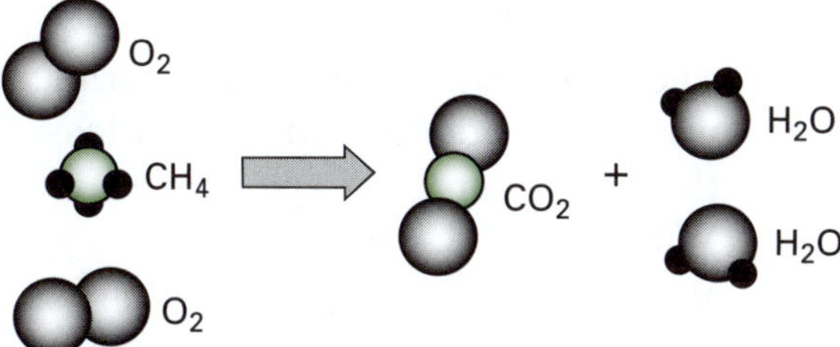

Figure T1.4 Combustion of methane

Which of the following statements is true?
A A total of three atoms are reacting.
B All the oxygen reacting ends up in the water.
C The carbon atom from the methane ends up in the CO_2.
D The number of atoms before the reaction does not equal the number of atoms after the reaction.

22. Which of these is an igneous rock?
A sandstone
B granite
C shale
D gneiss

23. Which of these is a non-renewable source of energy?
A solar
B hydroelectric
C wind
D coal

24. Basalt is usually grey to black, and is a fine-grained rock. The fine grains are due to
A the rock quickly cooling on the Earth's surface.
B being deposited in beds of slow-flowing streams.
C slowly cooling deep within the Earth's crust.
D being altered by high heat and pressure.

25. Which of these is the least important reason to recycle?
A It saves trees and protects biodiversity and wildlife habitats.
B It is popular with many people, organisations and councils.
C It helps curb global warming and pollution.
D It reduces the need for incinerators and landfill.

Part B: Restricted-response questions

(15 marks—1 mark for each question)

Complete the following restricted-response questions using the appropriate word.

26. Both plant and fungal cells contain cell walls. But one major difference is that fungal cells contain chitin in their cell walls while the cell walls of plants contain ....

27. Unlike animal cells, plant cells make their own food in a process called ....

28. Animal cells use food to extract the energy they need to reproduce and carry out basic functions. This process takes place in the .... of the cell.

29. The cell membrane is ...., allowing some substances to pass through while preventing others.

30. Most of the nutrient absorption that takes place during digestion occurs in the ....

31. The written notation used to represent the number of atoms of each element in a compound is called the ....

32. The lightest gas is ....

33. Burning a piece of wood is an example of a .... change.

34. .... is the capacity of a physical system to perform work.

35. .... rocks form from the alteration of existing rocks by either excessive heat and/or pressure, or through the chemical action of fluids.

36. Rocks are collections of ....

37. An .... is a metal-bearing mineral or rock, or a native metal, which can be mined at a profit.

38. .... include a number of elements, such as oxygen or phosphorus, that lack the physical and chemical properties of metals.

39. Hydroelectric power is produced by taking advantage of the gravitational .... energy stored in water.

40. How fast the atoms and molecules move determines the .... energy of a substance.

Part C: Knowledge, skill and processing data questions

(60 marks)

41. A fungus was grown on a number of culture plates. Anti-fungicide was added to some of these plates. Measuring the colony diameter monitored fungal growth. (See Figure T1.5.)

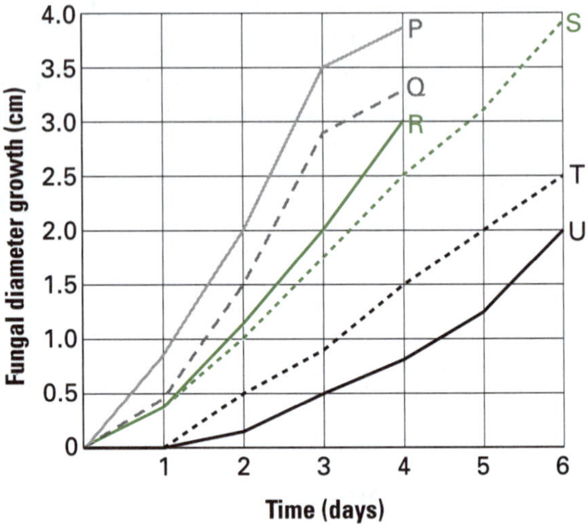

Figure T1.5 Fungus growth

a) Which of these results would most likely be the control? Explain. *(2 marks)*

b) How many anti-fungicides were tested? *(1 mark)*

c) How often were measurements taken? *(1 mark)*

d) Which is the most effective anti-fungicide? *(1 mark)*

e) Which colony had grown to a 2.0-cm diameter after 3 days? *(1 mark)*

f) The report states that 'the experiment was independently repeated at least three times and average data are shown'. Why was this? *(2 marks)*

g) Why would measuring the colony's diameter, rather than individual fungi, be appropriate? *(2 marks)*

42. Figure T1.6 shows steps during the mitosis of an animal cell, but the steps are out of sequence.

a) What is mitosis? *(1 mark)*

b) What is the correct sequence for these steps? *(2 marks)*

c) Which letter shows the phase of mitosis in which the chromosomes begin to separate? *(1 mark)*

d) Interphase is the stage where the cell is not undergoing mitosis. Does this mean nothing is occurring during this phase? Explain. *(2 marks)*

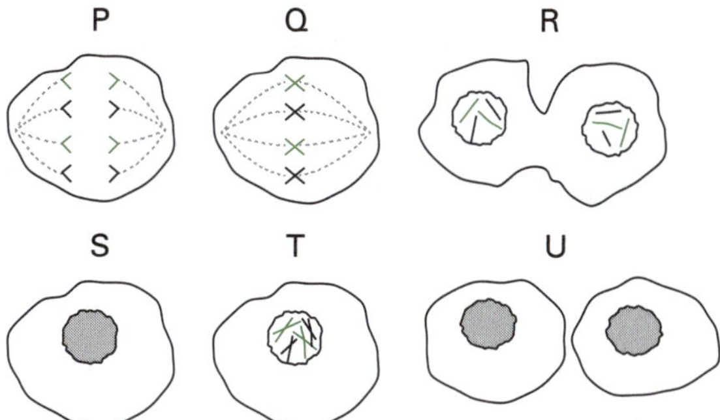

Figure T1.6 Mitosis

43. Some strains of fungus that grow in the soil can be harmful to agricultural crops. In order to test the effect of a certain bacterium in halting the growth of these fungi, some fungi were grown in a nutrient broth that had this bacterium added. At regular intervals some of the live fungal cells from the broth were taken out, dried and weighed. The results are shown in Figure T1.7.

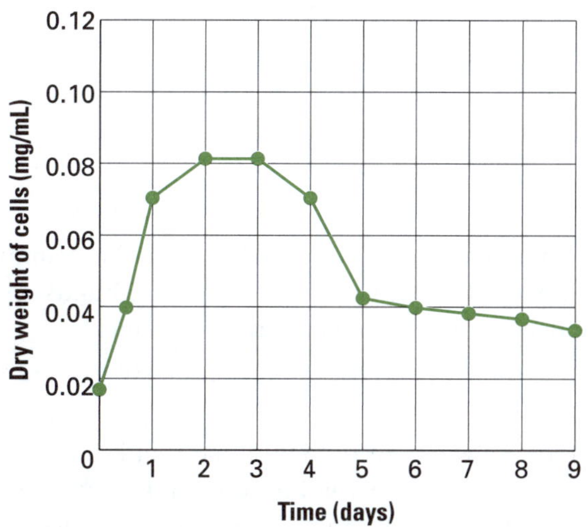

Figure T1.7 Weight of cells vs. time

a) Why was the dry weight of cells recorded? *(1 mark)*

b) Explain what the graph shows. *(3 marks)*

c) Is it certain that the fungal cells were slowed by the bacterium? Explain. *(4 marks)*

d) If a certain broth has a volume of 750 mL, what is the dry weight of the fungal cells it contains after 6 days? *(2 marks)*

e) Chemical fungicides are often used in agriculture. Explain why alternative methods for plant protection, which are less dependent on chemicals, are actively being researched? *(3 marks)*

44. Figure T1.8 provides a way to classify igneous rocks based on their mineral content of light- and dark-coloured minerals.

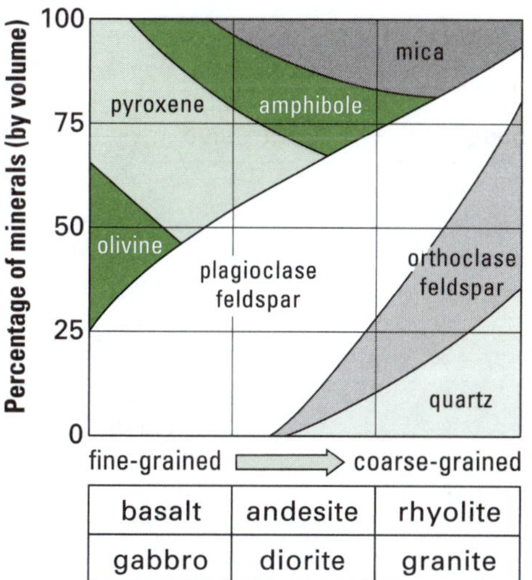

basalt	andesite	rhyolite
gabbro	diorite	granite

Figure T1.8 Igneous rock classification

Use the graph to determine whether each of the following statements is true (T) or false (F). *(1 mark for each part)*

a) Mica is not present in diorite.

b) Basalt does not contain any quartz.

c) Some andesite can contain over 50% feldspar.

d) If there is no olivine in the rock then it can't be gabbro.

e) Granite rocks contain at least 5% mica.

f) Orthoclase feldspar is less abundant than plagioclase feldspar in fine-grained igneous rocks.

g) The major component of andesite is the mineral plagioclase feldspar.

h) Samples of the basalt with the largest grains contain the following minerals in decreasing order by volume: plagioclase feldspar > pyroxene > amphibole > mica.

45. Figure T1.9 is a classification of rock types that gives one example in each category. Clastic rocks are rocks composed of fragments of older rocks. Foliated rocks refer to rocks that show a layered structure. For example, limestone is a sedimentary rock formed through chemical means. Use the information in Figure T1.9 to classify each of the rocks in Table T1.1. *(2 marks for each)*

46. Figure T1.10 shows how molten salts can be used to generate electrical power.

a) Descriptions showing the steps have been left off the diagram and are jumbled below. Determine the correct sequence to replace the letters (P, Q, R, S, T and U) in the diagram. *(6 marks)*

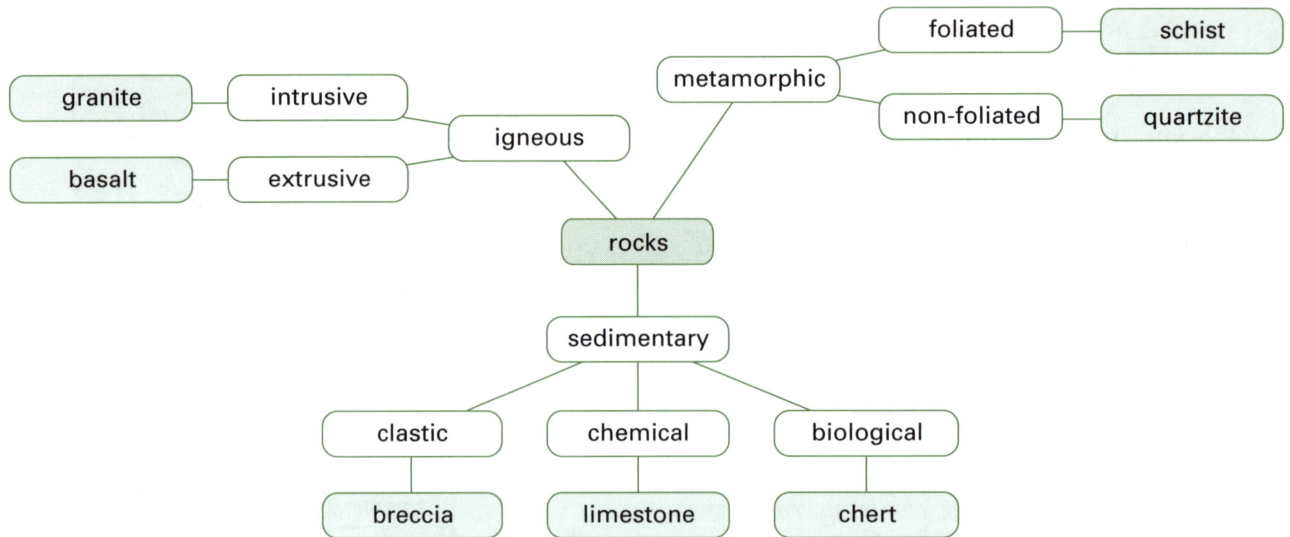

Figure T1.9 Rock classification

Table T1.1 Rock classification information

Rock	Description
coal	Forms from the build-up and decay of plant and animal material. This usually occurs in swampy regions where there is plenty of growing vegetation and little oxygen. The vegetation builds up so fast that new layers of plant matter buries the dead and decaying material very quickly.
gneiss	A high-grade rock that has been subjected to extensive heat and pressure. It has distinct banding with alternating layers composed of different minerals.
marble	A rock that has been altered from its parent rock of limestone or dolomite, and that is much harder. This rock has many different sizes of crystals and colour variations due to the impurities present at formation, but does not show banding. Some of the different colours include white, red, black, grey, pink, green, and mottled and banded.
sandstone	A rock formed from cemented grains that may either be fragments of a pre-existing rock or be single mineral crystals. Layers accumulate as sand settles out from suspension in a body of water and then are compacted by the pressure of overlying deposits.
rhyolite	This rock crystallises from silicate minerals at relatively low temperatures. The crystals are often too small to see, being due to the more rapid cooling at the surface.

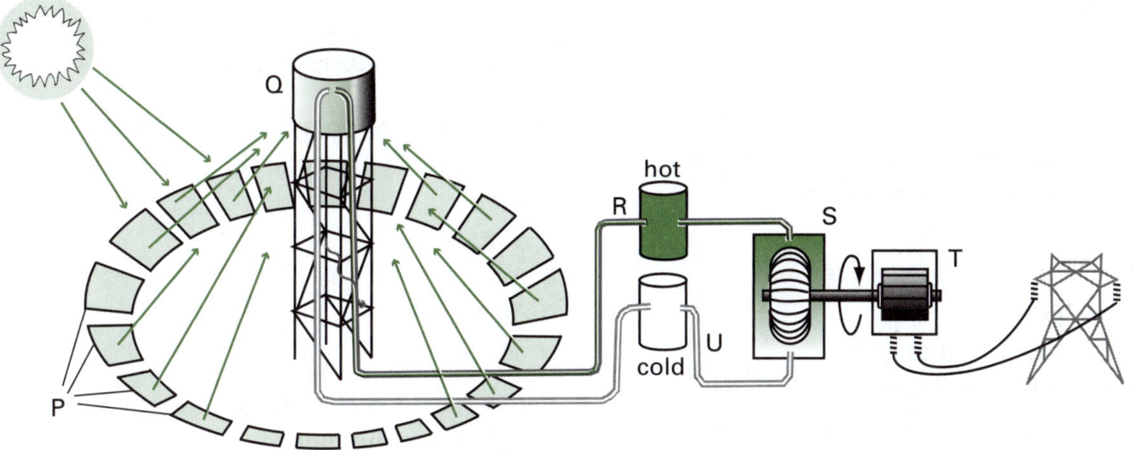

Figure T1.10 Molten salts generate electrical power

- **i)** The hot salt transfers heat to a steam generator.
- **ii)** The now cooled, but still liquid, salt is sent to another storage tank before repeating the cycle through the collection tower.
- **iii)** A large array of computer-controlled mirrors focuses sunlight onto a collection tower.
- **iv)** Steam is used to run turbines generating electricity, which moves to the electric grid.
- **v)** Molten salt in the collection tower is heated to over 500 °C.
- **vi)** The heated salt is now pumped to a storage tank.

b) Why is it important that the mirrors be computer controlled? *(2 marks)*

c) Suggest a reason for using molten salt rather than boiling water directly. *(1 mark)*

47. The reaction hydrogen + oxygen → water is shown in Figure T1.11.

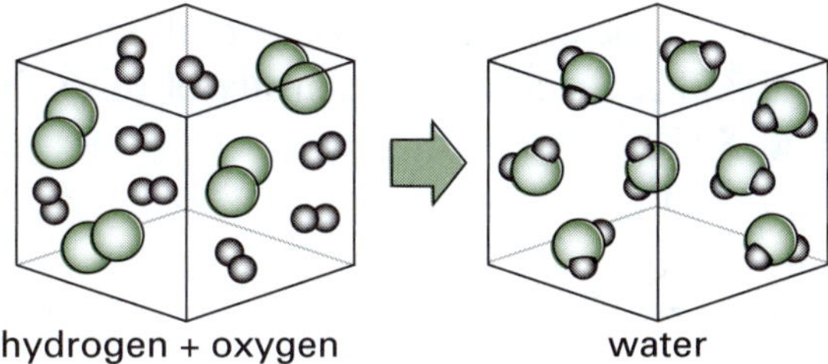

Figure T1.11 Chemical reaction

a) Both hydrogen and oxygen are described as diatomic molecules. What does this mean? *(1 mark)*

b) The diagram on the left shows a mixture, while the one on the right shows a compound. Explain. *(2 marks)*

c) What is the ratio of hydrogen atoms reacting to oxygen atoms reacting? *(1 mark)*

Go to pp. 211–213 to check your answers.

Course test 2

Go to p.v for *Tips for tests and examinations*

This test covers material from all the chapters in this book. It consists of the following.

- Section A: 25 multiple-choice questions *(1 mark for each)*
- Section B: 15 restricted-response questions *(1 mark for each)*
- Section C: 11 knowledge, skill and processing data questions *(total 60 marks)*

Total marks for this test: 100

Time: 2 hours

Part A: Multiple-choice questions

(25 marks—1 mark for each question)

1. The scientist who first observed cork cells under a simple microscope was
   - **A** Leeuwenhoek.
   - **B** Hooke.
   - **C** da Vinci.
   - **D** Janssen.
2. Select the true statement about microscopes.
   - **A** The objective lens has a lower magnification than the eyepiece lens.
   - **B** The iris reflects light onto the slide.
   - **C** The glass slide is placed on the stage for viewing.
   - **D** The fine focus knob is adjusted first and then the coarse focus knob is adjusted.
3. In plants the function of the xylem cells is to
   - **A** transport sugar and other nutrients to the leaves.
   - **B** store starch grains.
   - **C** produce sugars by photosynthesis.
   - **D** transport water from roots to the shoots and leaves.
4. What structure do many unicellular organisms use to move them through the water?
   - **A** flagella
   - **B** phagocytosis
   - **C** mitosis
   - **D** slime
5. Select the correct statement about exhalation in humans.
   - **A** The diaphragm relaxes upwards and the rib cage moves down and inwards.
   - **B** The diaphragm contracts down and the rib cage moves up and outwards.
   - **C** The lungs expand and air is drawn in from the outside.
   - **D** Air pressure in the chest cavity drops and is lower than the outside air pressure.
6. Select the true statement about asexual reproduction.
   - **A** There is less chance of harmful mutations in the species.
   - **B** There is a greater variation produced within a species.
   - **C** The offspring will be genetically stronger than the parent.
   - **D** The offspring are direct copies (or clones) of the original.
7. The scientist who developed the atomic theory in the early 19th century was
   - **A** Dalton.
   - **B** Democritus.
   - **C** da Vinci.
   - **D** Newton.
8. Fifty cubic centimetres of a solid had a mass of 350 g. Calculate the density of this solid.
   - **A** 50 g/cm^3
   - **B** 0.14 g/cm^3
   - **C** 7 g/cm^3
   - **D** 17 500 g/cm^3

9. The chemical symbol for the element iron is
   - **A** I.
   - **B** Ir.
   - **C** In.
   - **D** Fe.

10. The set that contains a metal, semi-metal and non-metal is
    - **A** Cu, B, I.
    - **B** Na, Ga, F.
    - **C** Co, Ge, Si.
    - **D** Ni, Cr, W.

11. A common use of carbon dioxide is
    - **A** bleach.
    - **B** fire extinguisher.
    - **C** rust removal.
    - **D** disinfectant.

12. Abrasion is a type of physical weathering of rocks. Abrasion typically occurs in
    - **A** cold regions where water freezes into ice.
    - **B** environments where volcanoes are active.
    - **C** regions where there are windblown sediments.
    - **D** environments where clay minerals absorb water.

13. Which one of the following is not a sedimentary rock?
    - **A** conglomerate
    - **B** shale
    - **C** sandstone
    - **D** pumice

14. Which one of the following is not a metamorphic rock?
    - **A** granite
    - **B** slate
    - **C** marble
    - **D** schist

15. The age of the Earth is about
    - **A** 2000 million years.
    - **B** 3000 million years.
    - **C** 4500 million years.
    - **D** 6500 million years.

16. The metal that can be extracted from pitchblende ore is
    - **A** uranium.
    - **B** aluminium.
    - **C** iron.
    - **D** zinc.

17. The kinetic energy of a body depends on its
    - **A** mass and speed.
    - **B** speed only.
    - **C** mass only.
    - **D** density and speed.

18. The gravitational potential energy of a body near the Earth's surface depends on its
    - **A** mass only.
    - **B** speed only.
    - **C** mass and height.
    - **D** speed and height.

19. In the late 18th century Benjamin Thompson noticed that boring out cannons with a drill generated large amounts of heat. Select the correct response.
    - **A** This observation could not be explained by the caloric theory of heat.
    - **B** Thompson's experiments showed that the caloric theory was correct.
    - **C** During the drilling of the cannons, elastic potential energy is converted to heat energy.
    - **D** Chemical energy is converted to heat energy in the boring of the cannons.

20. A student placed four rods of different materials in a fire and compared the rate at which heat was conducted along the rods. Which of the following rods has the slowest rate of heat conduction?
    - **A** copper
    - **B** gold
    - **C** silver
    - **D** pyrex glass

21. Select the true statement about electric cars at the present time.
    - **A** The lithium ion batteries are cheap to manufacture.
    - **B** Very few recharging sites have been constructed.

C Electric cars can travel 1000 km before recharging.

D Recharging back to 95% of full charge only takes 5 minutes.

22. Identify which one of the following sources of second-hand data on scientific theories and experiments is least reliable.
   A internet blogs
   B peer-reviewed science papers
   C university texts
   D online articles from *New Scientist*

23. A student wrote the following in his report on 12 different sodium compounds: 'Sodium compounds are white and crystalline'. What type of statement is this?
   A inference
   B generalisation
   C theory
   D prediction

24. A student investigated the rate of temperature increase when 50-g samples of steel that were painted different colours were placed in sunlight. The dependent variable in this investigation is the
   A mass of the steel samples.
   B time of day.
   C colour of the sample.
   D rise in temperature.

25. Various text types are used in science reports. One of these is a procedural recount. Which statement correctly identifies a procedural recount?
   A used to record the procedure of the experiment that has been previously performed
   B used to identify issues and presents points for and against
   C used to describe how an experiment is to be performed
   D used to present information on a particular topic

Part B: Restricted-response questions

(15 marks—1 mark for each question)

Complete the following restricted-response questions using the appropriate word.

26. A line-of-best fit on a graph is a smooth, straight or .... line that passes through as many points as possible.

27. When you label a science diagram, connect the labels to the diagram using ruled .... lines.

28. The .... of the measurements in an experiment depends on the equipment selected and how carefully the experiment is conducted.

29. The measurements are .... in a scientific investigation if repeated experiments (five or more) lead to consistent results.

30. A scientific report table commonly uses ...., scaffolds and diagrams to organise information.

31. .... reproduction is usually found in plants where new individuals are formed without producing seeds or spores.

32. The .... is the narrow passage between the vagina and the uterus in a human female.

33. .... are disease-producing microorganisms (microbes) such as bacteria, viruses, protozoa and some fungi.

34. Acids are secreted onto our skin and acids are produced in the stomach which kill many of the .... or slow down their reproduction.

35. In multicellular .... cells are specialised for different functions, carrying out extremely varied activities.

36. Many experiments have shown that solids and .... are incompressible whereas gases like oxygen can be compressed.

37. In general the atomic weight of each element .... as you move along each row of the periodic table.

38. The .... of a mineral is the way it splits into pieces with flat surfaces.

39. Traditional aboriginal hunters and gatherers used .... to encourage the regrowth of eucalypt trees and edible plant food such as bracken, from which the roots, young leaves and shoots could be eaten.

40. Adding drops of dilute acid to limestone produces a .... reaction.

Part C: Knowledge, skill and processing data questions

(60 marks)

41. Figure T2.1 shows different types of cells. Compare their similarities and differences and design a simple branching key to classify them. *(5 marks)*

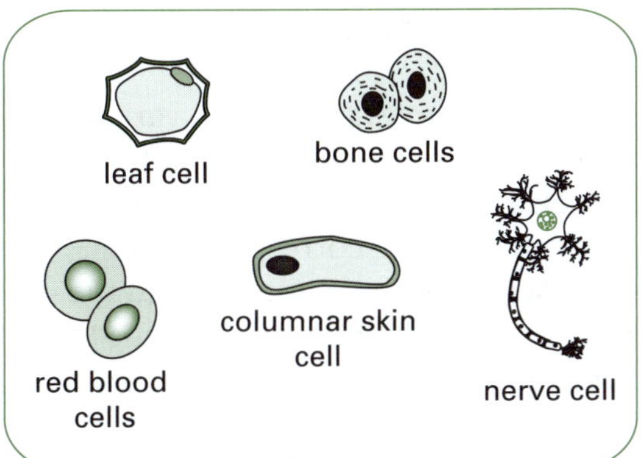

Figure T2.1 Cells

42. The iris is a common garden plant with long, thick leaves that emerge from an underground stem called a rhizome. In order to determine the effect of light and dark on the sucrose sugar content of iris leaves, a botanist conducted the following experiment.

Method

1. A small sample of leaf tissue was removed and weighed. The sucrose sugar was extracted and weighed. The sugar concentration in the leaf was calculated.
2. The plant was covered with a black cloth after taking the sample.
3. Every 6 hours a small sample of leaf tissue was removed and analysed for sucrose content. The cloth was replaced each time.

Results

The results are shown in Table T2.1.

Table T2.1 Sucrose concentration

Time of sampling (hours)	Sucrose concentration (µg/g leaf)
0	130
6	120
12	90
18	50
24	40
30	30
36	20

a) Identify the independent variable. *(1 mark)*

b) Identify the dependent variable. *(1 mark)*

c) Plot the data on a grid and draw the line of best fit. *(5 marks)*

d) How does the sucrose concentration change with time in the dark? *(1 mark)*

e) Suggest a reason why the sucrose concentrations change in this way as the plant continues to remain in the dark. *(2 marks)*

43. Helena has made a small key to help her identify some elements. Use the key on the next page to work out the name of the element on the card in Figure T2.2. Show how you used the key in determining your answer. *(2 marks)*

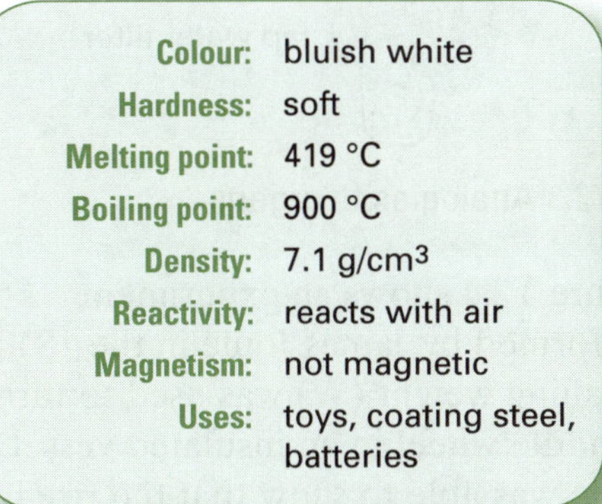

Figure T2.2 Description of an element

1A Density less than 7 g/cm^3 .... go to 2
1B Density more than 7 g/cm^3 .... go to 3
2A Melting point less than 100 °C .... sodium
2B Melting point more than 100 °C .... magnesium
3A Magnetic .... go to 4
3B Non-magnetic .... go to 5
4A Used in radioactive form to treat cancer .... cobalt
4B Used extensively in industry ..... iron
5A Melting point less than 500 °C .... zinc
5B Melting point more than 500 °C .... tungsten

44. Figure T2.3 shows some objects found and used around the home. Draw up a table that lists each object and the organs of the human body that complete the same job. *(3 marks)*

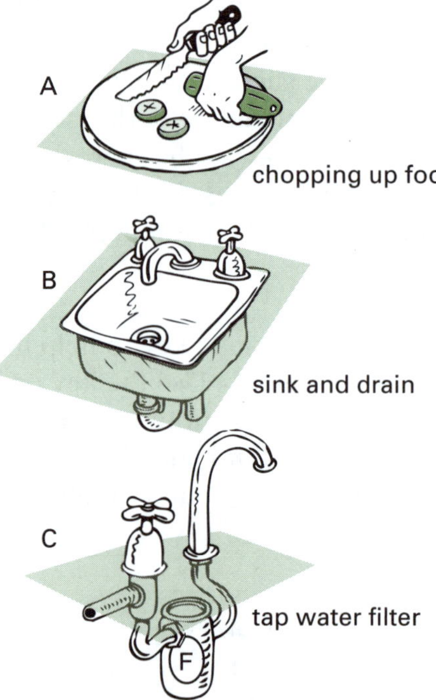

Figure T2.3 Analogies to organs

45. Figure T2.4 shows an experiment performed by James Joule in the 1840s. A falling weight (W) was used to turn a paddle wheel in an insulated vessel. Joule was able to show that the rise in temperature of the stirred water was related to the number of times the weight (W) fell. Analyse this experiment and describe all the energy transformations that occur. *(4 marks)*

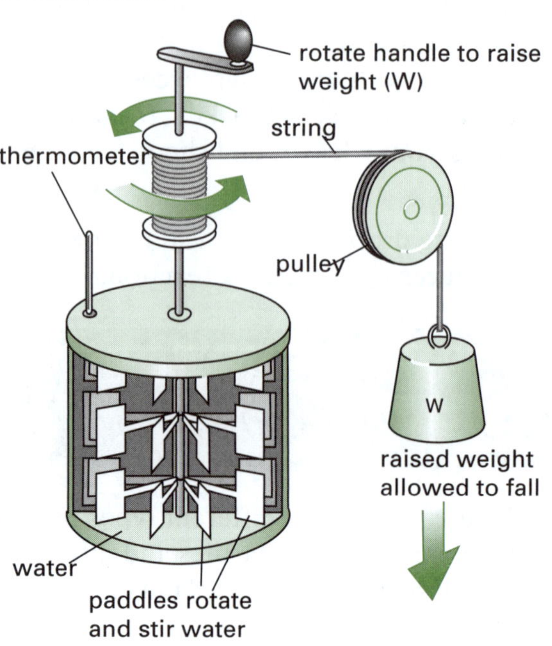

Figure T2.4 Joule's experiment

46. Table T2.2 shows the typical variation in blood pressure in the heart and blood vessels for a healthy adult. The high pressure reading corresponds to the contraction of the ventricles and the low pressure reading corresponds to the relaxation of the ventricles.

Table T2.2 Variation in blood pressure for a healthy adult

Chamber or vessel	High blood pressure reading (mmHg)	Low blood pressure reading (mmHg)
right ventricle	28	0
left ventricle	120	0
pulmonary vein	2	–2
capillaries (body tissues)	35	8
pulmonary artery	28	10
veins	8	–2
main body arteries	120	80

a) Examine the data. The information is not in a logical order. Arrange the data in the order in which blood flows through the circulatory system, starting with the right ventricle. *(3 marks)*

b) In which blood vessels is the blood pressure consistently high? *(1 mark)*

c) In which blood vessels is the blood pressure consistently low? *(1 mark)*

d) The right ventricle has a thinner muscular wall than the left ventricle. Does this have any effect on the blood pressure in the vessels that lead out of these chambers? Explain your answer. *(1 mark)*

e) Use the data to explain why the compression of veins by nearby body muscles is necessary to move blood through these vessels. *(2 marks)*

f) When you go to the doctor to have your blood pressure taken, the doctor is happy if your readings are in the range 120–80. In which vessels is the doctor measuring your blood pressure? *(1 mark)*

g) Predict what would happen to the capillary walls if the blood pressure was as high there as it is in the arteries. *(1 mark)*

47. Movement of air into and out of your lungs is due to pressure decreases and increases in the chest cavity. Figure T2.5 shows a simple model for inhalation. The tap is opened to allow air to enter and then is closed again.

The pressure and volume of the air initially in the lungs are as follows.

P_1 = initial pressure

V_1 = initial volume

The pressure and volume of the air finally in the lungs are as follows.

P_2 = final pressure

V_2 = final volume

Pressure is measured in hectopascals (hPa) and volume is measured in litres (L).

a) i) Calculate the final air pressure in the lung (P_2) by substituting the following values into the formula: *(1 mark)*

$$P_2 = P_1 \times V_1 / V_2$$

Values: P_1 = 1000 hPa; V_1 = 10 L; V_2 = 15 L

ii) How does the final pressure compare to the initial pressure in the above example? *(1 mark)*

iii) Make a copy of the final-state drawing (bottom of Figure T2.5) and draw in the air molecules using the initial state diagram as a guide. *(2 marks)*

iv) If the tap is now opened, what will happen? Use your answer for question ii) to explain. *(2 marks)*

b) i) Calculate the final air pressure in the lung (P_2) using the formula used in question a). *(1 mark)*

Values: P_1 = 700 hPa; V_1 = 15 L; V_2 = 12 L

ii) How does the final pressure compare to the initial pressure? *(1 mark)*

iii) If the tap is now opened, what will happen? Explain. *(1 mark)*

c) Which of questions a) and b) corresponds to the process of exhalation? *(1 mark)*

inhalation—chest cavity increases in volume

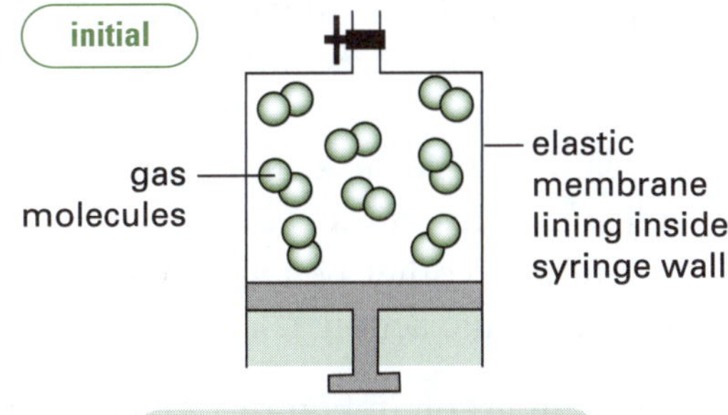

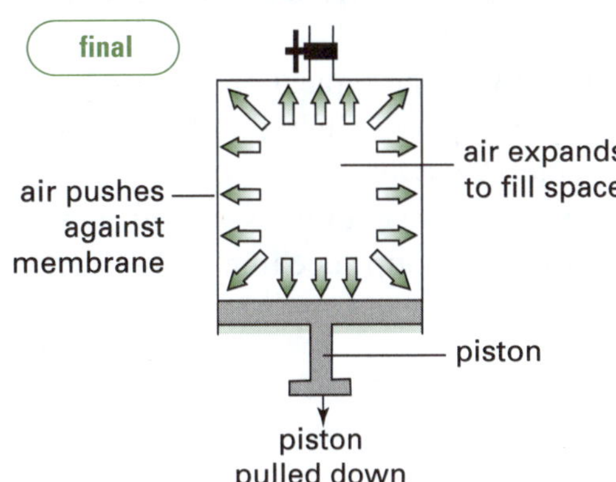

Figure T2.5 Chest cavity pressure

48. Read the following text about volcanoes and igneous rocks and answer the questions.

Uses of volcanoes

Although we often hear of the disasters caused by volcanoes, we rarely hear of the good things that have come from them. In some parts of the world, heat and power are generated by plants run by volcanic steam. This occurs, for instance, in Bolivia, Chile, Iceland, New Zealand and in parts of Italy.

Hot springs are used for medicinal purposes, laundry and bathing. Such springs form when subterranean water comes into contact with hot rocks and magma in volcanic regions. Some water is converted to steam and pushes the hot water to the surface to form the spring. The hot springs in New Zealand are famous with tourists as well as with the local Maori peoples.

Lava from volcanoes has many uses including road building. Blue metal or basalt is commonly added to cement to produce concrete. Even the lava beds themselves are useful because they are porous and allow fresh water to gather underground in huge cavities. These are important sources of drinking water in Hawaii.

a) Explain how power can be generated in volcanic areas of the world. *(2 marks)*

b) Identify two other benefits that derive from volcanic action. *(2 marks)*

49. The column graph in Figure T2.6 compares energy usage in different social groups over time.

a) What was the typical energy usage per day of a person living in the industrial society? *(1 mark)*

b) Explain why energy usage has increased as social groups became more advanced. *(4 marks)*

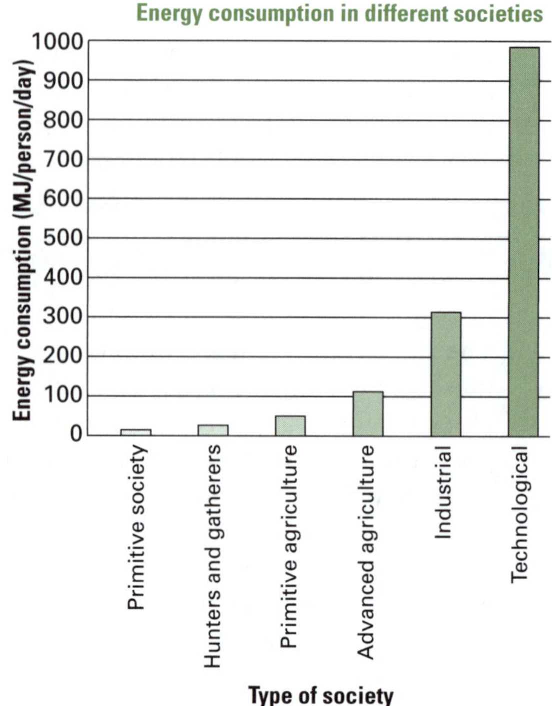

Figure T2.6 Energy consumption in different societies

50. Solve the following puzzle, called the alphabet code. The answer is a famous scientist. *(4 marks)*

Alphabet code

Examine each of the following questions and answers. The answers are written using an alphabet code. Use the answers to discover the solution to the alphabet code which has been used. Find the letters corresponding to the code letters K and H. Use this code to discover the name of a famous scientist associated with the study of energy.

Question	Answer
1. This form of energy makes a lot of noise.	TQXOI
2 This form of energy is waiting to happen.	CQSJOSZGP
3. This form of energy is deep inside the atom.	OXFPJGY

Famous scientist

K	G	H	J	T		K	Q	X	P	J

First name Surname

51. The stages of mitosis shown in Figure T2.7 are jumbled. Use the code letters to arrange the stages in the correct sequence. *(2 marks)*

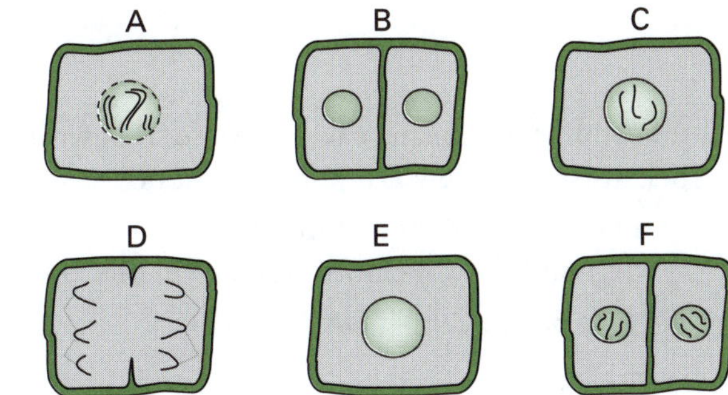

Figure T2.7 Mitosis

Go to pp. 214–217 to check your answers.

Chapter 1 answers

Experiment 1

Analysis: The onion epidermis shows long, narrow cells with a nucleus and cytoplasm and a cell wall. The iodine dye has coloured the nucleus and cell wall, making it easier to see.

Conclusion: The staining of the onion epidermis has allowed microscopic examination of the cells with greater clarity.

Experiment 2

Analysis: The pearly white colonies are bacterial and the furry irregular colonies are fungal.

Conclusion: Using the nutrient agar method, fungal and bacterial cells have been cultured.

Experiment 3

Conclusion: Seeds need water to germinate.

Test yourself 1

Part A: Knowledge

1. C ✓ Animal cells do not have a cell wall. A is wrong because both cells have a nucleus; B and D are wrong as both have a cell membrane and mitochondria.
2. A ✓ Cellulose is the main structural component of cell walls in plants. B is wrong as animal cells do not have cell walls; C is wrong as the nucleus is the control centre; D is wrong as plant cells may have large vacuoles.
3. C ✓ Some cells can be very large (e.g. a chicken egg). A, B and D are true statements.
4. B ✓ The genetic code is found inside the nucleus in structures called chromosomes. A is wrong as the cytoplasm is the fluid around the nucleus; C is wrong as mitochondria produce energy; D is wrong as the vacuole stores water and minerals.
5. C ✓ No flagella are present in these examples, so D is wrong; A is wrong as only some unicellular organisms cause disease; B is wrong because only a few are photosynthetic.
6. a) microscope ✓; b) membrane ✓;
   c) organelles ✓; d) chloroplasts ✓; e) blood ✓
7. A/H ✓; B/J ✓; C/F ✓; D/G ✓; E/I ✓

Part B: Skills

8. Total magnification = 15 × 40 = 600× ✓
9. e, b, g, ✓ a, d, ✓ c, f ✓
10. See Figure A.1.

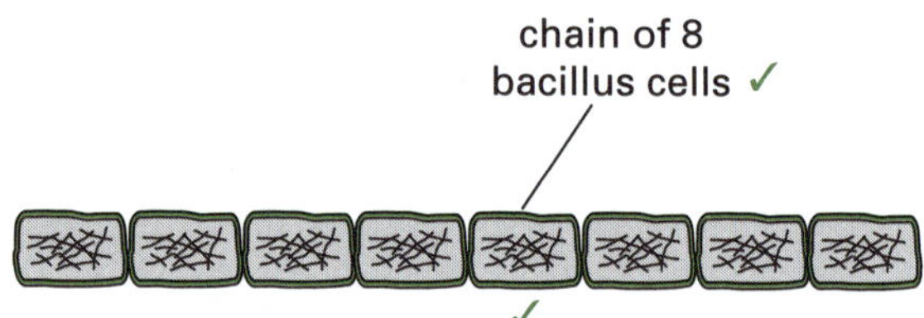

Figure A.1 Cells

11. C, F, G, ✓ D, A, E, ✓ I, H, B ✓
12. a) About four cells fit across the diameter. Thus, cell diameter = 1000/4 = 250 µm. ✓
    b) plant cells ✓ because they have a cell wall and a large vacuole ✓
13. a) plant cells ✓
    b) Its cells are dead and most contain only bubbles of air. ✓
    c) Each of its cells are closed, like small balls, and water can't get in. ✓
    d) The air inside the cells is compressible, and the cork's cells are not made of very rigid material. ✓
    e) Biologists have used the microscope to study the cellular structures that control the activities of the cell. This allows the function of each structure to then be investigated. ✓
14. Spontaneous generation theory says that somehow complex organisms such as worms and maggots, which are themselves made up of cells, came into existence by the action of non-cell substances. ✓ The cell theory states that all cells come from pre-existing cells.

Test yourself 2

Part A: Knowledge

1. A ✓ Enzymes in saliva start attacking food the moment it enters your mouth. B is wrong as the oesophagus is a tube that links the mouth and stomach; C is wrong as this is an intermediary site for digestion; D is wrong as the intestines are where digestion is completed.
2. B ✓ This method is used by more complex multicellular organisms. A, C and D are true statements.
3. C ✓ The marrow is the main source of blood cells to replace those that are damaged and removed. A is wrong because the heart pumps the blood; B is wrong because the plasma is the fluid of the blood; D is wrong because haemoglobin is the molecule that absorbs oxygen.
4. D ✓ Dentin is a bone-like material that lies under the enamel and makes up most of the tooth. A, B and C are all types of teeth in our mouths.
5. A ✓ B, C and D have the sequence in the incorrect order.
6. a) skull ✓; b) excretion ✓; c) aorta ✓;
   d) pollination ✓; e) digestive ✓
7. A/I ✓; B/F ✓; C/J ✓; D/G ✓; E/H ✓

Part B: Skills

8. a) asexual ✓ because there is no union of male and female parts ✓
   b) In sexual reproduction male and female sex cells join together in fertilisation. ✓ In asexual reproduction there is no fertilisation. ✓
   c) R ✓, T ✓, P ✓, S ✓, Q ✓, V ✓, U ✓
9. A = crown ✓; B = enamel ✓; C = dentin ✓; D = pulp ✓; E = root ✓; F = cementum ✓
10. a) 3.5 L ✓
    b) 17 years ✓
    c) 16 years and 45 years ✓
    d) 5.5 L ✓, occurring at about 25 years of age ✓
    e) Both males and females are growing ✓, and along with them so does their lung capacity.
11. a) day 16 ✓
    b) 6 days ✓
    c) No, the egg has not been released from the ovary. ✓
    d) No, the egg will have already died. ✓
    e) 32 days ✓
12. a) See Figure A.2.

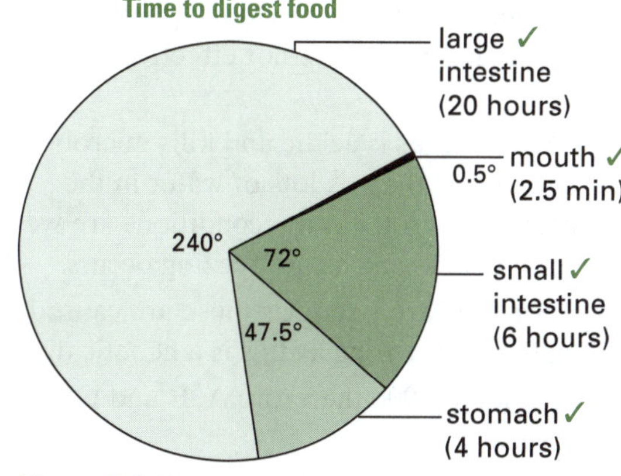

Figure A.2 Pie chart

    b) i) C ✓; ii) B ✓; iii) C ✓; iv) A ✓
13. a) i) There is no change in air pressure at A as the tube is connected to the outside air. ✓ (Note: the pressure at D decreases as the volume is now greater inside the bell jar.)
       ii) entering ✓ (as the space is now larger and the pressure lower than outside and so air enters)
    b) The balloon decreases in size ✓ as the pressure inside the jar (at D) increases. The pressure is now larger than the pressure at A and so the balloons deflate as air moves out of the T-piece. ✓
    c) rubber membrane = diaphragm ✓; balloons = lungs ✓; A = windpipe ✓; D = thoracic cavity (or chest cavity) ✓
    d) The model does not show how the ribs work in breathing. ✓ The balloons do not represent the lungs accurately. ✓ Lungs are not just spaces filled with gas. ✓
14. CO_2 levels are greater in exhaled air than inhaled air. ✓
    B is the inhalation side: the student breathes in and air pressure is reduced in B; air from the outside is being drawn in. As there is only a little carbon dioxide in inhaled air, the limewater flask B will be slightly white. ✓
    A is the exhalation side: the student breathes out and air bubbles through the limewater; this air is being tested for CO_2 content from the lungs. Flask A shows more carbon dioxide is present in exhaled air than inhaled air in B. The limewater in flask A will be milkier than in Flask B. ✓

Chapter test

Part A: Multiple-choice questions

1. **D** ✓ A is wrong as no red rash covers the body; B is wrong as HIV is caused by a virus; C is wrong as antibiotics are not effective for viral disease.
2. **B** ✓ The vinegar is acidic and kills microbes. A is wrong as there is lots of water in the vinegar; C is wrong as the conditions are wet not dry; D is wrong as no freezing occurs.
3. **B** ✓ A and C are wrong as these are caused by microbes; D is wrong as this is a genetic disease.
4. **D** ✓ $15 \times 60 = 900$ therefore A, B and C are incorrect.
5. **B** ✓ A is wrong as digestion occurs in the mouth, stomach and small intestine; C is wrong as waste is eliminated from the large intestine; D is wrong as the villi do not secrete acids and acids do not break down fats.
6. **A** ✓ B is wrong as only one ovary produces an egg each cycle; C is wrong as sperm are formed in the testes; D is wrong as the zygote is the fertilised egg.
7. **B** ✓ Fertilisation happens in the fallopian tube. A, C and D are correct statements about reproduction.
8. **B** ✓ A is incorrect as this is the function of the nucleus; C is wrong as this is the function of cell walls; D is wrong as this is the role of the vacuole.
9. **A** ✓ B is wrong as the objective lens is used for magnification; C is wrong as the mirror reflects light onto the slide; D is wrong as the eyepiece lens magnifies and focuses.
10. **C** ✓ The cells replicate and then divide into two identical cells. Therefore A, B and D are incorrect.

Part B: Short-answer questions

11. **a)** Each square of the grid is 1 mm × 1 mm. The field of view shows ~3 grid squares. Therefore, the diameter of the field of view is ~3 mm. ✓
    b) The image of the 'e' is inverted vertically and laterally. ✓
12. Approximately three cells in a line fit across a 1-mm square of the grid. Thus each cell has a diameter of $^1/_3$ mm (0.33 mm). ✓
13. **a)** mitochondrion ✓
    b) vacuole ✓
    c) cell membrane ✓
    d) nucleus ✓
    e) chloroplast ✓
14. **a)** scientist ✓; **b)** first ✓; **c)** nucleus ✓;
    d) functioning ✓; **e)** blood ✓; **f)** function ✓;
    g) replaced ✓
15. The correct sequence is B, A, D, C. ✓ The nucleus has to make a copy of itself before fission occurs. The two new daughter cells contain the same information inside their nuclei as the parent cell.
16. **a)** Cells V, X and Y will be in the most actively growing zone (1). ✓ The cells are undergoing mitosis as shown by the appearance of chromosomes and the parent cell dividing.
    b) Cell Z could be found in zone 2 where the cells are now elongating. ✓
    c) Cell W could be found in zone 3 as this cell is fully grown and mature. ✓
17. Antibiotic A is a very effective antibiotic for all three bacteria as colonies do not grow in the region around disc A. ✓
    Antibiotic B is effective against bacteria X but not effective against bacteria Y and Z. ✓
    Antibiotic C is effective against bacteria Y and Z but not effective against bacteria X. ✓
18. **a)** **i)** 8 cells ✓ ($2 \times 2 \times 2 = 2^3$)
    ii) 64 cells ✓ ($2 \times 2 \times 2 \times 2 \times 2 \times 2 = 2^6$)
    iii) $2^9 = 512$ cells ✓
    b) Bacterial cells do die and are food for other organisms. They are kept in a state of balance. ✓
19. Skeletal muscle cells move the bones to which they are attached. ✓ Their function is to produce force and cause motion. This is accomplished when muscle cells contract. Muscle cells contain contractile filaments that move past each other and change the size of the cell.
20. Root hairs are thin tubular outgrowths from single cells. They can be several millimetres long and are important in absorbing water and minerals. While water and minerals can be absorbed through other parts of the root, the root hairs are important as they offer a very large surface area. ✓
21. The cells of a multicellular organism become specialised and they cooperate. Hence the

cells combine to form a coordinated single organism with more capabilities than any of its component parts. ✓

22. A = skeletal ✓; B = digestive ✓; C = circulatory ✓; D = respiratory ✓

23. a) respiratory system ✓ (Her lungs take in air as she needs a greater oxygen supply.)
    b) nervous system ✓ (This is a reflex action that protects her from further harm.)
    c) nervous system ✓ (Eyes are part of the nervous system; the muscular/skeletal system is also involved in firing the arrow.)
    d) digestive system ✓ (Vomiting involves food moving back from the stomach and oesophagus to the mouth.)
    e) muscular system ✓ (A build-up of lactic acid in the muscles makes them sore and the muscles may cramp.)
    f) skeletal system ✓ (The collar bone is part of the skeleton.)
    g) circulatory system ✓ (Blood is transported by the circulatory system.)
    h) nervous/muscular/skeletal system ✓ (The ear collects information and relays the message through the nervous system; the muscles and skeleton are involved in the teacher falling forwards.)

24. Teeth are among the most distinctive (and long-lasting) features of mammal species. The shape of an animal's teeth is related to its diet. The function of teeth is to mechanically break down food into small pieces.
    - Incisors: teeth at the front of the mouth adapted for plucking, cutting, tearing and holding. ✓
    - Canines: lying on each side of the incisors; being pointed, they are used for cutting and tearing food, especially meat. ✓
    - Premolars: have a series of elevations, or cusps, that are used for breaking up particles of food. ✓
    - Molars: also have a series of elevations, used exclusively for crushing and grinding. ✓

25. The teeth of many vertebrates have been adapted for special uses. The following are some examples. *(1 mark each for any two examples)*
    - Rodents, hares and rabbits have curved incisors that are set deep in the jaws and grow continually throughout life.
    - The tusks of elephants are enlarged upper incisors.
    - In pigs the lower incisors lie close together and project forwards to form digging instruments.
    - Baboons have enlarged canines for defence and display.

26. Humans obtain their nutrients from a variety of sources. Cows obtain theirs from grasses and other vegetation. Cells of plant materials are mainly composed of cellulose and cows by themselves don't have any enzymes to digest this substance. So cows need a way to break it down. ✓ The rumen is the largest compartment of the stomach, containing billions of bacteria, protozoans, moulds and yeasts. The microbes present in the reticulum and rumen help in decomposing cellulose and other carbohydrates to produce various nutrients that are absorbed by thousands of 'finger-like' projections lining the bottom and sides of the rumen wall. ✓

27. The heart of mammals (and birds) is completely divided into two sides: left and right. Blood cannot pass directly from one side to the other. It has to leave the heart first. For example, blood from the right side is first pumped to the lungs, to be oxygenated, before returning to the left side of the heart. This oxygenated blood in the left side is pumped around the body before it returns to the right side of the heart. Blood returning to the right side is now deoxygenated, as oxygen has been removed as it circulated through the body. ✓ So each side of the heart works as a separate pump, one side pumping deoxygenated blood and the other pumping oxygenated blood. Because the two pumps are physically connected to form one organ, the heart is called a double pump. ✓

28. A = capillary ✓ (small diameter with very thin walls); B = vein ✓ (walls are thinner than arteries); C = artery ✓ (thick muscular walls present)

29. Circulating blood passes through the kidneys where used and unwanted water, minerals and a nitrogen-rich molecule (called urea) are removed. The kidneys filter wastes from the blood, producing a liquid called urine. ✓ If kidneys are diseased and not working properly, the build-up of wastes in a person's system will eventually kill them. Some kidney

diseases can be treated with medication. But severe kidney diseases require more intense treatment. One treatment is called dialysis. This is where the patient's blood is pumped through a dialysis machine that filters the waste from the blood and returns clean blood to the body. A dialysis patient has to spend nearly 60 hours each week attached to the machine. A more radical treatment for kidney disease is a kidney transplant. (Healthy people can live comfortably with only one kidney.) ✓

30. a) The difference comes from a number of small bones that fuse together during growth, such as the sacrum and coccyx of the vertebral column. ✓
    b) Without bones humans would be just like a beanbag of skin and organs on the floor. You couldn't stand up, let alone walk. Other bones protect the delicate, and sometimes soft, insides of your body. Your skull, a series of fused bones, acts like a hard protective helmet for your brain. ✓

31. It gives the impression there is one bone running along the length of your back. If that was so, you wouldn't be able to bend or twist. The spinal column consists of many vertebral bones, separated by flexible discs, allowing some movement to occur. ✓

32. a) Alveoli are where gas exchange occurs by diffusion. The diffusing capacity of the lung is limited by the size of the alveolar surface area and the thickness of the blood–gas exchange barrier. In order to be efficient, a sufficiently large surface area for gas exchange must be crammed into the limited space available in the chest. Hence, the lung's internal structure shows a gas exchange surface that is divided into a large number of small subunits connected to a branched conducting airway system. ✓
    b) The alveoli are the ends of the respiratory tree and act as the primary gas exchange units of the lung. Here the gas–blood barrier between the alveolar space and the pulmonary blood capillaries is extremely thin, allowing for rapid gas exchange. In order to reach the blood, oxygen must diffuse through a thin barrier to the blood; CO_2 removal follows the reverse course to reach the alveoli. ✓

33. a) Oxygen in the air slowly dissolves in the moisture on the mudskipper's skin, mouth and throat. This can then be absorbed into its blood stream. This is only possible when the mudskipper is wet, limiting mudskippers to humid habitats and provided that they keep moist. This mode of breathing is similar to that used by amphibians. ✓
    b) The large gill chambers close tightly when the fish is above water, keeping the gills moist and allowing them to function. Like a scuba diver's cylinders, they supply oxygen for respiration while on land. ✓

34. This is asexual reproduction and the new rhizome and shoots are identical to the older rhizome. This is an example of vegetative reproduction. ✓

35. a) Most birds need to remain light for flying. This is one way to reduce the weight they need to carry. ✓
    b) The bird lays eggs and the young develop outside the mother's body (external development). A uterus is useful for those animals that exhibit internal development, as this is where the egg implants as the embryo and foetus develops. ✓

36. a) Buffalo is hardy and resilient and suitable for growing in areas where full sun may not be available, such as along the shaded side of a house or under trees with thick canopies. Normal grass often dies off from lack of sun, leaving dirt patches. ✓
    b) The grass would require less watering since there would be less transpiration, being in a shady, cooler spot. ✓

37. a) He was testing to see whether the movement of the molecules into a cell depended on their size, not on how many there were (i.e. not on their number). He needs to keep this number constant. ✓
    b) A ✓
    c) The smaller the molecule, the less time it takes for a given concentration to be reached inside the cell. This indicates that smaller molecules can move into cells faster. ✓
    d) He should be cautious about coming to a general conclusion from only three chemicals. ✓ There may be factors other than size which determine the rate at which

chemicals are taken up by cells. More tests need to be done. ✓

e) No. These results only apply to the cells and chemicals he used. More experiments with different cells and a greater variety of chemicals are needed before he can determine whether there is a result which could be generalised. ✓

f) If the biologist's conclusion held for all chemicals, then the time reported for ethyl urethane should be somewhere between 6 and 40 hours. ✓ As it was not, the conclusion is not supported. ✓

g) There seem to be other factors that need to be considered, not merely the size of the molecule. ✓

38. A = nerve cell ✓; B = muscle cells ✓; C = skin cells ✓; D = red blood cell ✓

39. Transplanting brains is currently only science fiction. There are moral problems involved in either harvesting a brain-dead body, especially one deliberately created using human cloning, or otherwise obtaining a body (such as a criminal due to be executed for a crime), or an individual who is not dead but is soon to die of a brain-based illness. ✓ There is also the confusion and distress about personal identity that it would create. If, for example, a woman's living complete brain were successfully transplanted into the skull of a male's body (minus the male's brain-dead brain) it would be the woman who survived, though in an uncomfortable new guise. In her mirror, and to everyone else, she would appear to be the male donor. ✓ Then there is the issue of crimes being committed to obtain otherwise healthy bodies so that the brains of dying people could be transplanted into them. ✓

40. See Table A.1.

Table A.1 Differences between asexual and sexual reproduction

Asexual reproduction	Sexual reproduction
no sex cells produced ✓	sex cells produced ✓
offspring are identical ✓	offspring show differences to parents ✓
only one parent ✓	two parents required ✓

Chapter 2 answers

Experiment 1

Analysis: The dyed alcohol spreads or diffuses throughout the water. This has occurred as the particles of the liquid are in constant motion and they collide with the dye molecules and move them around until there is uniform mixing.

Conclusion: Dyes diffuse in water and continue to spread until there is a uniform distribution of the dye molecules.

Experiment 2

Analysis/calculation: Volume of glass = 45.5 – 40.0
= 5.5 mL

$$D = \frac{m}{V} = \frac{13.2}{5.5} = 2.4 \text{ g/mL}$$

This can be explained in several ways using the particle theory of matter. The particles that make up the glass have a greater mass or there are more particles in a fixed volume compared with water. Often both factors are important.

Conclusion: This experiment shows that glass is much denser than water. The density of the glass was 2.4 g/mL.

Experiment 3

Analysis: This experiment shows that the temperature of the solid wax rises at a steady rate for the first 10 minutes and then slows down. This corresponds to the time when the wax is starting to melt. Between 16 and 22 minutes the wax continues to melt and the temperature remains constant. Once all the wax is melted, the temperature begins to rise again. From these results we can see that the melting point of this wax is 44 °C. While the wax is melting, the added heat is used to separate the particles of the wax rather than heat them up.

Conclusion: The melting point of the wax was found to be 44 °C.

Test yourself 1

Part A: Knowledge

1. B ✓ In solids the particles vibrate when heated. In liquids and gases they move about with greater speed. A is incorrect as particles always move in some way; C is incorrect as the particles in a solid are close together; not all

matter is made up of pairs of particles so D is incorrect.

2. C ✓ The particles of a liquid touch one another but can slip and slide around each other. A is not correct as gases take the shape of the container; B is incorrect as the particles are not still; D is not correct as liquid particles are not far apart.
3. C ✓ $m = DV = 4 \times 40 = 160$ g; therefore, A, B and D are incorrect.
4. D ✓ Sublimation of a solid like iodine produces a vapour but no liquid. A, B and C do not refer to converting a solid into a gas.
5. D ✓ Zinc is a metal and nitrogen is a non-metal. In A both elements are metals; in B, uranium is a metal and silicon is a semi-metal; in C both elements are non-metals.
6. **a)** semi-metal ✓; **b)** ductile ✓; **c)** expand ✓; **d)** oxygen ✓; **e)** molecules ✓
7. A/H ✓; B/F ✓; C/J ✓; D/G ✓; E/I ✓

Part B: Skills

8. **a)** $V = 19 - 15 = 4$ mL ✓

   b) $D = \frac{m}{V} = \frac{32}{4} = 8$ g/mL ✓
9. Liquids take the shape of the base and sides of the container they occupy. ✓
10. The brown gas would gradually spread out and fill both gas jars. The oxygen will also spread out and fill both jars. The final appearance will be a pale brown colour in each jar. ✓ This is an example of diffusion. ✓
11. **a)** the thickness of one page $= \frac{5}{500}$ $= 0.01$ cm $= 0.1$ mm ✓

    b) size of 1 atom $= \frac{0.1}{1\,000\,000}$ $= 0.000\,000\,1$ mm ✓
12. X = gas ✓(as it can be easily compressed); Y = liquid ✓(takes the shape of the lower regions of the container.)
13. **a)** Br/non-metal ✓; **b)** B/semi-metal ✓; **c)** Co/metal ✓; **d)** K/metal ✓; **e)** Ga/metal ✓
14. **a)** non-metal ✓ (this is chlorine); **b)** metal ✓ (this is sodium); **c)** non-metal ✓ (this is carbon graphite)

Test yourself 2

Part A: Knowledge

1. A ✓ There is no change in the chemical composition on melting and the process is readily reversed. This is not true of B, C and D.
2. B ✓ The metal is named first and the non-metal ends in the letters 'ide'. A is not correct as the non-metal is named first; C is incorrect as water does not contain these elements; D is incorrect as there are not three atoms of chlorine in this compound.
3. D ✓ A, B and C are not heat-resistant polymers.
4. A ✓ B, C and D are not correct as nylon has good abrasion resistance, does turn yellow in sunlight and is not attacked by insects.
5. A ✓ The code 2 (B) is HDPE, 4 (C) is LDPE and 6 (D) is PS.
6. **a)** crease ✓; **b)** monomers ✓; **c)** physical ✓; **d)** chemical ✓; **e)** oxide ✓
7. A/I ✓; B/J ✓; C/F ✓; D/G ✓; E/H ✓

Part B: Skills

8. $H_2O = 2 \times 1 + 16 = 18$ units ✓
9. See Table A.2.

Table A.2 Chemical formuale

Formula	Name	Formula	Name
H_2SO_4 ✓	hydrogen sulfate	$CuSO_4$ ✓	copper sulfate
HCl ✓	hydrogen chloride	UO_2 ✓	uranium oxide
AlN ✓	aluminium nitride	CuO ✓	copper oxide

10. bubbles of gas produced ✓; a rise in temperature ✓
11. In this experiment: no bubbles were produced ✓; the taste of vinegar was still present ✓; the taste of salt was still present ✓. Therefore, there did not seem to be any change on mixing the two chemicals together.
12. The repeating unit is CHCl–CHCl ✓ (see Figure A.3).

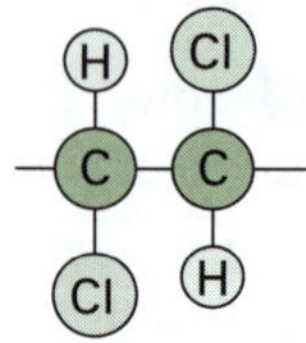

Figure A.3 Monomer

13. a) A, 8; B, 1; C, 8 ✓
b) structure A ✓
c) structure C ✓

14. See Figure A.4.

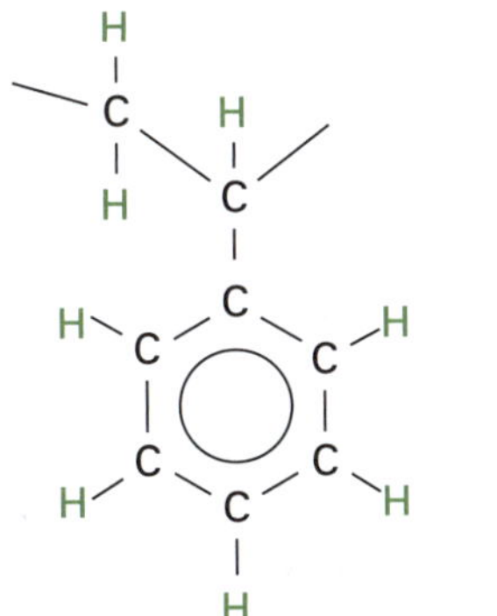

Figure A.4 Monomer

Chapter test

Part A: Multiple-choice questions

1. **B** ✓ Refer to the periodic table. A is incorrect as iron is not Ir and tin is not Ti; C and D are not correct as calcium is Ca and iron is Fe. No element has the symbol T.
2. **C** ✓ Metals are good conductors. A, B and D are characteristics of metals.
3. **B** ✓ A is incorrect as compounds are not elements; C is not correct as elements are pure substances; D is incorrect as alloys are mixtures of metals.
4. **C** ✓ Decomposition means new chemical substances are formed. A, B and D are all physical changes.
5. **B** ✓ Atoms cannot be divided into smaller particles according to the theory of Democritus. Thus A is incorrect. C and D are not relevant to the theory of atoms.
6. **A** ✓ The remaining responses are non-metals (B, C) or a semi-metal (D).
7. **D** ✓ A, B and C are all metals.
8. **A** ✓ B and D are metals and C is a non-metal.
9. **B** ✓ Oxygen is a molecule and water is a compound. A is incorrect because it contains two ionic compounds; C is incorrect as there are two molecules, both of which are elements; D is incorrect as there are two ionic compounds.
10. **C** ✓ Silver is a metal with the highest electrical conductivity. A and B are incorrect as they are non-metals; D is incorrect as germanium is a semi-metal.

Part A: Short-answer questions

11. a) temperature ✓; **b)** groups ✓; **c)** methane ✓; **d)** combustion ✓; **e)** mass ✓

12. A/G ✓; B/J ✓; C/F ✓; D/H ✓; E/I ✓

13. *(½ mark for each correct answer)*
U, Q, ✓ S, P, ✓ T, W, ✓ R, V ✓

14. Particles in air move in random motion, bumping into each other and other objects. They eventually spread out from one place and move to another. ✓ Particles from the cooking meat diffuse through the air to reach the front of the house. ✓

15. a) 5% ✓ (100 – 43.1 – 29.5 – 16.0 – 3.9 – 2.5)
b) See Figure A.5.

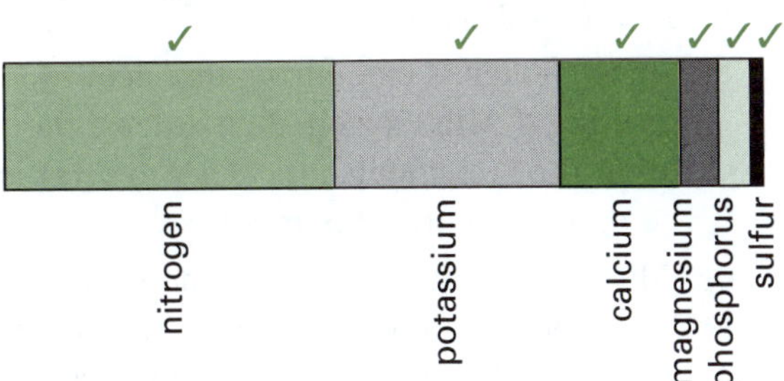

Figure A.5 Divided bar graph

16. See Table A.3.

Table A.3 Characteristics of elements

Element	Symbol	Metal, semi-metal or non-metal	Use	
tin	Sn	metal	tin plating of steel; pewter	✓
mercury	Hg	metal	amalgams	✓
iron	Fe	metal	steel structures	✓
carbon	C	non-metal	fuels	✓
silicon	Si	semi-metal	silicon chips	✓

17. See Table A.4.

Table A.4 Properties of metals and non-metals

Properties of metals	Properties of non-metals	
shiny surface	dull surface	✓
ductile	brittle	✓
malleable	non-malleable	✓
good conductor of heat	poor conductor of heat	✓
good conductor of electricity	poor conductor of electricity	✓

18. A ✓ and B ✓ because large changes in energy are typical of chemical changes.
19. A ✓ Copper sulfate has one atom of copper; one atom of sulfur and four atoms of oxygen.
20. a) four: carbon, hydrogen, nitrogen and oxygen ✓
    b) 12 ✓ (4 + 2 + 2 + 4)
    c) Since alloxan is made from uric acid by reacting it with oxygen, then at least the other three elements in it (C, N and H) must be present in urea. So urea is a compound. ✓
    d) Alloxan is a pure substance and so has a fixed composition. Therefore, a formula can be written for it. ✓ Petrol is a mixture of a number of different compounds. While we can indicate what substances are in it, the composition will vary; hence, no formula. ✓
21. a) the production of a gas ✓
    b) acid and laundry bleach ✓
    c) chlorine gas ✓
    d) Not knowing what the outcome might be, Jim should have taken more care. He should have done the experiment outside where the chlorine gas could be dispersed. ✓ He should have worn safety glasses. ✓
22. a) The mercury and sulfur must come from it because they are not present in air. ✓
    b) mercury ✓ and sulfur ✓
23. four different elements ✓
24. sample 3
    Workings:

    Sample 1: $D = \frac{m}{V} = \frac{29.70}{4.5} = 6.6 \text{ g/cm}^3$ ✓

    Sample 2: $D = \frac{m}{V} = \frac{48.18}{7.3} = 6.6 \text{ g/cm}^3$ ✓

    Sample 3: $D = \frac{m}{V} = \frac{42.77}{6.6} = 6.48 \text{ g/cm}^3$ ✓

    Sample 4: $D = \frac{m}{V} = \frac{53.46}{8.1} = 6.6 \text{ g/cm}^3$ ✓
25. a) aluminium ✓
    b) oxygen ✓
    c) silicon ✓
    d) sodium ✓ and potassium ✓
26. a) gas ✓ (the substance spreads out and fills the container)
    b) $\frac{100}{20} = 5$ times ✓
    c) The final appearance will be a much lighter brown. ✓
    d) diffusion ✓
27. The atoms of gold have a greater mass than the atoms of silver. ✓ The volume of each atom is almost the same. If samples of gold and silver have the same number of particles then the mass of the gold is greater as gold atoms are denser. ✓
28. $D = \frac{m}{V} = \frac{58.8}{75.0} = 0.78 \text{ g/mL}$ ✓
29. a) The alcohol contracts back into the bulb and then the column of alcohol returns to a lower value on the scale. ✓
    b) The volume of the liquid increases. ✓
    The volume of the solid glass increases. ✓
    However, the expansion of the glass is small, compared to the expansion of the liquid contained in it.
30. a) Take the reading of the position of the mercury bead while the test tube is in air. Then place the test tube into the liquid to be heated. The heat causes the air in the test tube to expand and push the bead of mercury up the glass tube, giving a new reading against the scale. The temperature difference gives a comparison of the hot liquid temperature and the air temperature. ✓
    b) Particles of hot liquid transfer heat to the particles of the glass test tube, which vibrate a little more vigorously. These vibrations are transferred to the gas particles of the gas in the tube. The gas particles move more rapidly. The gas particles hit the surface of the mercury bead, which moves up the

tube until the pressure on both sides of the mercury bead are equal once more. ✓

31. **a)** The mercury is liquid because room temperature is above its melting point but below its boiling point. ✓

b) **i)** At 400 °C the mercury is a vapour. As it cools the vapour will turn into a liquid at 357 °C. It will remain a liquid at 300 °C. ✓

ii) condensation ✓ (vapour is converted to a liquid)

32. *(There may be other answers than those provided.)*

a) Aluminium is used to make cooking foil ✓ as it is malleable ✓.

b) Zinc is used to coat iron roofs (galvanising) ✓ as zinc is more reactive than iron and protects the iron from rusting ✓.

c) Chlorine is used to sterilise drinking water ✓ as chlorine is toxic to microbes ✓.

33. **a)** **i)** magnesium ✓; **ii)** sulfur ✓; **iii)** silver ✓

b) **i)** Ti ✓; **ii)** Ar ✓; **iii)** Am ✓

34. Elements that are malleable are tin, nickel, silver and barium (they are metals). ✓ Sulfur is an element that is brittle (it is a non-metal). ✓

35. In elements all the atoms are the same. In compounds there are two or more different atoms joined chemically together in fixed proportions. A = element ✓; B = compound ✓; C = compound ✓; D = compound ✓; E = compound ✓; F = element ✓

36. A chemical change has occurred ✓, as the white powder obtained after evaporating the solution did not have the same properties as Fizzo when added to a new sample of water. The production of a gas is an indicator of a chemical change. ✓

37. **a)** The chemical change is the reaction of the zinc with the hydrochloric acid to form new products. ✓
The physical change is the evaporation of the solution to form a crystalline residue. The water evaporates into the air. This change is readily reversed by adding water to the crystals and allowing them to dissolve. ✓

b) zinc + hydrochloric acid → zinc chloride + hydrogen ✓

38. **a)** A fume cupboard is the safest place as any toxic fumes that may escape are removed rapidly. ✓

b) sulfur + oxygen → sulfur dioxide ✓

c) See Figure A.6.

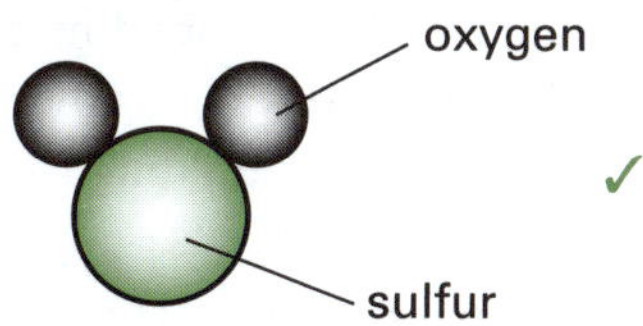

Figure A.6 Sulfur dioxide

39. Vinegar is used to pickle vegetables. ✓
The pickling process helps to preserve the vegetables as microbes such as bacteria and moulds are killed by the acidic solution. Acetic acid also adds taste to the food. ✓

40. Sample A: $D = \frac{m}{V} = \frac{34.5}{15.0} = 2.3 \text{ g/cm}^3$ ✓

Sample B: $D = \frac{m}{V} = \frac{33.6}{15.3} = 2.2 \text{ g/cm}^3$ ✓

Therefore, sample B is pure sodium chloride and sample A is impure. ✓

Chapter 3 answers

Experiment 1

Analysis: When water flow is slow then the water has only sufficient energy to dislodge and transport light sediment such as silt and some sand. As the water pressure increases, greater erosion results as the water has more energy. Some pebbles are now transported as well. In the shallow water, the lighter sediments are mainly deposited whereas the heavier particles may deposit in deeper channels in the coastal plain.

Conclusion: The stream tray model of the coastal river demonstrated that the faster the flow rate of the river, the heavier the sediment particles that could be transported.

Experiment 2

Analysis: Large or heavy particles (e.g. pebbles and sand) cannot be held by the water for the same length of time as clay particles, so they settle first. The lighter particles take longer to sediment. Some very small and light particles such as clay remain suspended in the water.

Conclusion: Heavy particles are deposited in still water before lighter particles. Very light or small particles remain suspended for long periods of time.

Experiment 3

Analysis: All the soils tested showed water-holding capacity. The loamy soil which contained humus proved to the best soil as it held water rather than letting it drain through rapidly or holding too much water which would kill the plant roots.

Conclusion: Clay soils have the greatest water-holding capacity and sandy soils have the least water-holding capacity.

Test yourself 1

Part A: Knowledge

1. **D** ✓ A is wrong as this is the innermost layer; B is wrong as this is the third layer; C is wrong as this is the outermost layer.
2. **B** ✓ A is wrong as the upper mantle is also part of the lithosphere; C is wrong as not all of the mantle is in the lithosphere; D is wrong as there is no core in the lithosphere.
3. **C** ✓ Diamond has a hardness of 10. A, B and D are wrong as their hardness is less than 10. Talc has a hardness of 1.
4. **A** ✓ Oxidation is a chemical reaction. B, C and D are all incorrect as these are physical changes.
5. **B** ✓ A is wrong as sand forms sandstone. C and D are wrong as these form conglomerate.
6. **a)** plants ✓; **b)** extrusive (volcanic) ✓; **c)** marble ✓; **d)** feldspar ✓; **e)** small (fine) ✓
7. A/F ✓; B/J ✓; C/G ✓; D/H ✓; E/I ✓

Part B: Skills

8. **a)** frost wedging (or ice cracking) ✓
   b) Ice has a greater volume than the liquid water from which it forms (i.e. water expands on freezing). ✓
9. **a)** **i)** cobble ✓; **ii)** pebble ✓
   b) All types of gravel—the smaller the particles the greater the transport. ✓
   c) **i)** clay ✓ (or between clay and coarse silt);
   ii) clay ✓ It needs the least energy to be carried, therefore is carried further out to sea.
10. Apatite is the hardest; ✓ calcite is the softest ✓.
11. The quartz is hard and will scratch the plastic lenses of the glasses. ✓
12. **a)** They are in horizontal layers. ✓
    b) E ✓ This is the lowest layer which is deposited first.
    c) A ✓ This layer has been deposited last.
13. D, H ✓, A, C ✓, G, E ✓, F, B ✓
14. See Figure A.7.

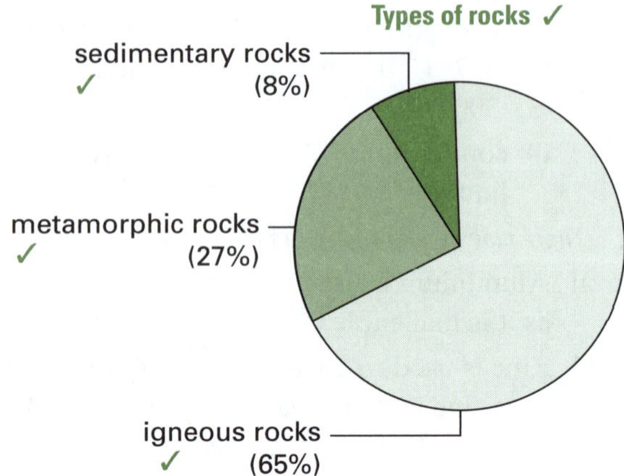

Figure A.7 Types of rocks

Test yourself 2

Part A: Knowledge

1. **D** ✓ The rock cycle diagram shows that all rocks can be weathered and eroded to form sediments. Thus A, B and C, which are specific rock types, are incorrect.
2. **B** ✓ Low grade means very little metallic mineral is present; therefore, A is wrong. C is wrong as coastal areas are not relevant and D is wrong as many different types of minerals (e.g. oxides, carbonates) may also be low grade.
3. **D** ✓ A is wrong as copper can come from many minerals such as chalcopyrite and chalcocite; B is wrong as iron comes from minerals such as magnetite; C is wrong as tin comes from cassiterite.
4. **C** ✓ Figure 3.20 shows most iron ore mines are in Western Australia. Thus A, B and D are wrong.
5. **A** ✓ B is wrong as no rotation of crops is involved. C is incorrect as fertiliser helps to grow plants and there are no plant changes involved; D is wrong as deep-rooted plants are alternated with shallow-rooted plants.
6. **a)** burning ✓; **b)** nutrients ✓; **c)** erosion ✓; **d)** energy ✓; **e)** aqueous ✓
7. A/J ✓; B/F ✓; C/I ✓; D/H ✓; E/G ✓

Part B: Skills

8. See Figure A.8.

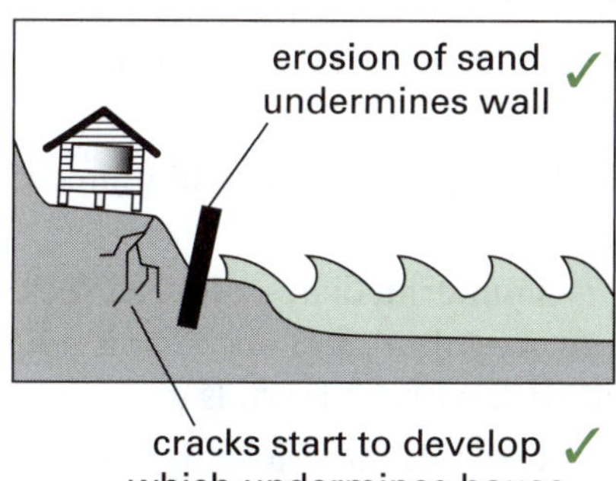

Figure A.8 Erosion

9. Q: How do coastal dunes form? ✓

   A: They form when sand dumped onto the shore dries and is blown further to the back of the beach. ✓

10. Marram has vertical as well as horizontal parts to its root system ✓ and its rapid growth means that the dune maintains its hilly shape rather than being smoothed out by the spinifex. ✓

11. **a)** 4.4 % ✓ (100 – 94.0 – 0.7 – 0.4 – 0.5)

    b) See Figure A.9.

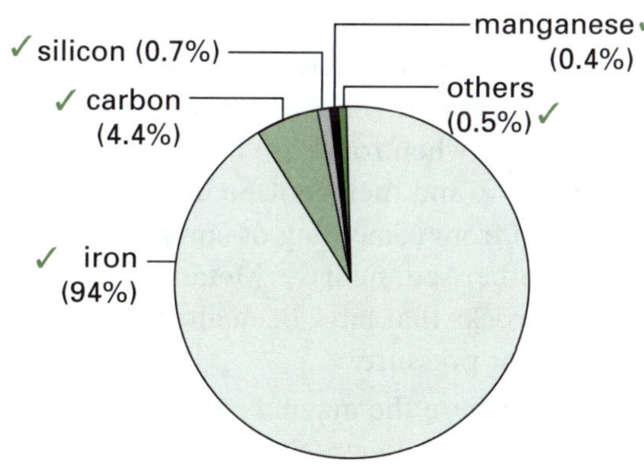

Figure A.9 Composition of steel

12. **a)** $\frac{2}{1000} \times \frac{100}{1} = 0.2\%$ ✓

    b) $50000 \times \frac{0.2}{100} = 100$ tonnes silver ✓

13. E, C, ✓ B, A, D ✓

14. Coal is burnt to provide the heat to smelt the iron ore into iron metal. ✓ Coal is burnt and the heat is used to boil water which turns turbines to generate electricity that is used in electrolysis to produce metallic aluminium from alumina. ✓

Chapter test

Part A: Multiple-choice questions

1. **D** ✓ Lustre describes the way light interacts with the surface of a crystal, rock or mineral. A is wrong as different colours can be shiny or dull; C is wrong as cleavage refers to the way the mineral splits when struck; B is wrong as streak is the colour of the powdered mineral.
2. **B** ✓ Most grains in limestone are skeletal fragments of marine organisms such as coral or foraminifera. A is wrong as sandstone is formed from sand; C is wrong as siltstone is formed from fine silt particles; D is wrong as mudstone is formed from compacted and cemented mud.
3. **C** ✓ Gabbro is a coarse-grained igneous rock. A, B and D are wrong as they are all minerals.
4. **B** ✓ The river slows around the inside of the bend and heavier minerals are deposited. Therefore A, C and D are wrong as the river is flowing faster in these locations.
5. **D** ✓ Aboriginal peoples used fire to encourage new growth. A is wrong as spears were used for hunting; B is wrong as many Aborigines moved from place to place and did not destroy the landscape at each location; C is wrong as traditional Aborigines did not plant crops as they were hunter-gatherers.
6. **A** ✓ B and C are wrong as they did not have the technology to make large structures and there was not sufficient iron available; D is wrong as iron was not used to make money.
7. **A** ✓ B is wrong as it refers to erosion; C and D are wrong as there are multiple of weathering.
8. **B** ✓ A is wrong as mineral crystals may have variable colours; C is wrong as it refers to cleavage; D is wrong as it refers to how shiny it is.
9. **D** ✓ A, B and C are incorrect as conglomerate consists of large pebbles and smaller grains cemented together.
10. **C** ✓ Granite forms underground. A is wrong as granite is igneous; B is wrong as it is intrusive not extrusive (formed from lava); D is wrong as the crystals are large.

Part B: Short-answer questions

11. **a)** mineral ✓; **b)** weathering ✓; **c)** erosion ✓;
    d) sedimentary ✓; **e)** fossils ✓; **f)** large ✓;

g) metamorphic ✓; h) bauxite ✓

12. A/K ✓; B/L ✓; C/J ✓; D/M ✓; E/H ✓; F/I ✓; G/N ✓

13. a) Yes, it is crystalline ✓, inorganic ✓ and solid ✓.

b) i) It is not naturally made. ✓

ii) It is formed from living things (plants which fell into ancient swamps). ✓

iii) It doesn't have a constant chemical composition. ✓ While many sands are made from silicon dioxide, there may be other particles present as well and still be called sand.

iv) These are dissolved in milk and not found in solid, crystalline form. ✓

14. A Weathering and erosion ✓; whether physical or chemical, weathering breaks down the rock particles and winds and rain cause erosion.

B Transportation ✓; the river transports the broken-down particles and carries them along through the flow.

C Deposition ✓; these particles are deposited when the river slows down, such as when it enters a lake.

D Sediments turn to rock ✓; as more layers are deposited, the lower layers are squeezed dry of moisture. Eventually over millions of years the particles are cemented together and new rock has formed.

15. a) Igneous rocks are rocks that have formed from molten rock material. ✓

b) Sedimentary rocks are formed from sediments ✓ being laid down or the organic remains of organisms ✓. At no stage in their formation were sedimentary rocks molten ✓, as were igneous rocks.

c) They cool over a long time and this allows the mineral crystals to grow. ✓

16. a) You can tell them apart by the size of their crystals. ✓ Basalt is very fine but gabbro has large crystals. ✓

b) No ✓, because intermediate rocks contain sodium-rich pyroxene feldspar ✓ while mafic rocks contain calcium-rich pyroxene feldspar. ✓

c) No ✓, because any fossils that may have been present in the parent rock are gone when material has melted. ✓

d) It would have formed on the Earth's surface ✓ as the crystals are non-existent, indicating rapid cooling.

17. a) Granite is light coloured ✓, gabbro is dark coloured ✓.

b) Granite has large crystals ✓, but pumice has no crystals ✓.

c) Diorite is an intermediate-coloured rock with large crystals ✓, and obsidian is a light coloured rock without crystals ✓.

18. a) Metamorphic rocks are those that have been changed by the application of heat and/or pressure. ✓

b) They could be all three ✓; adding more heat and pressure can further change even metamorphic rocks ✓.

c) Regional metamorphism operates over a large area ✓; pressure is an important aspect of regional metamorphism.

d) i) *(½ mark for each)* limestone ✓ metamorphoses to marble ✓; quartz sandstone ✓ metamorphoses to quartzite ✓; shale ✓ metamorphoses to hornfels ✓

ii) contact metamorphism ✓

19. a) Rocks are continually recycled. ✓ They are formed as igneous, sedimentary or metamorphic rocks. ✓ Igneous rocks are formed when rocks are heated to melting point ✓ and then cool. Sedimentary rocks form from cementing or compacting together sediments. ✓ Metamorphic rocks are rocks that have been altered by heat and/or pressure. ✓

b) D ✓ where the magma makes contact with the surrounding rock

20. a) topaz ✓ and diamond ✓

b) D ✓

c) B ✓

d) D ✓

21. a) Plants that did not have the ability to resist or survive these fires died off. ✓ Other plants had developed ways to survive these fires and could regenerate. These plants can then set seed and take over areas previously occupied by the plants that were burnt off. ✓ So fires favoured the plants that could withstand burning.

b) Fire suppression can lead to the build-up of inflammable debris and the creation of less frequent but much larger and destructive wildfires. ✓ Regular smaller burns would clean up the countryside from litter and debris and not provide fires that were so hot as to destroy everything in their paths. ✓

22. a) The more fertiliser applied to the ground, the higher is the grain yield per hectare. ✓

b) No, eventually too much fertiliser could be poisonous to the plant. Besides, there is a limit to how much a hectare of land can grow. ✓

c) The increased cost of fertiliser ✓; the excess runoff to streams and lakes can cause problems with native vegetation, ✓ and hence animal life.

23. a) i) ✓ and iv) ✓

b) The crown must have a sufficient store of nutrients (e.g. starch) to support a period of growth in the absence of photosynthesis. ✓

24. a) 40 + 20 + 3 + 2 = 65 kg per person ✓

b) $\frac{20}{65} \times 100 = 30.8\%$ ✓

c) less waste goes to the tip ✓; it costs less to recycle most materials ✓ than it is to form them from raw substances

25. a) The ore is dug out of the ground, loaded onto trucks and transported to where it will be crushed and mixed with other materials to help extract the metal from the ore. ✓✓

b) Although aluminium is extremely common and widespread, the common aluminium minerals are not economic sources of the metal. ✓ (Almost all metallic aluminium is produced from the ore bauxite. Australia is a major producer and exporter of bauxite ore.)

26. A = quartz crystal, B = sand grain, ✓
C = sandstone, D = sandstone strata ✓

27. X: Materials dissolved in water: b) mineral salts ✓
Y: Materials suspended in the water: c) clay, e) fine sand ✓
Z: Materials rolled/pushed along river bed: a) pebbles, d) boulders, f) very coarse sand ✓

28. The rock is conglomerate as it has particles of different colours and sizes cemented together and it does not fizz with acid. ✓

29. *(There is more than one possible answer.)*
1A. Mineral is white 2
1B. Mineral is not white 3
2A. Mineral has a dull lustre .... calcite ✓
2B. Mineral has a vitreous lustre....................................... fluorite ✓
3A. Mineral is yellow 4
3B. Mineral is not yellow .... magnetite ✓
4A. Mineral has a black streak ....chalcopyrite ✓
4B. Mineral has a green streak ... pyrite ✓

30. $\frac{40}{100} \times 1300$ kg = 520 kg ✓ per household
(over half a tonne!)

31. *(1 mark each for any three of the following points)*
These are some of the factors and many of them relate to the potential to make a profit.

- What is the size of the ore body?
- Where is it located (near the surface, under a city, under the ocean, in a national park and so on)? There may be government regulations prohibiting mining and also cost factors.
- What is the current and future market price for the metal?
- What are the set-up costs?
- What is the availability of staff (especially in remote locations)?
- How long can the mine be operated for?

32. A The continual pounding of waves attacks weak areas of rock and opens up a cave. ✓
B Through erosion, the cave widens and deepens, eventually forming an arch. ✓
C Eventually, weakened through further erosion, the roof of the cave collapses and an 'island' called a stack is left. ✓
D Even this stack eventually erodes away and so the headland retreats. ✓

33. a) cobbles ✓

b) 3 m/s ✓

c) clay, silt, sand, granules and pebbles ✓

d) At 100 cm/s (1 m/s) a stream can carry up to medium-sized pebbles. As the stream slows to 45 cm/s (0.45 m/s) larger particles are dropped, so it can now not carry any pebbles at all, while still carrying clay, silt, and sand. ✓

34. a) P ✓ since it is beneath Q
b) Q is harder. ✓ It does not show as much erosion. The water in the bay shows significant erosion and the undercutting of P rocks at the shore.
c) Sedimentary rock P was laid down first, then Q later. ✓ Over millions of years the particles in the rocks were cemented together as pressure pushed them together. Erosion occurred, exposing the rocks to the surface. ✓ The layers were then tilted. Either the bay-side rocks were pushed up, or the lake-side rocks subsided, or maybe both. Water in the bay eroded P ✓ while Q formed part of a depression allowing a lake to form. As rock Q is harder, it resists erosion more, thus standing higher in the middle and protecting part of rock P below. ✓ (Other possible answers may also score 4 marks.)

35. Gabbro is plutonic, formed when molten magma is trapped beneath the Earth's surface and cools slowly into a crystalline mass. ✓ Basalt has very fine or no crystals due to rapid cooling of lava at the surface. ✓ Dolerite forms in shallow intrusions, such as dykes (which cut across the rock strata) and sills (which push between beds of sedimentary rock). ✓

36. The rock is slate. ✓ It does not fizz when acid is added. It is grey-black, layered and does not burn.

37. The unknown mineral is sphalerite. ✓ This mineral can be black and its streak is white. It has a resinous lustre and its hardness of 3.8 fits the range of 3.5–4. ✓

38. He saw a cyclical nature to geological processes. He thought of the rock cycle as being endlessly repetitive. Rocks do not remain in equilibrium and are forced to change as they encounter new environments. The rock cycle explains how the three rock types are related to each other, and how processes can change them from one type to another with time. ✓

39. a) about 190 million tonnes ✓
b) 2007–08 ✓
c) between 350 and 400 million tonnes, ✓ assuming the increased trend for more exports continues ✓

40. a) paper, glass, plastic bottles, plastic bags ✓
b) plastic bags, motor oil ✓
c) motor oil ✓ Many home handymen change the oil on their cars themselves. This waste oil often ends up in garbage bins. Take this oil to a garage or repair workshop which can send it off to be recycled. ✓

Chapter 4 answers

Experiment 1

Analysis: See Figure A.10.

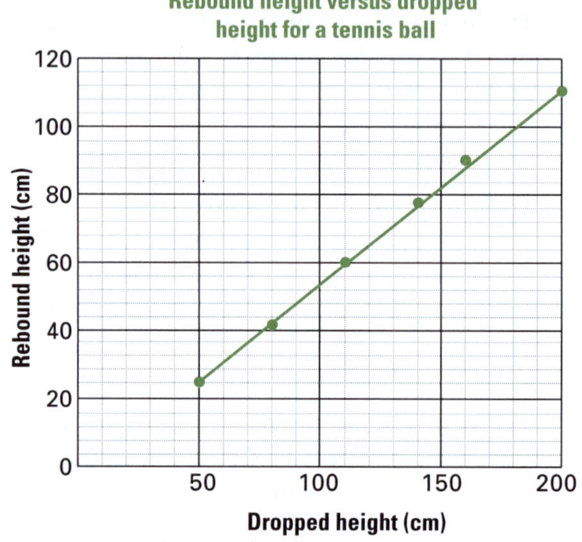

Figure A.10 Rebound height

As the ball falls it loses gravitational potential energy and gains kinetic energy. The tennis ball hits the surface and deforms. Some of this deformation energy is converted to heat and on rebound only a proportion of the kinetic energy is regained and the ball rises to a lower height than it was when dropped.

Conclusion: The height of rebound of a tennis ball is directly related to the height from which it is dropped.

Experiment 2

Analysis: The water reacted with the plaster of Paris and heat was produced. The chemical energy of the plaster and the water was partly converted into heat energy which caused the temperature of the mixture to rise.

Conclusion: In the chemical reaction between water and plaster of Paris, the chemical energy of the system was partly transformed into heat energy.

Experiment 3

Analysis:

1. polytherm, because the temperature rose more slowly
2. conduction through the plastic
3. 210 seconds = 3.5 min; at that time the temperature is 35.5 °C
4. The variables that should be controlled are:
   - temperature of the original liquid
   - depth at which the temperature was taken
   - volume of hot liquid
   - type of probe used.

Conclusion: Polytherm is the better heat insulator.

Test yourself 1

Part A: Knowledge

1. **A** ✓ Energy causes change. B is wrong as heat is only one form of energy; C is wrong as forces are pushes or pulls or twists; D is wrong as adding energy can heat things up.
2. **C** ✓ The stored elastic potential energy can be released and do work. A is wrong as there is no change in the chemicals in the rubber band; B is wrong as no electricity is involved; D is wrong as the rubber band is not raised above the Earth's surface.
3. **B** ✓ The chemicals inside the torch are converted to electrical energy as well as some heat energy. A is incorrect as no sound is involved; C is incorrect as the reverse is true; D is wrong as elastic potential energy is not produced.
4. **A** ✓ The moving particles of air have kinetic energy and as the blades rotate, electrical energy is then produced. B is wrong as the wind does not have potential energy; C is wrong as solar energy is not directly involved; D is not correct as a gravitational potential energy is not involved.
5. **B** ✓ The chemicals in the battery store chemical potential energy. A is wrong as electricity is only produced by chemical reactions; C is wrong as heat energy is not involved in this instance; D is wrong as kinetic or gravitational potential energy are not involved.
6. **a)** photons ✓; **b)** height ✓; **c)** elastic ✓; **d)** mechanical ✓; **e)** kinetic or heat ✓
7. A/G ✓; B/H ✓; C/I ✓; D/J ✓; E/F ✓

Part B: Skills

8. GPE = $9.8mh = 9.8 \times 10 \times 10 = 980$ J ✓
9. 300 J ✓ On the Moon there is no air and so the gravitational potential energy should be completely converted to kinetic energy as it falls. No energy should be converted to heat in this case. ✓
10. **a)** kinetic energy ✓
    b) The molecules of water are much closer together than air molecules are. Moving water molecules collide with their neighbours and pass the energy of the sound onwards very rapidly. In air, the air molecules have to travel further before they collide with other air particles. Energy can be transferred on collision. ✓
11. The two chemicals react together to form new substances. ✓ In this reaction some chemical potential energy is converted to heat energy and so the temperature of the mixture rises. ✓
12. **a)** gravitational potential energy ✓
    b) kinetic energy ✓
    c) Kinetic energy is transformed into elastic potential energy as well as heat energy and some sound energy as the plasticine is deformed. ✓
13. People sweat when their bodies are hot. Sweat glands release sweat onto the skin surface. Heat energy from the body is transferred to the sweat and increases the evaporation rate. ✓ Heat energy is converted to kinetic energy and the energised water molecules can more readily escape from the skin surface. ✓ Therefore, the body loses heat and becomes cooler.
14. A = light energy ✓; B = heat energy ✓; C = chemical potential energy ✓

Test yourself 2

Part A: Knowledge

1. **B** ✓ Cooling involved loss of caloric. A is wrong as this is the opposite answer to B; C is wrong as these different materials are absorbing rather than losing caloric; D is wrong as this does not explain cooling.

2. **B** ✓ The water vapour at 100 °C has absorbed additional energy to change state and so has a higher heat content. A is wrong as the 5 kg will have a greater heat content; C is wrong as ice at 0 °C has absorbed heat and turns into liquid water and so the liquid water has a higher heat content; D is wrong as different materials can absorb different amounts of heat.
3. **D** ✓ All the other materials (A, B and C) are metals and are good heat conductors.
4. **B** ✓ The molecules in cool air occupy less space and so it is more dense. A is wrong as warm air rises as it is less dense; C is wrong as convection is responsible; D is wrong as they have a longer wavelength.
5. **A** ✓ The curtains absorb the heat and prevent its loss through the glass. B is wrong as these structures should be on the north and west sides; C is wrong as some energy is wasted; D is wrong as aluminium is lighter than steel and will save more energy.
6. **a)** rechargeable ✓; **b)** efficiencies ✓; **c)** insulated ✓; **d)** hot ✓; **e)** vibrating ✓
7. A/J ✓; B/F ✓; C/H ✓; D/G ✓; E/I ✓

Part B: Skills

8. **a)** an electric bar heater (radiator) ✓
   b) No, not all the energy is changed to electrical energy. ✓ As the filament glows, another form of energy that is produced is light (both visible and infra-red). ✓
9. A = 3 ✓; B = 1 ✓; C = 4 ✓
10. W = 2 ✓; X = 4 ✓; Y = 3 ✓; Z = 1 ✓
11. A = 2 ✓; B = 4 ✓; C = 5 ✓; D = 1 ✓; E = 3 ✓
12. **a)** See Figure A.11.

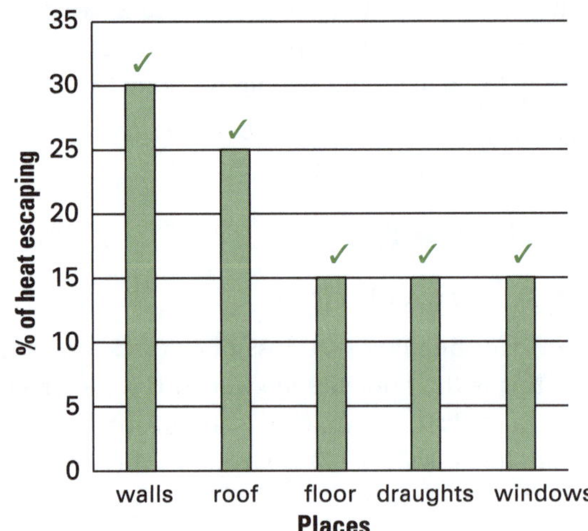

Figure A.11 Heat losses

   b) No; ✓ houses vary and this data is only for a typical house. ✓
13. **a)** 5 × 6 = 30 J/s lost ✓
   b) Use curtains or double glazing. ✓
14. See Figure A.12.

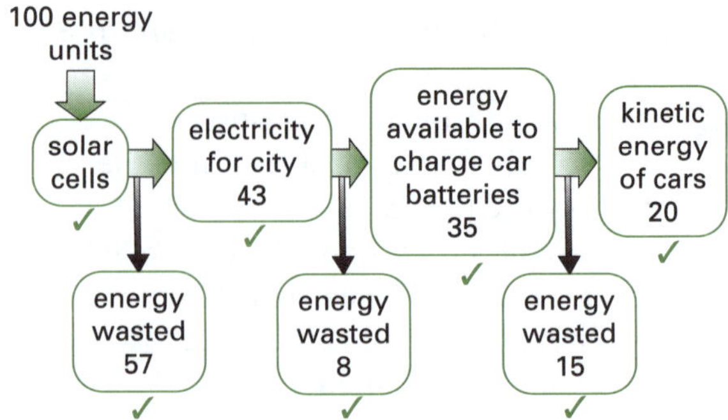

Figure A.12 Solar energy uses

Chapter test

Part A: Multiple-choice questions

1. **B** ✓ Electrical motors are connected to the electrical power supply. The rotor turns and so the machine produces kinetic energy. A is wrong as the reverse is true; C is wrong as motors do not produce electricity; D is wrong as producing sound is not the main function of a motor.
2. **A** ✓ The gas burns to release heat and some light (you can see the flame). B is wrong as gas ovens do not use sound; C is wrong as heat is not the only form of energy produced (some light is also formed); D is incorrect as no electricity is formed.
3. **C** ✓ An electric current is a form of kinetic energy. A, B and D are forms of elastic potential energy.
4. **C** ✓ The battery provides the electrical energy which flows into the bulb and is converted to light and heat. A is wrong as a bicycle does not use electrical energy; B is wrong as the main function of mobile phones (particularly smart phones with an internet connection) is to produce sound energy and light energy; D is wrong as stereo speakers do not produce light.
5. **A** ✓ The chemicals in the wax store the chemical potential energy in their bonds. B is wrong as the wax does not use its heat content as a source of energy; C is wrong as there is no light energy in the wax; the particles in the wax

do not move about and so no kinetic energy (D) is stored.

6. C ✓ The air is compressed and so stores elastic potential energy (PE) and the air molecules are in constant motion so they possess kinetic energy (KE). A and B are wrong as they are only part of the answer; D is wrong as no sound is stored.
7. B ✓ Gravity slows the ball down as it rises and so its KE decreases with time. A is wrong as its GPE increases as it gains height; C is wrong as its KE will be zero at the highest point as it comes to a stop before falling; D is wrong as it has no elastic PE.
8. A ✓ PV cells are very costly. B is wrong as sunlight does not damage the cells; C is wrong as the displays are heavy; D is wrong as there is not much grease on the road to cause this problem.
9. A ✓ Drag is reduced if the vehicle is aerodynamic. B is wrong as vehicles need to be made from lightweight alloys; C is wrong as ethanol does not increase fuel efficiency; D is wrong as heavy cars use up more fuel.
10. A ✓ If warm air gets behind the curtain, it is cooled by contact with the cold glass and then sinks down to produce a draught that re-enters the room. B is wrong as pelmets do not change the amount of solar energy; C is wrong as they minimise cold currents; D is wrong as they cannot prevent hot air accumulation in the summer.

Part B: Short-answer questions

11. a) insulators ✓; b) heat ✓; c) convection ✓; d) photovoltaic ✓; e) fossil ✓
12. A/G ✓; B/J ✓; C/F ✓; D/H ✓; E/I ✓
13. The heat from your hand is conducted into the ice block. This heat causes the ice to melt. ✓
14. a) The coloured water will move up the narrow-bore tube. ✓ The water is expanding as it is heated. ✓
    b) by conduction through the glass into the water ✓; by convection in the water and again by conduction into the inner flask ✓
    c) convection currents ensure good mixing ✓
    d) radiation = heat rays move out in all directions ✓; convection = heated gas (air moves away from the hot flask) ✓
15. • Insulating stopper: prevents heat from leaving the flask by conduction. ✓
    • Vacuum between glass wall: prevents heat loss by conduction and convection because there are no particles present. ✓
    • Silvered glass surfaces: reflect radiant heat rays. ✓
    • Glass, felt: they are poor conductors. ✓
16. a) *(1 mark each for any four of the following)*
    • wool or synthetic tea-cosy
    • type of material (knitted/woven)
    • type of teapot (metal/ceramic/porcelain/glass)
    • starting temperature of the water
    • teabags/tea leaves/no tea
    • amount of water in the teapot
    • length of time before temperature is recorded

    b) a teapot without a tea-cosy ✓
    c) *(1 mark each for any three of the following)* teapots; thermometers; tea-cosies; stopwatch; kettle; water; measuring cup or jug; tea
    d) 1. Boil water. ✓
       2. Measure out 500 mL of water and record its initial temperature. ✓
       3. Place the hot water into three identical teapots (one with a woollen tea-cosy, one with a synthetic tea-cosy and one control pot). ✓
       4. Take the temperature after 10 minutes and 20 minutes. ✓

    e) The teapot covered with the woollen tea-cosy should lose less heat than the uncovered teapot and the water in it should remain hotter. ✓
    f) It traps a layer of hot air next to the teapot, which acts as an insulation layer (air and wool are poor conductors). ✓
17. a) See Figure A.13. (next page)
    b) the black can ✓
    c) room temperature (28 °C) ✓
18. a) radiation ✓; b) conduction ✓; c) convection ✓; d) radiation and convection ✓
19. The air next to the skin and in the foam bubbles is a poor conductor of heat, so the heat from the diver's body is not readily lost to the surroundings. ✓
20. a) the uneven heating of the Earth's surface and atmosphere by the Sun ✓

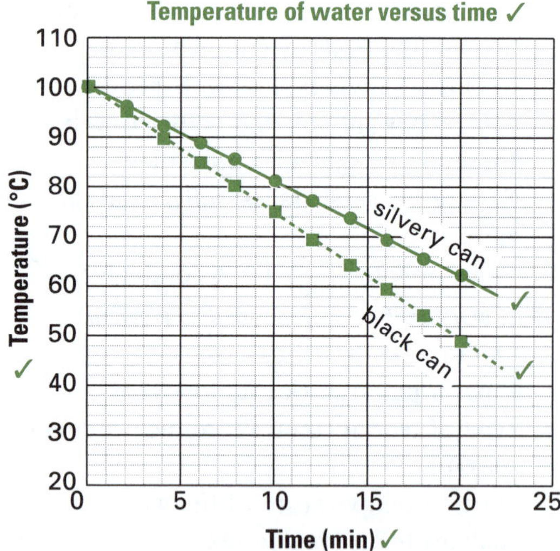

Figure A.13 Silver and black can graph

b) *(1 mark each for any two correct examples)* sailing, flying a kite, drawing water from a well, windmills for grinding grain, hang-gliding ✓✓

c) kinetic energy ✓ (wind) ✓ kinetic energy ✓ (or mechanical energy) ✓ electrical energy ✓ (This last step is not in every windmill.)

d) It can produce a power output of 50 kilowatts (50 000 watts). ✓

e) 60 times (3 000 000 watts/50 000 watts) ✓

21. a) Electricity is used ✓ to turn blades and this makes wind in an electric fan. Wind turbines use wind to make electricity. ✓

b) i) Power can be generated ✓ from the wind, when it is blowing. Excess energy can be stored ✓ in batteries. On days when the wind is not blowing, diesel engines can 'kick in' ✓ to generate electricity.

ii) in remote locations, where a connection to the electricity grid is not available ✓; where there is sufficient wind to make them worthwhile ✓

c) i) There is not enough power output to generate sufficient electricity. At these low wind speeds the blades may not turn sufficiently. ✓

ii) The generators could overheat at very high speeds. ✓

22. a) the ears receive energy ✓

b) the food being cooked ✓

c) the ocean or land surface/trees/buildings ✓

23. a) The person uses his weight to compress the spring in the pogo stick. The spring rebounds, pushing the person up. ✓

b) when the pogo stick is fully compressed ✓

c) when the pogo stick is at its maximum height above the ground and the spring is fully extended ✓

d) when moving up or down, whenever the spring is half-compressed ✓

e) when the person is at the top or bottom of the bounce ✓

24. a) 40 °C ✓ (1200/60 = 20; 20 + 20 = 40 degrees)

b) i) temperature increases with depth (50 °C per 1000 m) ✓

ii) 150 °C ✓

iii) 5 km ✓

c) between H and I ✓, as the temperature gradients are closer to the surface here and closer together ✓

25. a) the flask will begin to spin ✓

b) chemical PE ✓; heat energy ✓; elastic PE ✓; KE ✓

c) increase the size of the Bunsen flame ✓

26. Fred is correct. ✓ Energy is neither created nor destroyed, only changed to other forms. ✓

27. Ball B has the greater kinetic energy as it has the greater speed. ✓

28. Ball Y has the lower kinetic energy as it has less mass. ✓

29. The 2.5-t truck ✓ will be more damaged as it has a greater kinetic energy when it collides with the wall. In order to bring the truck to rest, a greater reaction force is applied to the truck and this will cause more damage. ✓

30. a) Mass (A) at 3 m has a greater gravitational potential energy. ✓ More energy was used to lift it off the floor to a height of 3 m. This amount of energy expended by our muscles is now stored in the mass prior to release.

b) Mass (A) will register the greater impact force ✓ as it has fallen from a greater height and will have a greater kinetic energy on impact. In order to bring it to rest on the force meter, a greater reaction force must be applied.

31. The string does not stretch. As the archer pulls on the middle of the string, the forces are applied to the ends of the bow causing the bow to bend/stretch and store elastic potential energy. ✓ When the archer lets go of the string, the stretched bow contracts back to its rest state and transmits the stored energy to the arrow which is attached to the string. The stored energy is converted into the kinetic energy of the arrow. ✓

32. *(1 mark each for any two correct examples)*
heat, movement of pistons, overcoming friction in engine parts and tyres, air resistance, resistance between tyres and road, wearing of brakes when slowing down and stopping ✓✓

33. $5(F - 32) = 9 \times 40$
$5(F - 32) = 360$ ✓
$F - 32 = 360/5 = 72$
$F = 72 + 32 = 104$ °F ✓

34. **a)** $5(F - 32) = 9 \times 0$
$(F - 32) = 0$
$F = 32$ °F ✓

b) $5(F - 32) = 9 \times 100 = 900$
$F - 32 = 900/5 = 180$
$F = 180 + 32$
$F = 212$ °F ✓

35. When a car brakes, a brake pad is pushed against a spinning disc attached to the wheel. Friction between the pads and the disc slows the disc down and therefore slows the car down. Kinetic energy is being converted into heat energy. ✓ If the car is travelling very fast then the braking force has to be applied over a longer time and so more heat is produced. Brake pads are made of various materials including ceramic compounds and copper fibres. ✓

36. A = kinetic energy ✓; B = heat energy ✓; C = sound energy ✓ (B and C are interchangeable)

37. A metal gauze is a good conductor of heat. ✓ The gauze quickly spreads the heat from the flame, preventing it from setting the gas below the gauze alight. For the gas to burn, it has to reach a certain temperature, called the ignition temperature, but the gauze prevents this from being reached. ✓

38. The tungsten wire glows red hot as electrical energy is converted to heat energy. The heat cannot be conducted or convected away as there is a vacuum in the bulb. The hot wire radiates energy into the vacuum and then through the glass blub into the environment. Visible light is being radiated as we can see the wire is glowing. Infra-red radiation is also being emitted. Although these rays are invisible, they travel through the glass and are absorbed by the outside air and the hand, and are converted to heat energy. ✓ In addition, some of the infra-red rays are absorbed by the glass and it gets hot. The air in contact with the hot glass will absorb some of this heat and the heated air will move via a convection current into the local environment. Therefore, the hand gets hot due to radiation and some convection. ✓ (Tungsten is used because, of all the metals on the periodic table, it is the one with the highest melting point and therefore does not melt and break.)

39. The cylinder is much more massive and contains more heat energy than the nail. ✓ The nail may have a higher temperature but its mass and heat content are much lower. ✓ A bathtub of warm water contains more heat than a kettle of boiling water.

40. In many parts of Australia the Sun can contribute 60 to 70% of the power required to heat water. ✓ Conventional hot water systems use electricity or gas to heat the water. By using solar hot water, the use of electricity or gas will be reduced. There will also be a reduction of greenhouse gases. ✓

Chapter 5 answers

Experiment 1

Analysis:

1. See Table A.5.

Table A.5 Similarities and differences between *Eucalyptus* flower buds, nuts and leaves

Similarities	Differences
all produce woody nuts	length of leaves
all have at least seven flowers per cluster	width of leaves
all have a cap on the flower buds	shape of nuts

2. See Figure A.14. on the next page.

Conclusion: The five different *Eucalyptus* species had flower buds, nuts and leaves of different sizes. These observable features can be used to create a classification key.

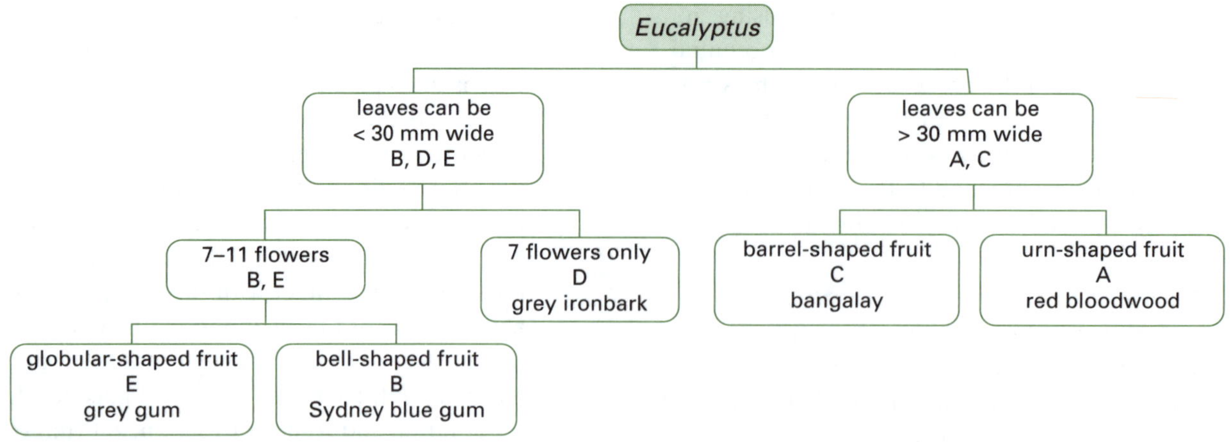

Figure A.14 Diagrammatic key

Experiment 2

Analysis:

1. A: between the edges and centre of the wet eucalyptus forest or the centre of the dry woodland

   B: anywhere in the open, dry eucalyptus woodland

   C: centre to half-way out to the edge of the dry woodland, or from the edge of the wet eucalyptus forest
2. Many answers are possible, such as different ages of leaves, therefore different amounts of chlorophyll.

Conclusion: The graph does seem to support the hypothesis at this stage as most of the results for the wet eucalyptus forest show a darker green than the dry forest. However, many more samples must be taken to confirm the hypothesis.

Experiment 3

Analysis:

1. See Table A.6. Column 3 is calculated by taking an average of the readings in column 2.
2. loamy soil: average height = 32 cm
   clayey soil: average height = 26 cm
   sandy soil: average height = 21 cm
3. The best soil for growing these seeds is loamy soil and the poorest is sandy soil. Loamy soil has greater nutrient value than sandy soil. Sandy soil has a poor water-holding capacity.

Conclusion: Plants grow better in loamy soils than in sandy or clayey soils.

Table A.6 Height of each plant after 6 weeks

Pot	Plant height (cm)	Average height (cm)
L1	30 35 34	33
L2	29 33 36	33
L3	28 30 35	31
L4	29 34 36	35
C1	24 26 28	26
C2	27 25 29	27
C3	26 25 27	26
C4	28 25 22	25
S1	20 20 23	21
S2	17 19 21	19
S3	20 18 22	20
S4	21 23 22	22

Test yourself 1

Part A: Knowledge

1. **D** ✓ A, B and C are wrong as they all refer to responsibilities of teamwork.
2. **B** ✓ Algae are not a flowering plant and therefore are not relevant. A, C and D are wrong as all these pieces are about water movement in flowering plants.
3. **A** ✓ The parent did not observe how the blood got on the shirt. It could have got there by a nose bleed. Thus she has inferred the reason for the blood being on the shirt. B is wrong as she did not make an observation when the event occurred; C is wrong as generalisations are made after extensive investigations and similar

patterns emerge; D is wrong as no prediction is being made.

4. C ✓ A is wrong as this is the dependent variable; B is wrong as this is a controlled variable; D is wrong as this variable should be controlled.
5. C ✓ A and B are wrong as these are controlled variables; D is wrong as this is the dependent variable.
6. **a)** independent ✓; **b)** plagiarism ✓; **c)** fair ✓; **d)** constant ✓; **e)** reliable ✓
7. A/I ✓; B/H ✓; C/G ✓; D/J ✓; E/F ✓

Part B: Skills

8. **a)** High levels of trace metals in fertilisers make plants grow faster. ✓
   b) the species of plant ✓; the size of the pot ✓; the frequency and amount of watering ✓; the types of soil ✓ (others may be correct, such as the amount of sunlight)
9. Rose's idea can be tested scientifically. ✓ Plants don't have feelings, nor can they think. These are not attributes we can measure scientifically. ✓ (It is not uncommon to attribute human form or personality to things or objects. This is called anthropomorphism. 'My garden will be happy to see me again' is an anthropomorphic statement.)
10. **a)** 18 ✓
    b) temperature ✓ (or other correct controlled variables such as the amount and quality of food)
    c) See Figure A.15.

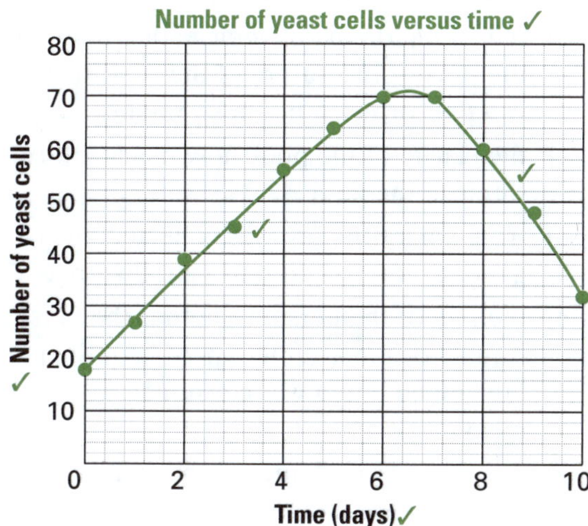

Figure A.15 Graph yeast cells versus time

11. **a)** 6 ✓
    b) after 7 days ✓
    c) One possible answer is that the number of yeast cells produced were exactly balanced by the number of yeast cells dying. ✓
12. **a)** The growing tips of grass seedlings always grew towards the light. The bending began several millimetres below the tip and extended to 20 mm down the stem. ✓
    b) Light caused the production of a chemical in the tip of the seedling that moved down the stem and caused it to bend. ✓
    c) He cut off portions of the tips of the plants before exposing them to light to determine whether the stems still bent towards the light. He also covered the tips with tinfoil caps. ✓ The variables to be controlled were: same intensity light; same temperature; plants of the same age and height; plants watered the same; same distance from light source. ✓
    d) Yes. ✓ The removal of the top 4 mm of the stem stopped the plants from bending towards the light. ✓
    e) No. ✓ He has not shown that a chemical substance was transferred. More experiments are needed. ✓
13. A = 3 ✓; B = 4 ✓; C = 1 ✓
14. Hypothesis: Different metals conduct heat at different rates. ✓
    Metallic rods are being heated at one end and the initial temperature of each rod is recorded. ✓

Test yourself 2

Part A: Knowledge

1. D ✓ Inferences must be logical explanations. A is wrong as inferences may not necessarily be correct; B is wrong as this is the definition of a hypothesis; C is wrong as a theory is only developed after considerable testing.
2. A ✓ The independent variable is a condition that is deliberately manipulated by scientists to test the response in an experiment. B is wrong as this would be the dependent variable; C is wrong as the control does not change;
   D is wrong as the experimental variable does change.

3. A ✓ A scientific hypothesis is a proposal intended to explain certain facts or observations. Statement A is an opinion. B, C and D are all hypotheses which can be tested scientifically.
4. B ✓ A prediction is a statement or claim that a particular event will occur in the future. A is wrong as a new observation must occur which may result in a new prediction; C is wrong as much testing needs to occur before a theory can be developed; D is wrong as it does not refer to results of a future experiment.
5. D ✓ A truly objective experiment means accepting all valid results and not only those that support your ideas. A, B and C are incorrect as they are all factors which must occur in developing an experimental procedure.
6. a) indicate ✓; b) true ✓; c) temporary (or provisional) ✓; d) variables ✓; e) independent ✓
7. A/G ✓; B/H ✓; C/I ✓; D/J ✓; E/F ✓

Part B: Skills

8. a) The river slows down at this point so it can drop its heavy load. ✓
   b) There must be plenty of nectar for them to collect. ✓
   c) Water is heating the solid CO_2 and it sublimes, releasing carbon dioxide gas. ✓
   d) The density of oil is less than that of water, and the two don't mix. ✓
   e) The rate at which objects fall doesn't depend on their masses. ✓
9. The fastest rate of reaction occurs with the least time. The table only gives temperatures with 10-degree intervals, so you need to interpolate. This temperature will be near 40 °C, and possibly somewhere between 35 °C and 40 °C. ✓
10. a) the quantity (volume) of soup in the pot ✓
    b) the temperature of the handle ✓
    c) the nature and size of the pot ✓; the type of soup ✓; the length of time the ladle is in the soup ✓ when the temperature is measured; the size ✓ and type of metal ✓ the ladle is made from
11. a) lengths of the different wires ✓; a variety of weights ✓; a ruler or other measuring tape ✓; hook to attach wires from ceiling ✓ or clamp and stand to attach wires from
    b) the initial lengths of wire ✓; the thickness of wire ✓; the mass of weights attached to the wire ✓
    c) Wires will stretch to a different extent when masses are attached to them. ✓ Suppose a 2-m length of wire stretched 1.0 cm when a given weight was attached. Then this wire would be expected to stretch 2.0 cm with the same mass when the length of wire is 4 metres. ✓
    d) Using wires of the same thicknesses ✓, measure out exactly 2-m (or the same) lengths of wire ✓ and have them attached to a hook ✓ or other point at least 2.5 metres above ground. Attach a weight carrier at the other end. ✓ (You may need to experiment to determine the best weight to use for this experiment: you don't want something so light that the extension is very small and barely measurable, but on the other hand you don't want anything so heavy that it snaps the wire. (Safety tip: wear goggles in case the wire does snap.) Attach the weight and measure the length of wire. ✓ The difference gives the amount of stretching that has occurred. ✓ Repeat with the other wires.
12. There are a number of advantages (see p. 138) such as sharing the workload ✓, discussing the issues involved ✓.
13. *(two marks each for any two of the following)*
    a) the size, shape and weight of the balloon; how much helium is placed in the balloon; cross-winds; updrafts or downdrafts ✓✓
    b) the size and type of vehicle; the tread on the brakes; the nature of the road surface; whether or not it has been raining ✓✓
    c) the clarity of the pool; how much chlorine/salt is in the pool ✓✓
    d) from where the ore was collected; the gangue material present with the ore ✓✓
    e) the size of the person; whether the person is male or female; the type/size of the meal; how much liquid was consumed with the meal ✓✓
14. a) the quantity (volume) of the solution ✓; the concentration of the blue dye ✓; the quantity of bleaching agent added ✓; the rate of stirring/mixing and so on

b) See Figure A.16. Notice that temperature is the independent variable, and needs to go on the horizontal axis.

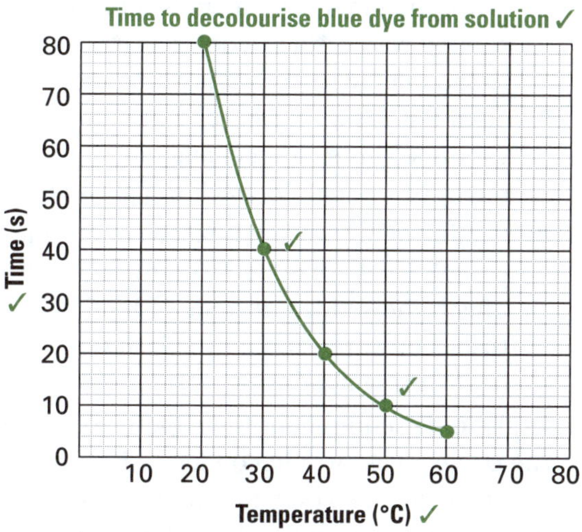

Figure A.16 Time versus time

c) time for discolouration halved for each 10-degree rise in temperature ✓

d) Yes, since for every 10-degree rise it only took half as long to decolourise the blue colour, so the rate of reaction must be twice as fast. ✓✓

Chapter test

Part A: Multiple-choice questions

1. **A** ✓ This is the logical procedure that should be followed in all scientific experiments. B, C and D are not in the correct sequence.
2. **C** ✓ This is what a hypothesis is. A, B and D do not refer to tentative explanations.
3. **A** ✓ The term scientific method is a misnomer because it can apply to other fields of study where evidence is required. It often involves much experimentation (so B is wrong), and wouldn't be of much use if it was random and haphazard (so C is wrong). While it can include D, it is not limited to it.
4. **C** ✓ Before you even pose a question, or come up with a hypothesis, or even plan an idea, you make a certain observation that challenges you to investigate it. Therefore, answers A, B and D are incorrect.
5. **D** ✓ The other words (A, B and D) are much easier to answer; 'why' is the difficult one. For example, why are positive protons found in the nucleus, while negative electrons whiz around it? Why are protons and neutrons much heavier than electrons?
6. **B** ✓ The researcher needs to be able to compare the effects of the experimental drug with a group of people who don't take the drug (control) to establish whether it actually has an effect. The experimental group is actually taking the drug. A, C and D are not related to this information.
7. **A** ✓ A hypothesis is a proposed explanation made on the basis of limited evidence as a starting point for further investigation. A theory is a system of ideas intended to explain something, especially one based on general principles independent of the thing to be explained. Hence, a theory is the next step along. It is not a conclusion, and is a long way from becoming a law. B, C and D are not part of the definition of a hypothesis.
8. **B** ✓ This is the method or procedure. A, C and D do not outline the steps that need to be performed.
9. **D** ✓ Scientific laws are not final (A) and can change with further evidence (B) . They originate from hypotheses (C) and result from accumulated observations over time.
10. **C** ✓ Coming up with a method to examine a novel problem is difficult and is often associated with some advanced pieces of technology and equipment. A, B and D are not as difficult.

Part B: Short-answer questions

11. a) problem ✓; b) information ✓;
    c) hypothesis ✓; d) experiment ✓; e) variable ✓;
    f) independent ✓; g) dependent ✓;
    h) controlled ✓; i) control ✓; j) record ✓;
    k) data ✓; l) conclusion ✓
12. a) Liquids on the floor can make it slippery and therefore dangerous. Also, many chemicals can damage the floor or footwear, or release foul odours. ✓
    b) Hair has a habit of falling into chemicals or brushing against a lit Bunsen burner. ✓ Short hair is best, or you should use a hair net or tie your hair back.
    c) Anything can potentially fall on your feet. Solid footwear can lessen damage. ✓
    d) Running increases the risk of tripping or knocking something off a bench. ✓

e) Directions have been developed over time through experience and are there for safety reasons. ✓ They are not there to hinder you.

f) As in cooking, you would not cook in a dirty pan or serve food in an unwashed plate. Also, it's good manners to clean your equipment after use. ✓

13. not wearing protective eyewear ✓; holding the hot test tube rather than using a test tube holder or tongs ✓; using a cracked beaker ✓; not wearing a lab coat or protective garment ✓; listening to music so attention is not completely focused on the task at hand ✓

14. Correct: hair tied back ✓; wearing goggles ✓. Incorrect: equipment too close to the edge ✓; bringing paper too close to the lit candle ✓.

15. a) Corrosive chemicals. They can burn or eat through material, including skin. ✓ So handle with care.

b) No smoking. ✓ This sign is often seen around flammable liquids such as petrol.

16. a) Bunsen burner ✓; b) test tube ✓;
c) retort stand ✓; d) wash bottle ✓;
e) measuring cylinder ✓; f) beaker ✓;
g) pipeclay triangle ✓; h) wire gauze ✓;
i) goggles ✓; j) conical flask ✓;
k) evaporating basin ✓; l) crucible (with lid) ✓;
m) spatula ✓; n) eyedropper ✓;
o) mortar and pestle ✓; p) tongs ✓;
q) test tube rack ✓; r) funnel ✓

17. Many chemicals, such as acids, react with metals. ✓ It is not uncommon for test tube racks to be wetted with spilt chemicals. This can cause corrosion. Also, metals react with acids producing hydrogen, which is flammable and so is dangerous when a naked flame is nearby. ✓ Wooden racks can be quickly rinsed and allowed to dry.

18. a) distillation ✓

b) A = Bunsen burner ✓; B = wire gauze ✓; C = distillation flask ✓; D = distillation head ✓; E = thermometer ✓; F = condenser ✓; G = water out ✓; H = water in ✓; I = receiving flask ✓ containing distillate

c) cold water circulating around the inner tube ✓ cools the vapour and they liquefy ✓

d) The arrows indicate the in and out directions of cooling water as it circulates around the inner tube. ✓ Just before exiting, the warmer cooling water meets the hot vapour and cools it down. ✓ As the vapour moves down the inner tube it cools further, and the cooling water is also cooler as it is closer to the inlet. ✓ If the inlet were near the distilling flask, cold water would hit the hot vapours. This sudden change in temperature might be enough to crack the glassware. ✓

e) Boiling chips are small pieces of marble gravel. They are added to prevent bumping. This is the sudden gushing of liquid that rushes upwards and may splash out of the flask during heating. ✓

19. In the older equipment the vapour touching the cooled glass would condense. But this had a large volume so only a small portion of it would cool sufficiently. ✓ Much of the vapour could also escape. ✓ In the modern distillation apparatus, the tube is long and narrow so the vapours would almost all touch the cool glass and condense. ✓ As this vapour condensed to liquid, it allowed other vapour to come into contact with the cool glass and also condense. ✓ Additionally, by using the thermometer you could control the rate at which the cooling waters circulated to prevent any vapours escaping. (See Figure A.17.)

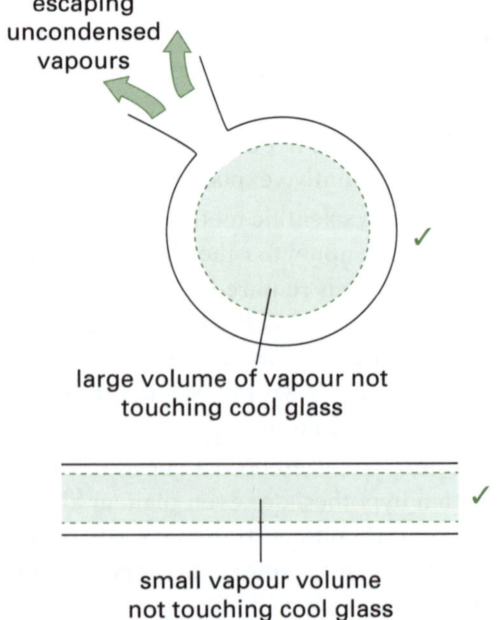

Figure A.17 Condensation

20. a) Who are these dentists? Were they paid for their endorsements (in which case they

would not be unbiased)? How many dentists were approached before nine of them would endorse the toothpaste? ✓✓

b) Which university conducted these tests? Using the word 'university' sounds impressive but adds little weight if payment depended on finding a favourable outcome. And 22% whiter … whiter than what? Mud? How was this measured? ✓✓

c) 'Leading' is misleading. Is a retailer defined as leading if they sell their product? It is there to give the impression that high quality stores wouldn't sell anything cheap and nasty, so it follows that the product must be good. ✓✓

21. a) Inference ✓: coughing can be caused by many conditions. A virus is just one possibility.

b) Generalisation ✓: examination of a wide variety of copper compounds shows that most of them are blue and so this is a useful generalisation.

c) Observation ✓: the colour of the different minerals can be observed.

d) Prediction ✓: the statement suggests that the student has already tried the dilute acid and is now predicting that a more concentrated solution will react even faster.

22. a) This is a hypothesis. ✓ It is suggesting an answer to a problem and it can be experimentally tested. It predicts the outcome of the research. ✓

b) This is NOT a hypothesis. ✓ It does suggest an answer to a problem. It cannot, however, be tested directly. While this is one way they could have started fires, it could have been a later development. ✓

c) This is NOT a hypothesis. ✓ It involves personal opinions and this will vary from one person to another. ✓

23. independent variable = time period ✓; dependent variable = average size (possible measuring diameter) ✓
The size of the crystals depends on the time allowed for them grow. Time does not depend on the crystal's size. The student can make measurements over regular time intervals. For instance, he/she can measure the size of the crystals every second day for a fortnight. Consider the absurd situation if time depended on the crystal's size. When the crystal reached its maximum size and stops growing (because there is no more alum salt in the solution), does this mean that time now stands still?

24. a) fire safety blanket ✓ (Know where the safety blanket is in the laboratory, and how to use it in case of emergencies.)

b) no running ✓ (There should never be any running in a laboratory.)

c) wear safety goggles ✓ (Eye protection is necessary in many experiments. Foresight is better than no sight!)

25. Student B will obtain a more accurate answer as she is avoiding parallax error. ✓ Student A will obtain an answer that is too low due to parallax error. ✓

26. The measuring cylinder is more accurate than the beaker. ✓ The scale on the measuring cylinder is divided into 0.1-mL units whereas the beaker's scale is divided into 10-mL units.

27. The volume of liquid is 96 mL. ✓ The volume is measured at the base of the curved meniscus.

28. See Figure A.18. *(take off 1 mark for each incorrect drawing or labelling)*

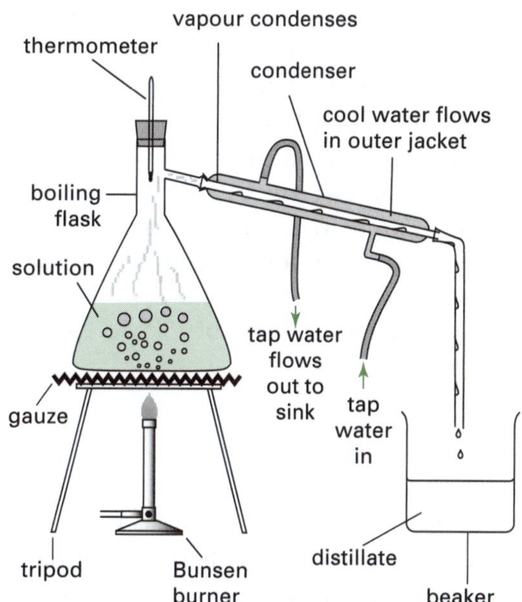

Figure A.18 Distillation

29. a) R ✓

b) 4.4 cm ✓

c) $Rf = 2.6/6.0 = 0.43$ ✓✓ (Measure the distance component Q has moved from where the original mixture was spotted. Measure the distance from where the original mixture was spotted to where the solvent front finally reached. Now divide the first number by the second. Your answer should be close to that given here.)

30. **a)** T ✓

b) The container is covered to make sure that the atmosphere in the beaker is saturated with solvent vapour. ✓ This stops the solvent from evaporating as it rises up the paper. ✓

c) Measurements should be taken from the 2 cm mark. ✓ This is where the solvent makes contact with the spotted substance and begins to move it up the paper. ✓

d) component P ✓; $Rf = 9.5/14 = 0.68$ ✓

31. **a)** less dense than water ✓; does not react with water ✓; does not dissolve in water ✓

b) no ✓ (flammable gases, such as butane or hydrogen, can be collected like this)

32. **a)** natural gas = 23%, coal = 22%, petroleum = 40%: total = 85% ✓

b) $\frac{8}{100} \times 360 = 28.8°$ ✓✓

c) Renewable energy is energy generated from natural resources such as sunlight, biomass, wind, rain, hydroelectric, tides, and geothermal heat, which are renewable (naturally replenished). ✓

d) There are several components to it and each would be only a sliver if these replaced the 7% renewable energy sector. ✓ It also emphasises the components of renewable energy. ✓

e) You need to find 36% of 7%:

$\frac{36}{100} \times 7 = 2.52\%$ ✓

f) $\frac{53}{100} \times 7 = 3.71\%$ ✓

33. **a)** **i)** gas Q ✓; **ii)** gas P ✓; **iii)** gas R ✓

b) The gas that builds up in the flask from the reaction needs to have only one path out: the shorter tube on the right. ✓ If the tube on the left was shorter too, gas could force its way through.

34. The correct sequence is: A ✓, J ✓, K ✓, H ✓, B ✓, G ✓, I ✓, L ✓, F ✓, C ✓, E ✓, D ✓.

35. R will last the longest. ✓ If you discourage evaporation, the film should last a long time. Placing a large beaker, moist inside, over the frame carrying the film will provide a moist environment that slows evaporation. ✓

36. **a)** **i)** The metal sphere would rotate faster ✓ because the increased steam would issue faster from the outlet tubes ✓.

ii) The extra boiling water adds weight and will initially reduce the rotational speed. As the water boils, the speed of the aeolipile will increase. ✓ The extra will allow the aeolipile to rotate for a longer time. ✓

iii) A larger ball would be heavier ✓ and therefore not rotate as fast ✓.

iv) The metal sphere would rotate faster ✓ as the steam would issue faster from the outlet tubes ✓.

b) Yes ✓, but possibly not as fast.

c) No ✓, as each outlet tube would be acting against each other.

d) Yes ✓, the shape of the ball should not make any difference.

37. When reading the graph, your values could be slightly different to the ones given here.

a) true ✓

b) 11.3 L/100 km ✓

c) 15 miles/gallon ✓

d) From the graph, it would use 7 L/100 km ✓. So in 500 km it would use 7 × 5 = 35 L ✓.

e) 20 miles/gal = 19 L/100 km ✓; 30 miles/gal = 9.5 L/100 km ✓; I could save 19 – 9.5 = 9.5 L ✓ for each 100 km

f) 112.5/7.5 = 15 L/100 km ✓; from the graph this is 19 miles/gal ✓

38. **a)** See Figure A.19 on the next page.

b) column graph ✓

c) No. ✓ The insects are discrete entities (organisms). There is no continuous relationship between one insect and the next. ✓

39. **a)** U ✓

b) No. ✓ If anything, there seems to be a negative correlation between wing beat frequency and speed. That is, the smaller the wing beat frequency, the slower the speed. ✓

c) Only 10 insects are shown in this dot plot. If he plotted the results of several dozen more, ✓ he could see whether there is any substance to his hypothesis.

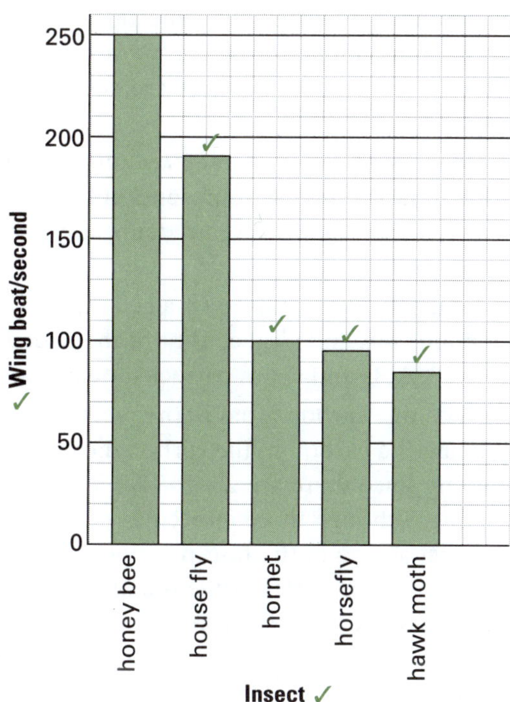

Figure A.19 Insect wing beats

d) the weight of the insect ✓; the shape of the insect ✓ (is it aerodynamic?); the size of its wings ✓

40. a) decreasing wing area ✓

b) See Figure A.20.

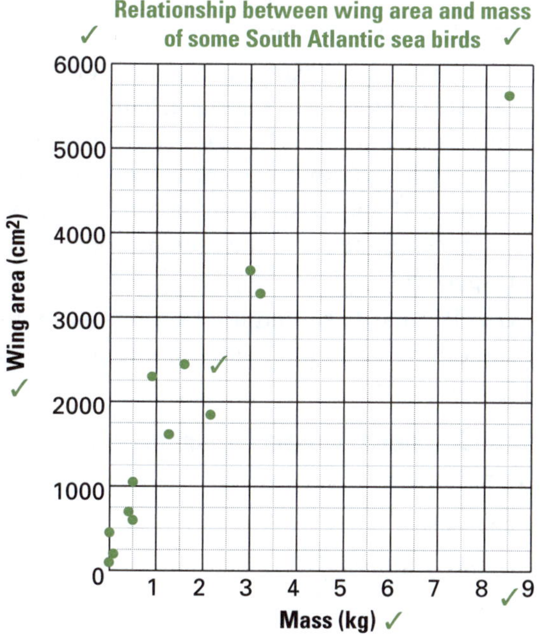

Figure A.20 Wing area and mass

c) The outlier is the result for seabird A. ✓ Its mass and wing area are away from the cluster of other results. (An outlier is an observation that is numerically distant from the rest of the data.)

d) Yes. ✓ It appears that the larger the wing area, the heavier is the sea bird. ✓

Test 1 answers

Part A: Multiple-choice questions

1. **D** ✓ A, B and C are examples of a fungus.
2. **B** ✓ Reactions take place in the cytoplasm, in the presence of organelles in it. A, C and D are incorrect as, while they are part of any cell, they are not the jelly-like part referred to in the question.
3. **A** ✓ This is sexual reproduction. B, C and D are examples of asexual reproduction.
4. **D** ✓ Fertilisation involves the sperm and egg joining together to form a zygote. Gestation is the time a baby spends in the womb. Chickens do not have a womb , so A is incorrect. Pollination is the transfer of pollen from the anther to the stigma of a plant, so B is also incorrect. Mitosis is used to duplicate existing cells so, C is incorrect.
5. **C** ✓ Therapeutic cloning is a procedure where cells are taken from a patient and used to create stem cells from it. Scientists can then use those cells to grow tissue that is a perfect genetic match for the patient. These cells can be potentially transplanted into the patient to treat a disease from which the patient suffers. Hence, alternatives A, B and D are incorrect.
6. **D** ✓ A belongs to the digestive system; B belongs to the circulatory system; C belongs to the muscular system.
7. **C** ✓ This is a major blood vessel taking oxygenated blood to the lower part of the body. A is the kidney; B is the ureter; D is the urinary bladder.
8. **A** ✓ This is what allows muscles to contract. B, C and D do not need to contract.
9. **B** ✓ Cartilage is rather rare in adult humans, but it is very important during the developing foetus because of its firmness and its ability to grow rapidly. In embryos, cartilage precedes most of the bones of the skeleton. The bones

of fish are made mostly of calcium, but sharks don't have any bones. Its skeleton is made of cartilage, so A is incorrect. Collagen is a fibrous protein found in flesh and used to connect and support other body tissues, so C is incorrect. The protein keratin is found in fingernails, so D is incorrect.

10. **C** ✓ The respiratory system includes the lungs where gaseous exchange occurs. The urinary system gets rid of liquids (urine) and substances dissolved in it, so A is incorrect. The circulatory system takes dissolved gases (e.g. CO_2) to the lungs, but it is the lungs which gets rid of them. At the end of the digestive system solids (faeces) are removed, so D is incorrect.
11. **A** ✓ Elements are composed of atoms. While C and D are part of an atom, they do not define the element and, besides, protons are identical in any atom. B is incorrect because molecules are larger and may consist of several elements.
12. **A** ✓ Solids have a definite shape. Liquids (B) have particles that are somewhat packed tightly but don't have definite shape, while gases (C) are neither packed tightly nor have definite shape. Plasma (D) is a state of matter, similar to gas, where a certain portion of the particles are ionised.
13. **C** ✓ A physical property can be discerned without changing the structure of the substance. A, B and D are all examples of chemical properties.
14. **D** ✓ A, B and C could all indicate that a chemical change has taken place.
15. **B** ✓ Chemical energy in the battery produces electrical energy that flows through the wires, to be converted to heat and light within the bulb. A, C and D are incorrect as they are in the wrong sequence and/or contain energy forms not referred to. Chemical energy is a special form of potential energy.
16. **C** ✓ Hydrogen reacts with oxygen in the air to form water (H_2O). The chemicals mentioned in A, B and D are not formed (produced).
17. **D** ✓ The number of protons in the nucleus of an atom distinguishes one element from another. A is completely wrong as elements are not distinguished by how many atoms they have; B and C do not distinguish one element from another.
18. **D** ✓ There are four carbon atoms and 10 hydrogen atoms. B and C are incorrect because the amounts are wrong. The question asked for the total number of atoms, not how many different kinds of atoms there are, so A is incorrect.
19. **A** ✓ Kinetic (from the Greek *kinesis*) energy is due to movement. B represents potential energy, while C and D do not define any form of energy.
20. **D** ✓ All cells are enclosed by a cell membrane, but plant cells and viral cells have the added cell wall. A, B and C are correct statements.
21. **C** ✓ The one carbon atom in the methane ends up being the carbon in the carbon dioxide. A is wrong since there are $2 + 5 + 2 = 9$ atoms reacting, although three molecules react. B is incorrect as two of the four reacting oxygen atoms end up in water, while the other two are in carbon dioxide. D is wrong as atoms are not created nor destroyed in a reaction. There are four O atoms, one C atom and four H atoms reacting. In the products there are also four O atoms, one C atom and four H atoms.
22. **B** ✓ Granite is igneous, while sandstone (A) and shale (C) are sedimentary, and gneiss (D) is metamorphic.
23. **D** ✓ Coal was formed many millions of years ago and, while there are currently sufficient reserves, it will eventually run out. A, B and C are all renewable sources.
24. **A** ✓ Basalt is an extrusive igneous rock, cooling rapidly on the Earth's surface, hence it has very fine grains. If it had cooled deep within the Earth, as C suggests, the crystals would be large. B and D are both incorrect as they refer to sedimentary and metamorphic rocks respectively.
25. **B** ✓ Popularity should not be a reason to do something. A, C and D are important reasons for recycling.

Part B: Restricted-response questions

26. cellulose ✓
27. photosynthesis ✓
28. mitochondrion ✓
29. semipermeable ✓
30. small intestine ✓

31. chemical formula ✓
32. hydrogen ✓
33. chemical ✓
34. energy ✓
35. metamorphic ✓
36. minerals ✓
37. ore ✓
38. non-metals ✓
39. potential ✓
40. heat ✓

Part C: Knowledge, skill and processing data questions

41. a) P ✓ Since this is the set-up without any anti-fungicide added, it should grow the fastest. Hence, the curve would be the steepest. ✓
    b) 5 ✓
    c) every day, once a day ✓ (supposedly at the same time of day)
    d) U ✓ (the one which caused the fungus to grow the slowest)
    e) R ✓
    f) There could be aberrations that can alter the results ✓ from trial to trial. So by repeating the experiment several times, you increase the reliability ✓ of the data.
    g) Fungi grown in a culture plate are too small for individual counting ✓, and it would be too fiddly. But how fast the colony grows (diameter expands) gives a good and easy indication ✓ of the size of the colony.
42. a) Mitosis is the duplication and division of a plant's or animal's cell nucleus and the nuclear material (chromosomes) contained in it. ✓
    b) *(½ mark for each correct answer)*
    S, T, Q, P, R, U ✓✓
    c) P ✓
    d) While this is sometimes called the resting phase, because no mitosis is occurring, the cell still undergoes the chemical reactions and growth it requires. ✓ Also, there is a period during interphase of the cell's life cycle where the chromosomes are replicated. ✓
43. a) The dry weight gives a truer indication of the amount of fungus present. ✓ Otherwise you are just measuring water.
    b) Over a period of 9 days ✓ the number of fungal cells increased and then decreased ✓, stabilising to a level around 0.02 to 0.04 mg/mL ✓.
    c) No ✓, the fungal cells could have died off naturally for any number of reasons, besides the presence of the bacterium ✓; for example, overcrowding or being killed by their own wastes. ✓. In order to be certain, a control is needed ✓ with broth, fungi and no bacterium.
    d) At day 6, the dry weight measured is around 0.04 mg/mL. ✓ So in 750 mL there would be $750 \times 0.04 = 30$ mg. ✓
    e) Excessive use of chemical fungicides in agriculture has led to problems in human health ✓, environmental pollution ✓, and fungi developing resistance to fungicides ✓. Hence, a serious search is needed to identify more environmentally friendly solutions to the problem.
44. a) F ✓; b) T ✓; c) T ✓; d) F ✓; e) T ✓; f) F ✓; g) T ✓; h) T ✓
45. coal: sedimentary ✓, biological ✓
    gneiss: metamorphic ✓, foliated ✓
    marble: metamorphic ✓, non-foliated ✓
    sandstone: sedimentary ✓, clastic ✓
    rhyolite: igneous ✓, volcanic ✓
46. a) P, iii ✓; Q, v ✓; R, vi ✓; S, i ✓; T, iv ✓; U, ii ✓
    b) It is important to maximise the amount of sunlight falling on the tower. ✓ A computer can turn the mirrors ever so slightly as the Sun moves across the sky so that the beam is always focused on the tower. ✓
    c) Super-heated steam builds up high pressures and can rupture containment vessels. ✓
    So pipes and tanks would need to be built stronger, which adds to costs. Other factors may include enhanced rusting.
47. a) Each molecule consists of two atoms ✓ chemically bonded together: H_2 and O_2.
    b) There are different kinds of particles mixed in together, which are not chemically bonded ✓, in the reacting vessel. The right-hand container contains only one kind of particle, H_2O. ✓ While consisting of different atoms, they are chemically bonded together.
    c) Ratio = H:O = 2:1 ✓

Test 2 answers

Part A: Multiple-choice questions

1. **B** ✓ A is not correct as Leeuwenhoek developed single-lens microscopes 11 years after Hooke; C is wrong as da Vinci only used a magnifying glass; D is wrong as the Janssen brothers did not view cork cells.
2. **C** ✓ A is wrong as the eyepiece lens has the lower magnification; B is wrong as the iris adjusts the amount of light; D is wrong as the coarse focus is used first.
3. **D** ✓ A is wrong as sugars are transported in the phloem from the leaves; B is wrong as starch is stored in other cell types; C is wrong as this is the function of cells in the leaf that contain chloroplasts.
4. **A** ✓ Flagella are whip-like tails that help to propel the cell. B is wrong as phagocytosis involves engulfing food; C is wrong as this is cell division; D is wrong as slime is not used to move about.
5. **A** ✓ B, C and D are wrong as these occur during inhalation.
6. **D** ✓ A, B and C are all wrong as these statements refer to sexual reproduction.
7. **A** ✓ B is wrong as Democritus lived in ancient Greece and he proposed the existence of atoms; C is wrong as he had no knowledge of atoms; D is wrong as Newton was a physicist and he believed in the atomic nature of matter but he did not propose that theory.
8. **C** ✓ $D = \frac{m}{V} = \frac{350}{50} = 7 \text{ g/cm}^3$, thus A, B and D are wrong.
9. **D** ✓ The symbol for iron is Fe as the Latin name is ferrum. Therefore A, B and C are wrong.
10. **A** ✓ Cu is a metal; B is a semi-metal and I is a non-metal. B is wrong as Ga is a metal and so is Na; C is wrong as Ge and Si are both semi-metals; D is wrong as they are all metals.
11. **B** ✓ Carbon dioxide smothers a fire as it excludes oxygen. A, C and D are wrong as carbon dioxide is not used for any of these.
12. **C** ✓ Windblown sand can abrade rocks very rapidly. A is wrong as ice cracking is not abrasion; B is wrong as volcanic action does not lead to abrasion; D is wrong as this is a chemical weathering process.
13. **D** ✓ Pumice is a volcanic igneous rock. A, B and C are wrong as they are all sedimentary.
14. **A** ✓ Granite is an igneous rock. B, C and D are wrong as they are all metamorphic.
15. **C** ✓ Various radiometric dating techniques suggest the Earth is about 4.5 billion years old. Thus A, B and D are incorrect.
16. **A** ✓ Pitchblende contains uranium oxide. Thus B, C and D are wrong.
17. **A** ✓ KE depends on both the mass and speed of a body. Thus B, C and D are all wrong.
18. **C** ✓ GPE = mgh and so the GPE depends on mass and height. On Earth $g = 9.8 \text{ m/s}^2$ and so GPE = $9.8mh$. Thus A, B and D are wrong
19. **A** ✓ The caloric theory cannot explain this observation as no other body got cooler as a result of the drilling. B is incorrect as the experiment did not support the caloric theory as the cannons got hotter and hotter and the caloric (if it existed) seemed to be inexhaustible; C is wrong as the cannons did not possess elastic potential energy; D is wrong as kinetic energy is converted to heat, not chemical energy converted to heat.
20. **D** ✓ Glass is an insulator, whereas A, B and C are all metals and good conductors.
21. **B** ✓ The number of recharging sites will need to increase significantly for electric cars to be useful. A is wrong as lithium ion batteries are expensive; C is wrong as their range is much less (~200 to 300 km); D is wrong as a quick recharge takes much longer and only gets to about 80%.
22. **A** ✓ Internet blogs are just different people's opinions and ideas and they are not reliable as they are not peer-reviewed by experts. B, C and D are all wrong as these information sources are all peer reviewed by experts in the field.
23. **B** ✓ A is wrong as an inference refers to a possible explanation of one observation; C is wrong as a theory is based on considerable collection of evidence and the review of many experimental results; D is wrong as a prediction gives a possible outcome of a future experiment.
24. **D** ✓ The rate of temperature rise will depend on the different colours that the steel objects are painted with. A and B are wrong as these are controlled variables; C is wrong as this is the independent variable.

25. A ✓ When you recount the experiment you are referring to a past event. B is wrong as this refers to a discussion text; C is wrong as this is a procedure text; D is wrong as this is a report text type.

Part B: Restricted-response questions

26. curved ✓
27. straight ✓
28. accuracy ✓
29. reliable ✓
30. graphs ✓
31. vegetative ✓
32. cervix ✓
33. pathogens ✓
34. microbes (or bacteria or fungi) ✓
35. organisms ✓
36. liquids ✓
37. increases ✓
38. cleavage ✓
39. fire ✓
40. fizzing ✓

Part C: Knowledge, skill and processing data questions

41. See Figure A.21 (other keys are possible).
42. a) time in darkness ✓
    b) sucrose content of the leaf ✓
    c) See Figure A.22.
    d) The concentration of sucrose decreases with time in the dark. ✓

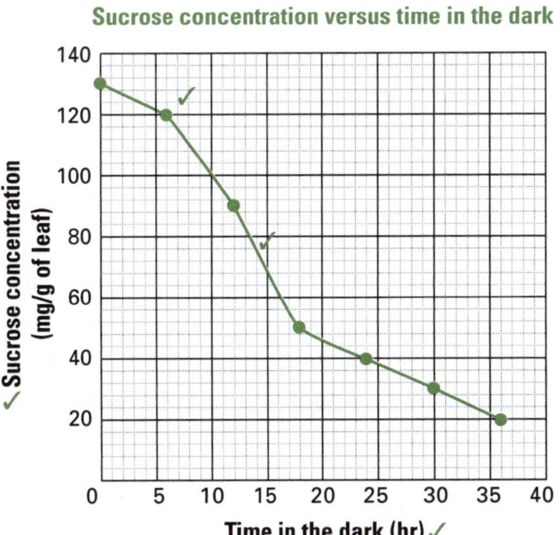

Figure A.22 Sucrose concentration

    e) In the dark the plant respires to obtain energy. ✓ The sucrose is removed from storage and its concentration decreases as it continues to respire. ✓

43. 1b; 3b; 5a ✓ metal = zinc ✓
44. See Table A.7.

Table A.7 Organs that complete the same job as household objects

Object	Part of body
A	teeth/mouth ✓
B	bladder/urethra ✓
C	kidney ✓

45. Muscles in the arm turn the handle: chemical potential energy is transformed into kinetic

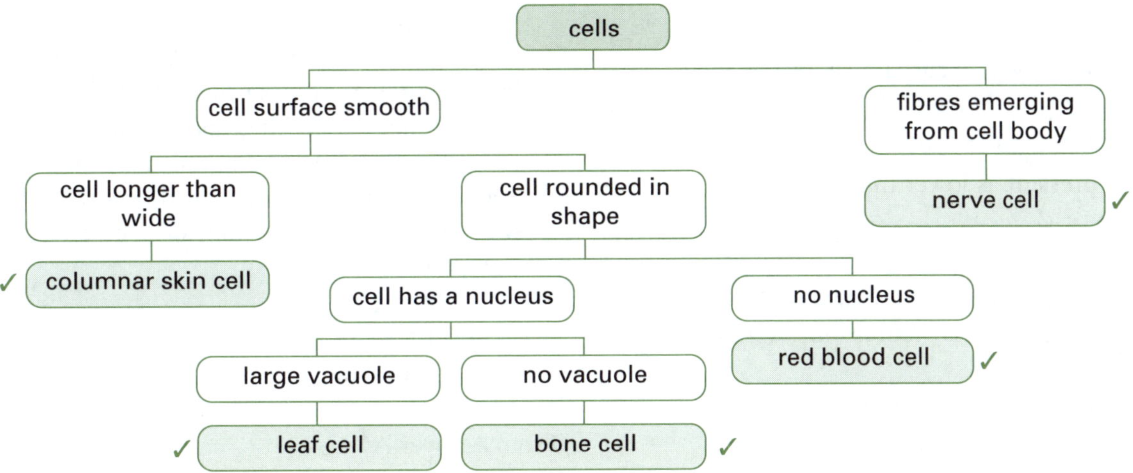

Figure A.21 Key to cells

energy. ✓ Rotating the handle and raising the weight: kinetic energy transformed into gravitational potential energy. ✓ Weight falls down: gravitational potential energy is transformed into kinetic energy. ✓ Paddles turn and water warms up: kinetic energy is transformed into heat energy. ✓

46. a) See Table A.8. *(1 mark for the first three correct ✓; 1 mark for the next two correct ✓; 1 mark for the last two correct ✓)*

Table A.8 Variation in blood pressure for a healthy adult, according to the order in which it flows through the circulatory system

Chamber/vessel	Blood pressure range (mm of mercury) (high–low)
right ventricle	28 to 0
pulmonary artery	28 to 10
pulmonary vein	2 to –2
left ventricle	120 to 0
main arteries of the body	120 to 80
capillaries in body tissues	35 to 8
veins	8 to –2

b) main arteries ✓

c) pulmonary vein/vein ✓

d) Yes, it exerts less force on the blood so blood going to and from the lungs is at a lower pressure than that of the body. ✓

e) The blood pressure in the veins is low ✓ so the blood is moved back to the heart via nearby muscle contractions. ✓

f) the arteries ✓

g) The capillaries would rupture (burst) as they are thin. ✓

47. a) i) $P_2 = 1000 \times 10/15 = 666.6$ hPa ✓

ii) Final pressure is lower than the initial pressure. ✓

iii) See Figure A.23.

iv) Air will move into the apparatus ✓ because pressure is greater outside than inside. ✓

b) i) $P_2 = 700 \times 15/12 = 875$ hPa ✓

ii) It is greater. ✓

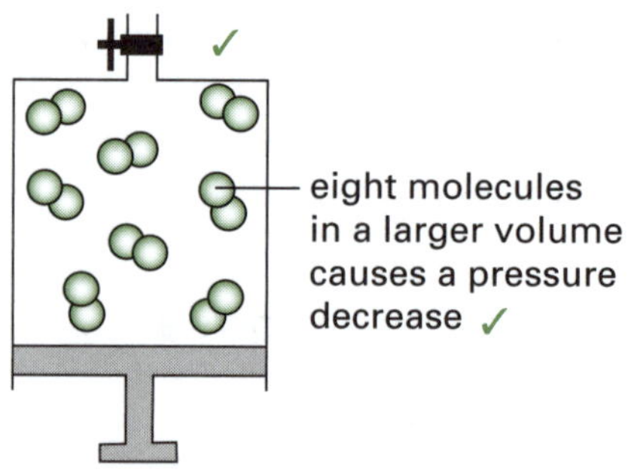

Figure A.23 Model of lung

iii) Air will move out as the pressure is higher. ✓

c) In question a) the volume V_2 is greater than V_1. This is equivalent to inhalation as the lungs expand. In question b) the volume V_2 is less than V_1 and so this is what happens as the lungs deflate in exhalation. Therefore b) ✓ is the process of exhalation.

48. a) When subterranean water comes in contact with hot magma, it is converted to steam. ✓ The steam can be used to turn turbines and generate electrical power. ✓

b) Basalt is used to make concrete. ✓ Fresh water can be collected in porous volcanic rocks in Hawaii and this is an important source of drinking water. ✓

49. a) ~325 MJ/person/day ✓

b) Primitive societies and hunter-gatherers used little energy as they did not have machines to do their work. ✓ When the agricultural revolution began, more energy was expended in planting and harvesting crops and keeping animals. ✓ As people moved into more advanced agriculture and then into the industrial revolution, more machines were being used. ✓ In the technological age, information transfer is increasing. This occurs using television, mobile phones, satellites and the internet, as well as by books and newspapers. In addition, there is greater travel occurring. More people are doing this, so more energy is expended. ✓

50. A1 = SOUND ✓
A2 = POTENTIAL ✓
A3 = NUCLEAR ✓
Code:
From Answer A1:
T = S; Q = O; X = U; O = N; I = D (sound)

From Answer A2:

C = P; Q = O; S = T; J = E; O = N; Z = I;
G = A; P = L (potential)

From Answer A3:

O = N; X =U; F = C; P = L; J = E; G = A;
Y = R (nuclear)

Famous scientist: The codes for K and H has not been found; leave a blank.

First Name = _ A _ E S

Surname = _ O U L E

So missing code must be K = J and H = M to give JAMES JOULE. ✓

51. E, C, A, D, F, B ✓✓

Index

Page numbers in **bold** are definitions of terms.

D

E

F

G

H

N

O

P

Q

R

S